Ecology of Sonoran Desert Plants and Plant Communities

Ecology of Sonoran Desert Plants and Plant Communities

Edited by Robert H. Robichaux

The University of Arizona Press
Tucson

The University of Arizona Press
www.uapress.arizona.edu

© 1999 by The Arizona Board of Regents
All rights reserved. Published 1999
Century Collection edition 2016

Printed in the United States of America
21 20 19 18 17 16   7 6 5 4 3 2

ISBN-13: 978-0-8165-1869-2 (cloth)
ISBN-13: 978-0-8165-3540-8 (Century Collection paper)

Publication of this book was supported by a generous grant from the Advisory Committee of the Desert Laboratory at Tumamoc Hill, Tucson, Arizona.

Library of Congress Cataloging-in-Publication Data
Ecology of Sonoran Desert plants and plant communities / edited by Robert H. Robichaux.
p.   cm.
Includes bibliographic references and index.
ISBN 0-8165-1869-9 (cloth : acid-free paper)
1. Desert plants—Ecology—Sonoran Desert. 2. Plant communities—Sonoran Desert. I. Robichaux, Robert Hall.
QK142.4.E36   1999
581.7'54'097917—ddc21
98-40107
CIP

British Library Cataloguing-in-Publication Data
A catalogue record for this book is available from the British Library.

∞ This paper meets the requirements of ANSI/NISO Z39.48-1992 (Permanence of Paper).

Contents

	List of Abbreviations	vii
	Introduction	3
1	Diversity and Affinities of the Flora of the Sonoran Floristic Province *Steven P. McLaughlin and Janice E. Bowers*	12
2	Vegetation and Habitat Diversity at the Southern Edge of the Sonoran Desert *Alberto Búrquez, Angelina Martínez-Yrízar, Richard S. Felger, and David Yetman*	36
3	The Sonoran Desert: Landscape Complexity and Ecological Diversity *Joseph R. McAuliffe*	68
4	Population Ecology of Sonoran Desert Annual Plants *D. Lawrence Venable and Catherine E. Pake*	115
5	Form and Function of Cacti *Park S. Nobel and Michael E. Loik*	143
6	Ecological Genetics of Cactophilic *Drosophila* *William J. Etges, William R. Johnson, Garry A. Duncan, Greg Huckins, and William B. Heed*	164

7	Ecological Consequences of Agricultural Development in a Sonoran Desert Valley *Laura L. Jackson and Patricia W. Comus*	215
8	Deep History and a Wilder West *Paul S. Martin*	255
	List of Contributors	291
	About the Editor	293
	Index	295

Abbreviations

AMOVA	Analysis of Molecular Variance
ARIZ	Arizona Upland
BP	before present
CAM	Crassulacean acid metabolism
CGC	Central Gulf Coast
CIW	Carnegie Institution of Washington
HBAR	average heterozygosity
IBP	International Biological Program
INEGI	Instituto Nacional de Geografía Estadística e Informática
LC	Lower Colorado Valley
NM	National Monument
NSF	National Science Foundation
PLS	Plains of Sonora
PPFD	photosynthetic photon flux density
RR	relative richness
STS	Sinaloan thornscrub
TMR	average annual temperature range
USDA	U.S. Department of Agriculture
USGS	U.S. Geological Survey

Ecology of Sonoran Desert Plants and Plant Communities

Introduction

The Sonoran Desert region of the southwestern United States and northwestern Mexico harbors a rich diversity of plant life. Though best known for its large columnar cacti, such as the saguaro (*Carnegiea gigantea*), cardón (*Pachycereus pringlei*), and organ pipe cactus (*Stenocereus thurberi*), the Sonoran Desert includes an array of plant forms that is unrivalled elsewhere in North America. In addition to stem and leaf succulents, the plants include a wide variety of evergreen and deciduous shrubs and trees, which differ greatly in leaf size, shape, color, and duration. Among the more unusual plants are the drought-deciduous trees with large, swollen trunks, such as the elephant trees (*Bursera microphylla* and *Pachycormus discolor*) and boojum tree (*Fouquieria columnaris*). Other perennial elements are the grasses, root perennials, and arborescent monocots, such as yuccas (e.g., *Yucca valida*) and palms (e.g., *Sabal uresana*). The plants also include a large group of annuals, which complete their ephemeral life-cycles in close association with either winter or summer rains. In wet years, winter annuals may carpet the desert floor with spectacular floral displays.

The diversity of plant life in the Sonoran Desert, and the ecological patterns and processes that underlie it, have been the focus of a major scientific research effort since the turn of the century. Of paramount importance to early research in the Sonoran Desert was the establishment of the Desert Laboratory on the slopes of Tumamoc Hill in Tucson, Arizona. Founded by the Carnegie Institution of Washington in 1903, the Desert Laboratory served as a wellspring of pioneering research on desert plants.

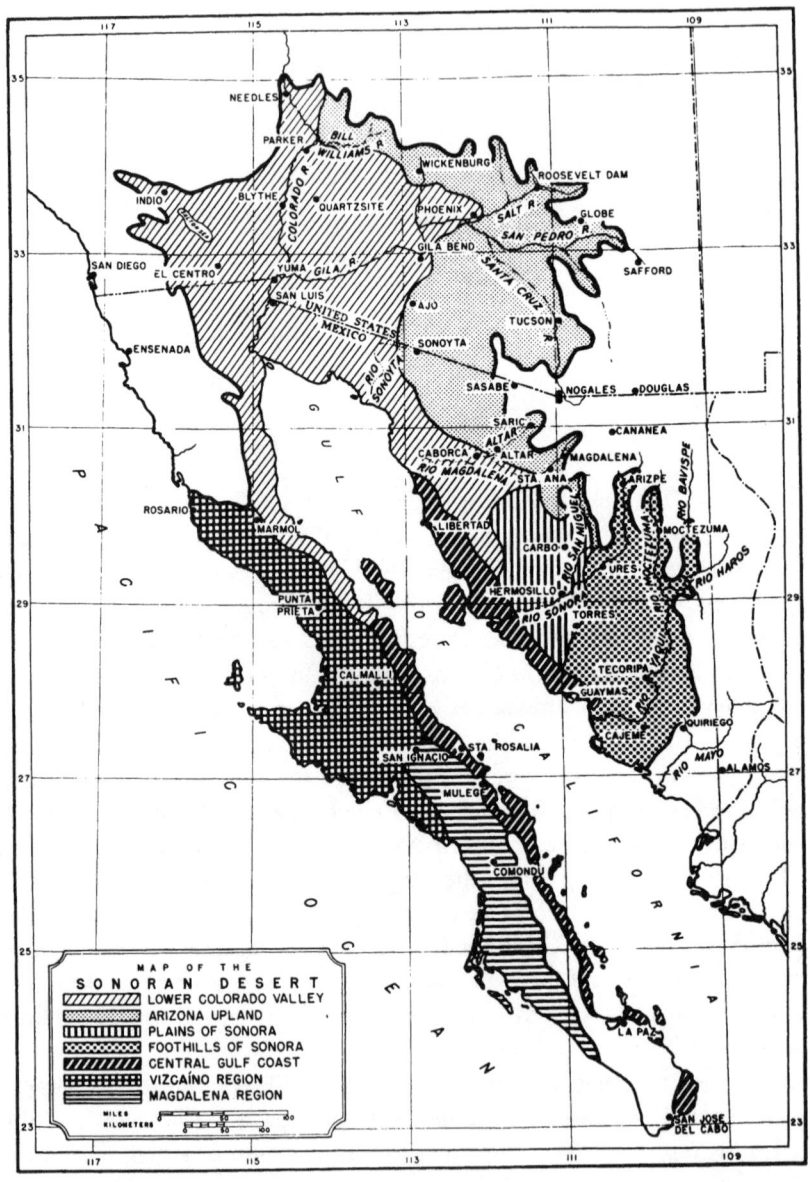

Figure I.1 The Sonoran Desert and its vegetational subdivisions (Shreve 1951).

As detailed in many of the chapters in this volume, much of our current understanding of ecological phenomena in the Sonoran Desert is based in part on the research of early Desert Laboratory scientists, including W. A. Cannon, B. E. Livingston, D. T. MacDougal, E. B. Shreve, F. Shreve, and V. M. Spalding, among others.

Of the many publications deriving from early research at the Desert Laboratory, Forrest Shreve's classic *Vegetation of the Sonoran Desert* (1951) stands out for its comprehensive assessment of the plants and plant communities of the region. Shreve (1951) identified the Sonoran Desert as the region immediately surrounding the upper portion of the Gulf of California. The extent of the region and the location of its boundaries were determined using criteria related only to the distribution of plants and plant communities. Based on biological criteria, the Sonoran Desert encompassed about 310 000 km^2, and included significant parts of Arizona, Sonora, California, Baja California Norte, and Baja California Sur (fig. I.1). Of the various topographic, climatic, and edaphic features that were assessed, the amount and seasonal distribution of precipitation were considered to be the most important in terms of governing the distribution of plants and plant communities in the region. The number and duration of drought periods were considered to be particularly significant. Also important on a more local scale were the texture, depth, and surface characteristics of the soil.

Shreve (1951) recognized that the Sonoran Desert was distinguished by its great diversity of life-forms, with plants differing in "height, bulk, manner and amount of branching, character of stem, size, duration, color, and seasonal behavior of leaves, degree of succulence, and time of flowering." Differences in these traits formed the basis for distinguishing 25 principal life-forms, excluding those of minor importance in the vegetation. The appearance of the landscape and the physiognomy of the vegetation were considered to be determined in large measure by the varied grouping of plants exhibiting the different life-forms.

Shreve (1951) identified seven major subdivisions of Sonoran Desert vegetation. The subdivisions were based solely on the character and organization of the plant communities, with the designations reflecting the dominant character of the vegetation. The microphyllous desert, for example, was named for the low, open stands dominated by small-leaved shrubs, such as creosote bush (*Larrea tridentata*) and white bursage (*Ambrosia* [=*Franseria*] *dumosa*), whereas the sarcocaulescent desert was named for the abundance of drought-deciduous trees and shrubs with large, swollen trunks, such as elephant tree (*Bursera microphylla*) and sangrengado (*Jatropha cinerea*). Corresponding names were also provided for each subdivision that reflected the most distinctive genera, and the location or outstanding geographical feature. (The latter names for the subdivisions appear in figure I.1.) Thus, the microphyllous desert was also designated the *Larrea-Ambrosia* region and the Lower Colorado Valley subdivision, whereas the sarcocaulescent desert was also designated the *Bursera-Jatropha* region and the Central Gulf Coast subdivision.

Shreve (1951) provided detailed descriptions of the perennial plants and plant communities of each subdivision. The influences of climatic factors and soil conditions on the composition and structure of the vegetation were emphasized. Special attention was given to water availability, and its relation to the physiological and morphological adaptations of the plants. Patterns of variation in the vegetation that were associated with major topographic features were also highlighted. In the Arizona Upland subdivision, for example, several "physiographic units" were identified with which the vegetation showed close correlation: plains and lower bajadas, upper bajadas, slopes of hills and mountains, and streamways. Within some subdivisions, geographical patterns of variation were distinguished. The Foothills of Sonora subdivision, for example, was further divided into northern, central, and southern regions, with the latter region representing a transitional zone between desert and "thorn forest" vegetation.

In addition to analyzing perennial vegetation, Shreve (1951) recognized the importance of "ephemeral herbaceous," or annual, plants in the Sonoran Desert, noting that annual plants accounted for about half of the flora and played a conspicuous role in the vegetation in the appropriate seasons. Annual plants were grouped into winter ephemerals and summer ephemerals, depending on whether germination and growth were associated primarily with winter or summer rains, respectively. Important regional patterns in the distributions of annual plants were identified, as were the roles that dormancy, dispersal, and competition might play in influencing the temporal and spatial distributions of annual plants on a more local scale. The crucial importance of seed banks, for example, was anticipated by the observation that the relative abundances from year to year of the regularly recurring species did not bear a constant relation to the extent of the population and seed crop of the immediately preceding two or three years.

The major subdivisions of Sonoran Desert vegetation were based on the character and organization of the plant communities. Yet, Shreve (1951) recognized that the great variation in form among Sonoran Desert plants was accompanied by variation in ecological behavior and physiological response, thus making "almost every species a distinct entity requiring separate investigation." Key ecological, morphological, and anatomical features of 26 of the more common plants in the Sonoran Desert were thus provided, as were maps of their geographical distributions where possible. In highlighting the "individuality" of the plants, the descriptions served as an important supplement to the description of the vegetation.

The seminal insights in *Vegetation of the Sonoran Desert* (Shreve 1951) have been extended and refined by subsequent research. Three publications during the last 48 years serve as excellent examples. A direct extension was the monumental *Vegetation and Flora of the Sonoran Desert* (Shreve and Wiggins 1964). The two-volume treatise, in which Shreve's earlier analysis of the vegetation was combined with Wiggins' description of the flora, became the standard reference on plant life in the Sonoran Desert. A modified version of the major subdivisions of Sonoran Desert vegetation was proposed by Turner and Brown (1982). Six of the subdivisions in Shreve (1951) were retained (Lower Colorado Valley, Arizona Upland, Plains of Sonora, Central Gulf Coast, Vizcaíno Region, and Magdalena Region), with the Foothills of Sonora subdivision being recognized as "thornscrub" rather than desert vegetation. Based on more extensive data on the composition and structure of the vegetation, Turner and Brown (1982) refined the boundaries of the six subdivisions and identified various "series," or plant communities, contained within them. In *Sonoran Desert Plants: An Ecological Atlas*, Turner et al. (1995) expanded directly on the earlier efforts of Shreve (1951) by providing detailed maps of the geographical distributions of 339 perennial plants of the Sonoran Desert. For each plant, Turner et al. (1995) also provided an "elevational profile," showing the distribution as a function of latitude and elevation, and identified key ecological, morphological, and physiological attributes, based in part on an extensive review of the published literature.

The present volume further explores the ecology of Sonoran Desert plants and plant communities, building on the foundation established by Shreve (1951) and others. McLaughlin and Bowers (chapter 1) analyze the diversity and affinities of the flora of the Sonoran Desert region. Understanding patterns of floristic diversity at a variety of scales is an important complement to understanding patterns of variation in the vegetation. McLaughlin and Bowers provide an overview of floristic research in the Sonoran Floristic Province, assess previous phytogeographic treatments, and review and extend recent numerical studies. They also estimate total species diversity and characterize the flora in terms of both its major elements and its largest families and genera. In addition, they analyze patterns in species diversity among local floras, providing a quantitative assessment of the relative importance of area, elevation, and geographic location to landscape diversity.

Búrquez et al. (chapter 2) examine the physical environment and vegetation of the Sonoran Desert, with a focus on the central and southern desert regions. They provide overviews of important climatic, geomorphic, and hydrologic patterns within the region, then examine vegetation and habitat diversity at a variety of scales, ranging from large-scale ecoclines associated with regional variation in temperature and precipitation to small-scale patterns associated with the microhabitats under tree canopies. They also analyze ecosystem dynamics in terms of primary productivity and standing crop biomass. In addition, they discuss the geographical limits of the Sonoran Desert, with a special focus on the transition to thornscrub in the southern region, and describe in detail the gradient in vegetation encountered along a transect from the Gulf of California to the foothills of the Sierra Madre.

McAuliffe (chapter 3) analyzes the relationship between landscape complexity and ecological diversity in the Sonoran Desert. He focuses on local variation in the composition of vegetation and its relationship to physical processes shaping the landscape. He examines the geological evolution of alluvial piedmonts (or bajadas) flanking mountain ranges in the Sonoran Desert, demonstrating that piedmonts are mosaics of alluvial deposits of different ages. He assesses the direct impacts of recent geological changes, such as alluvial aggradation and erosion, on populations and communities of long-lived desert plants. He also examines the complex mosaic of different soils that develop on the alluvial surfaces of different ages and discusses the influence of soil characteristics on the timing, quantity, and vertical distribution of available soil moisture. Using examples from the Arizona Upland, Vizcaíno Region, and Lower Colorado Valley subdivisions, he then analyzes how the temporal and spatial distributions of soil moisture in different kinds of soils exert a strong control over the predominance of different kinds of plants in the landscape.

Venable and Pake (chapter 4) examine the population dynamic properties of Sonoran Desert annual plants. Annual plants are highly responsive to the environmental variation characteristic of the desert, particularly the high level of variation in the amount and seasonal distribution of rainfall. Venable and Pake document long-term patterns of germination, survival, and reproduction of annual plants at a site in the Arizona Upland subdivision. They analyze the relationships between variance in reproductive success, germination fractions, and seed size. In addition, they examine mechanisms of species coexistence mediated by temporal variance in the environment, assessing in particular the roles of persistent seed banks and predictive germination. Using experimental manipulations, they also assess patterns of seed dispersal, and evaluate the relative importance of seed dispersal, seed banks, and *in situ* reproduction from the previous year.

Nobel and Loik (chapter 5) analyze the morphological and physiological mechanisms by which cacti are adapted to the characteristic droughts and thermal extremes of the Sonoran Desert. The diversity of cacti is a distinguishing feature of the region, with species differing greatly in form. Nobel and Loik assess the influence of stem morphology on tissue temperatures (especially near the apex), the ability to compete for light, and water storage capacity. They also examine the effects on stem temperature and light interception of features such as stem orientation, spines, and pubescence. In addition, they highlight the trade-offs that ribs and tubercles represent in terms of water storage, stem temperature, light interception, and photosynthetic CO_2 uptake. With respect to the belowground environment, they examine patterns of root growth and branching in the context of water uptake and assess the mechanisms that reduce loss of water from the roots during soil drying. They also document the influence of "nurse plants" on the thermal and light environments experienced by seedlings.

Chapters 1–5 discuss the plant life of the Sonoran Desert across spatial scales ranging from the landscape level to the organismal level, paralleling the broad range of scales in Shreve (1951) and Shreve and Wiggins (1964). In Chapters 6–8, the intimate link between plant and animal life in the Sonoran Desert is highlighted, including the special role of humans in altering the landscape.

Etges et al. (chapter 6) explore the relationship between Sonoran Desert plants and the animals that use them as feeding and breeding resources. The diversity of cacti in the Sonoran Desert has shaped a wide variety of animal-plant associations, including many insect species that have become highly dependent on living or dead plant tissues. One of the more intriguing associations involves pomace flies (*Drosophila*), which feed on bacteria, yeast, and cell sap contained in the decaying tissues of host cacti. Etges et al. analyze patterns of karyotypic (or genetic) variation in two species of *Drosophila* in relation to major subdivisions of Sonoran Desert vegetation. They also analyze karyotypic variation in relation to host cacti, given that multiple hosts are used over the geographical ranges of the flies. In addition, they examine the correlation between heterozygosity and climate, specifically annual temperature range. Using genetic distance trees, they assess the evolutionary history of the flies in relation to host cacti and major vegetation subdivisions.

Jackson and Comus (chapter 7) analyze the ecological consequences of agricultural development in the Sonoran Desert. The changes wrought by intensive modern agriculture on lowland desert vegetation have been profound. Focusing on the large Santa Cruz Valley in the northern Sonoran Desert, Jackson and Comus examine the characteristics of native vegetation prior to modern agricultural development. They also analyze the timing and extent of clearing for agriculture and the pattern of subsequent abandonment of farmland, using data from aerial photos, topographic maps, agricultural censuses, and field visits. With additional insights from soil surveys and personal interviews, they assess the hydrologic and ecological effects of agricultural development in terms of altered surface water flow patterns and habitat fragmentation. They conclude by discussing the implications of the profound changes in terms of ecosystem recovery.

Martin (chapter 8) examines the major imprint that 40 000 years of change in climate, vegetation, megafauna, and ancient cultures have left on the modern biota of the Sonoran Desert region. He reviews data on patterns of climate change in the late Quaternary, providing insight into the aberrant nature of Holocene climatic conditions. Capitalizing on a wealth of fossil data obtained from ancient packrat middens, much of which derives from research at the Desert Laboratory, he analyzes changes in the vegetation of the region that accompanied the changes in climate. In addition, he analyzes the catastrophic late-Quaternary extinction of large mammals, especially large herbivores, in the region and assesses the central role played by early human hunters. In a provocative conclusion, he examines the possibility that a clearer understanding of deep history may alter our approach to some traditional conservation issues in the Sonoran Desert region.

By design, the present volume covers a broad range of spatial and temporal scales in exploring the ecology of Sonoran Desert plants and plant communities. It thus samples the diversity of research actively being pursued in the region. At almost any one scale, contributions from other scientists would be sufficient to fill additional edited volumes. Through their collective efforts, contributors to this volume and other scientists are providing great insight into the diversity of plant life in the Sonoran Desert and into the ecological patterns and processes that underlie it, thereby continuing the great tradition established by the early Desert Laboratory scientists.

Publication of the present volume closely follows the 95th anniversary of the founding of the Desert Laboratory. The volume celebrates that anniversary, paying tribute to an institution with a long and distinguished history. As editor, I gratefully acknowledge the support given to the Desert Laboratory and to this volume by a host of individuals. May our lives, and those of our children, continue to be enriched by the exquisite diversity of plant life that surrounds us in the Sonoran Desert.

Literature Cited

Shreve F. (1951) *Vegetation of the Sonoran Desert.* Carnegie Institution of Washington Publication no. 591, Washington, D.C.

Shreve F., Wiggins I. L. (1964) *Vegetation and Flora of the Sonoran Desert.* Stanford University Press, Stanford.

Turner R. M., Bowers J. E., Burgess T. L. (1995) *Sonoran Desert Plants: An Ecological Atlas.* University of Arizona Press, Tucson.

Turner R. M., Brown D. E. (1982) Sonoran desertscrub. *Desert Plants* 4: 181–221.

1 Diversity and Affinities of the Flora of the Sonoran Floristic Province

Steven P. McLaughlin and Janice E. Bowers

This chapter is concerned with floristic features of the Sonoran Desert region. We first contrast flora and vegetation and discuss the scales appropriate for floristic work. We then review floristic work in the Sonoran Desert region and discuss the various phytogeographic treatments that have been applied to the Sonoran Floristic Province, which includes the warm deserts of California, Nevada, Arizona, Sonora, and Baja California (fig. 1.1). We provide a preliminary tabulation of the total species diversity of the Sonoran Floristic Province and describe the flora in terms of its major elements and most important families and genera. Finally, we investigate patterns in species diversity at the landscape scale using 25 local floras from the Sonoran Floristic Province.

Flora and Vegetation

The plant life of an area can be described in terms of its vegetation or its flora. Vegetation refers to attributes of the plant community such as dominant species, physiognomy, cover, density, and biomass. Flora refers to the total species composition; that is, all vascular plant species occurring in a particular plot, area, or region of interest. Botanists are occasionally lax in making this distinction. Benson (1982), for example, referred to vegetation types as "floras or floristic associations."

Phytogeographic treatments are most often based on vegetation. Shreve's (1951) subdivisions of the Sonoran Desert, for example, are based primarily on the relative dominance of different life-forms. Shreve

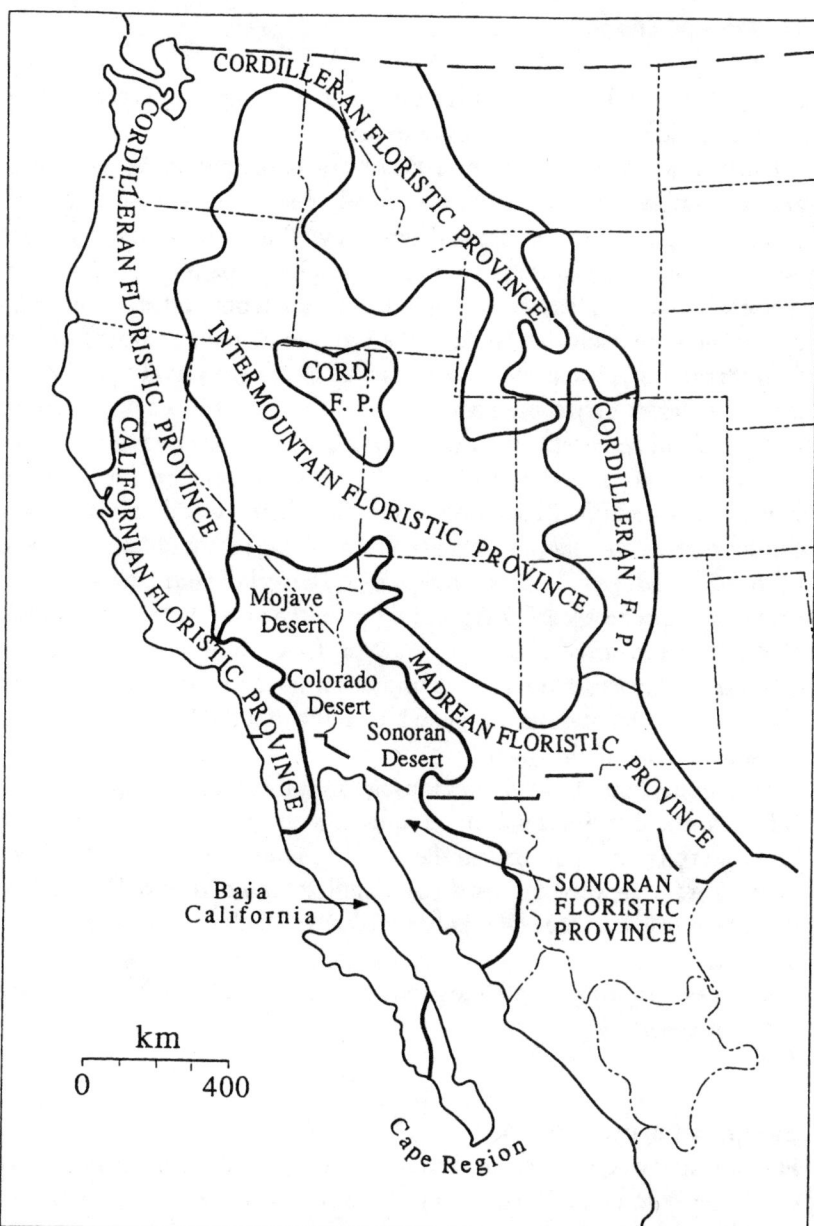

Figure 1.1 Approximate boundaries for the Sonoran Floristic Province, with geographic place names as traditionally applied to various desert areas within the province. Floristic provinces adjacent to the Sonoran Floristic Province in the United States are also identified.

thought that subdivisions of a region based on geography, vegetation, or flora "coincide to a greater or lesser extent."

Floristic areas and vegetation types are both mappable, phytogeographic entities. Floristic areas are usually delineated as continuous regions having a characteristic and distinctive flora. Vegetation types are discontinuous regions with characteristic physiognomy and dominant species. Boundaries between floristic areas are seldom sharply defined; an exception is the boundary between the Sonoran Floristic Province and the Cordilleran Province at the eastern base of the Sierra Nevada in California. Boundaries between adjacent vegetation types (for example, forest and grassland or woodland and desert) are often sharp enough to allow detailed mapping at fairly small scales. Within a floristic province, various vegetation formations may occur; pinyon-juniper woodland and even conifer forests, for example, occur within the higher mountain ranges in the northern Mojave Desert (Clokey 1951; DeDecker 1984), which is part of the Sonoran Floristic Province. Likewise, the vegetation characteristic of a particular floristic province may be found in adjacent provinces; Sonoran Desert vegetation, for example, occurs in the Rincon Mountains (Bowers and McLaughlin 1987), which lie within the Madrean Floristic Province (McLaughlin 1992).

Traditional boundaries of floristic regions sometimes are drawn to coincide with the distributional limits of dominant species. MacMahon and Wagner (1985) maintained that the northern limits of creosote bush (*Larrea tridentata*) marked the northern boundary of the Mojave Desert, that the range of Joshua tree (*Yucca brevifolia*) was indicative of the Mojave Desert (except in parts of Arizona), and that the ranges of blue paloverde (*Cercidium floridum*) and ironwood (*Olneya tesota*) closely defined the Sonoran Desert.

Scale Considerations

Floristic studies generally are conducted at a broader spatial scale than vegetation studies. The basic unit of floristic research is the local flora — a checklist for a particular circumscribed area, generally smaller than an average-size county. Examples include both artificially delineated areas such as wildlife refuges, national parks and monuments, and regional parks, and natural areas such as small mountain ranges, large watersheds, and topographic basins. Regional floras span a much broader scale and usually encompass political entities such as states or provinces; Wiggins' (1964) binational flora of the Sonoran Desert is a notable exception.

Vegetation studies sample a small fraction of the local and regional flora, generally no more than 10 to 100 species. These species constitute a highly biased sample of the regional flora, weighted heavily to the most abundant woody and herbaceous plants, which also tend to be widespread species. Local floras typically range from about 100 to 1000 species, and regional floras usually have from 1000 to 10 000 species.

Observations regarding species diversity are highly scale-dependent, and the terminology for describing diversity (Whittaker 1977) reflects this. Alpha diversity refers to the number of species within a habitat (in practice, often a research plot), while beta diversity refers to the variation in species composition between habitats. Gamma diversity refers to the total number of species in a landscape; conceptually, it is the product of alpha and beta diversities. In floristic research, gamma diversity corresponds closely to the scale of local floras. Delta diversity, analogous to beta diversity, refers to the variation in species composition along climatic gradients, between landscapes, or between local floras. Epsilon diversity refers to regional diversity, the total number of species in a floristic province or larger area (Whittaker 1977).

Floristic Research in the Sonoran Floristic Province
Early Exploration

Floristic work in the Sonoran and Mojave desert regions of the United States essentially began with Frederick Coville's 1891 expedition to Death Valley, California (Kurzius 1981; Schramm 1982). Ever since, botanists have been fascinated with the taxonomic diversity and morphological complexity of desert plant life in the southwestern United States and northwestern Mexico. Coville's catalog of plants from Death Valley and the surrounding mountain ranges (Coville 1893) was the first local flora from the Sonoran Floristic Province. Other early floristic studies treated the Desert Laboratory grounds in the Tucson Mountains of Pima County, Arizona (Thornber 1909), and the Salton Sink in southeastern California (Parish 1914). Shreve (1915) included a plant list for the Santa Catalina Mountains, which have desert vegetation on their lower slopes, in his classic study of their vegetation.

Several of these early floristic projects were connected in one way or another with the Desert Laboratory (Bowers 1988, 1990), a tradition that continues today with the floristic research of Janice Bowers, Raymond Turner, and Paul Martin. The scope of floristic work undertaken at the Desert Laboratory has extended well beyond the boundaries of the Sonoran Floristic Province. Shreve planned comprehensive treatments of the Sonoran, Mojave, Great Basin, and Chihuahuan Deserts (Bowers 1988). Howard Gentry was based at the Desert Laboratory while working on his flora of the Rio Mayo area in southern Sonora (Gentry 1942); Paul Martin and his collaborators have continued work in this subtropical area. Bowers' research interests extend well into the Apachian District of the Madrean Floristic Province, including surveys of the Rincon Mountains (Bowers and McLaughlin 1987) and Huachuca Mountains (Bowers and McLaughlin 1996).

Regional Floras

Major portions of the Sonoran Floristic Province are treated in the floras for Arizona (Kearney and Peebles 1960), California (Munz 1935, 1974; Munz and Keck 1959; Hickman 1993), and northern and southern Baja California (Wiggins 1980). Wiggins' flora of the Sonoran Desert covers the desert portions of Sonora, Baja California, Arizona, and extreme southeastern California, excluding the most northern portions of the Sonoran Floristic Province (Wiggins 1964). DeDecker (1984) provides a regional treatment of the flora of the northernmost portion of the Sonoran Floristic Province. Felger (1992) gives a checklist for northwestern Sonora, a large area with a relatively small flora. Portions of the Sonoran Floristic Province are also covered in the floras for Kern County, California (Twisselmann 1967); San Diego County, California (Beauchamp 1986); and central-southern Nevada (Beatley 1976).

Local Floras

Local floras have been published for many sites within the Sonoran Floristic Province of the United States (table 1.1, fig. 1.2). A few from the Mexican portion of the province have also been published (Felger and Lowe 1976; Felger 1980; Moran 1983a, 1983b). In addition to the published floras listed in table 1.1, there are many unpublished reports and checklists; some of the latter are listed in Bowers 1982. Local flora projects conducted in the Sonoran Floristic Province mostly have been surveys of small, isolated mountain ranges.

Table 1.1 Published local floras from the Sonoran Floristic Province of the southwestern United States. Map numbers refer to figure 1.2.

Map No.	Area[a]	Reference(s)
1	Grapevine Mountains	Kurzius (1981)
2	Funeral Mountains	Annable (1985)
3	Cottonwood Mountains	Peterson (1984)
4	Death Valley NP	Coville (1893); Norris (1982)
5	Black Mountains	Schramm (1982)
6	Nevada Test Site	Beatley (1976)
7	Spring Range	Clokey (1951)
8	Beaver Dam Mountains	Higgins (1967)
9	Lower Grand Canyon	Phillips (1975)
10	Muddy Mountains	Swearingen (1981)
11	McCullough Mountains	Bostick (1973)
12	Newberry Mountains	Holland (1982)
13	Kingston Range	Thorne et al. (1981)
14	Clark Mountain Range	Thorne et al. (1981)
15	New York Mountains	Thorne et al. (1981)
16	Providence Mountains	Thorne et al. (1981)
17	Granite Mountains	Thorne et al. (1981)
18	Joshua Tree NP	Adams (1957); unpublished updates
19	Salton Sink	Parish (1914)
20	Eastern Imperial County	McLaughlin et al. (1987)
21	Castle Dome Mountains	Russo (1987)
22	Cabeza Prieta NWR	Simmons (1966); Russo (1987)
23	Buckeye Hills Recreation Area	Pierce (1979)
24	White Tank Mountains	Keil (1973)
25	Lake Pleasant RP	Lehto (1970)
26	McDowell Mountain RP	Lane (1981)
27	South Mountains	Daniel and Butterwick (1992)
28	Sierra Estrella RP	Sundell (1974)
29	Tonto NM	Burgess (1965)
30	Organ Pipe Cactus NM	Bowers (1980); Pinkava et al. (1992)
31	Tucson Mountains	Bowers and Turner (1985); Rondeau (1991)

[a] NM = National Monument; NP = National Park; NWR = National Wildlife Refuge; RP = Regional Park.

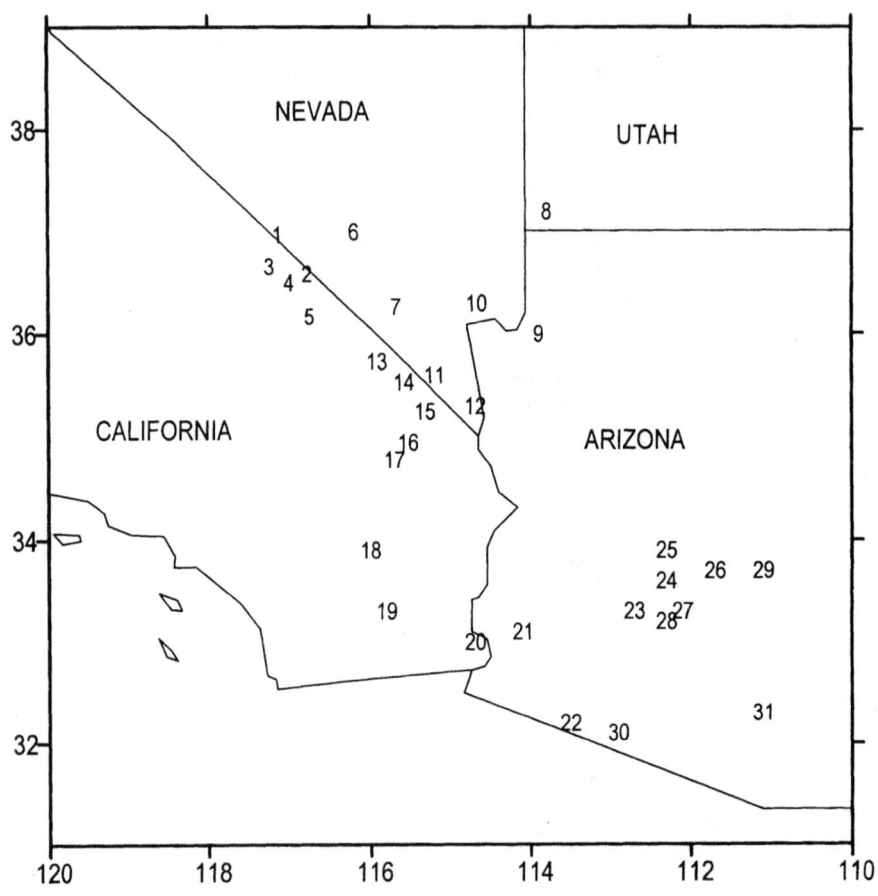

Figure 1.2 Locations for published local floras from the Sonoran Floristic Province of the southwestern United States. Map numbers are explained in table 1.1.

Not unexpectedly, since local floras have long been regarded as a suitable topic for master's theses, centers of local flora research are associated with major universities, most notably Arizona State University (Lehto 1970; Keil 1973; Sundell 1974; Pierce 1979; Lane 1981; Russo 1987) and the University of Nevada at Las Vegas (Bostick 1973; Kurzius 1981; Swearingen 1981; Holland 1982; Schramm 1982; Peterson 1984; Annable 1985). There thus has been a concentration of floristic work around Phoenix, Arizona, and in southern Nevada and around Death Valley (fig. 1.2). The Colorado Desert area of southeastern California, the western portions of the Mojave Desert, and the desert portions of northwestern Arizona are poorly sampled by local floras.

Modern Floristic Patterns
Chorology
Previous Phytogeographic Treatments
California botanists have long subdivided the desert region of southeastern California into a northern Mojave Desert and a southern Colorado Desert (Parish 1930; Munz 1935). Parish (1930) placed the boundary between the two deserts at the Chuckwalla Mountains while Munz (1935) put it somewhat farther north along the Little San Bernadino, Cottonwood, and Eagle Mountains, which together form an eastward extension of the San Bernadino Mountains. Shreve (1942) placed this boundary approximately at the Riverside–San Bernadino County line.

The geologist William Blake first proposed the name "Colorado Desert" in 1853 but applied it only to the Salton Sink and adjacent dunes, specifically excluding the adjacent arid mountain slopes (Blake 1914). Nevertheless, botanists traditionally include all of southeastern California directly east of the peninsular ranges in the Colorado Desert; this area lies mostly at low elevations, from below sea level up to about 600 m. The Mojave Desert, which lies east of the Sierra Nevada and north of the transverse ranges, occurs at higher elevations, mostly between 600 and 1500 m; the highest ranges in the northern Mojave Desert exceed 3000 m in altitude.

Parish (1930) estimated that about half the xerophytic flora of the Mojave Desert enters the Colorado Desert, while approximately one third of the flora of the Colorado Desert does not reach the Mojave Desert. Parish (1930) and Munz (1935) suggested that affinities of the Colorado Desert flora are with the deserts of southwestern Arizona and northern Mexico, while those of the Mojave Desert flora are with the Great Basin.

Forrest Shreve, the premier student of desert vegetation in North America, had several changes of opinion regarding the chorology of the deserts. In 1924 he described the Sonoran Desert, a region he considered quite distinct from the Colorado and Mojave Deserts in California, as a natural province bounded on the west by the Colorado River and the Gulf of California. A year later, he suggested that the distinction between the Colorado and Mojave Deserts was more geographic than botanical, but that the desert north of Death Valley was distinctive botanically (Shreve 1925). In this same paper, he also noted that the desert vegetation changed rapidly in going from the west to the east side of the Colorado River into Arizona. Later still, he recognized the Mojave Desert as a separate region and added the Colorado Desert and the deserts of Baja California to the Sonoran Desert (Shreve 1942). His final assessment (Shreve 1951) divided the Sonoran Desert into seven subdivisions, only two of which occur in the United States — the Lower Colorado Valley, extending east to the vicinity of Gila Bend and Ajo, Arizona, and the Arizona Upland, in the eastern and northern portions of the area.

Dice (1939) initially included both the Sonoran and Mojave desert regions in his Sonoran Biotic Province, but later he recognized Sonoran, Mohavian, and Sanlucan (Baja California) Biotic Provinces within the area considered here to be the Sonoran Floristic Province (Dice 1943). Biotic provinces, as defined by Dice, are continuous areas with characteristic vegetation, flora, and fauna. Dice's Sanlucan Province covered all of the peninsula south of the 30° lat., including the Cape Region, which Shreve (1942, 1951) excluded from the Sonoran Desert. Dice (1943) stated that the Mohavian Biotic Province differed from the Sonoran Biotic Province in climate, vegetation, and fauna. Regarding the flora, Dice noted only that cacti were largely lacking in the Mohavian Biotic Province because of its extreme climate.

Stebbins and Major (1965) included the Sheephole Mountains of San Bernadino County and the Granite Mountains of Riverside County in the Colorado Desert. They also recognized a third desert region in California north of the Mojave Desert, which they designated the "Inyo" region because it coincides closely with the desert portions of Inyo County. The Inyo subdivision of Stebbins and Major (1965) is roughly equivalent to the northern Mojave Desert of DeDecker (1984).

Diversity and Affinities of the Flora   21

Takhtajan (1986, treatment prepared by Arthur Cronquist) broadly defined the Sonoran Province to include the Sonoran and Mojave Deserts as defined by Shreve (1942), all of Baja California (including the Cape Region), the Chihuahuan Desert, and the Tamaulipan area of northeastern Mexico. Baja California is treated as a separate subprovince; the deserts of Sonora, Arizona, California, and southern Nevada constitute a second subprovince.

Early botanists were not in agreement regarding the affinities of the desert flora. Coville (1893) noted that "the arid plants of . . . [Death V]alley, almost without exception, are species which extend far southward into Mexico through the States of Sonora and Chihuahua, and which are there confined to the arid interior plateau of that country." Parish (1914) commented that "the one prominent fact in the distribution of the xerophytic flora of the desert is the successive dropping out of species as one proceeds from the southeast to the northwest." Shreve (1915), working in the Santa Catalina Mountains, stated that the affinities of the desert flora of southeastern Arizona lie mostly with the Sonoran/Sinaloan regions to the south and with the Chihuahuan region to the southeast, with little relationship to the Mojave region to the northwest.

Recent Numerical Studies

Local floras sample the regional flora in the same sense that plots or transects sample vegetation; such floras can be combined into a database for numerical analysis of regional floristic patterns (McLaughlin 1986, 1989, 1992, 1994). As samples of the regional flora, local floras are subject to several forms of sampling error, including overlooked taxa and misidentifications. So long as such sampling errors are random, they do not invalidate the use of local floras in quantitative, regional studies. Nonrandom or systematic errors include inadequate collecting in certain habitats, at certain times of year, or of certain families, any of which can result in the under-representation of particular floristic elements in the checklist.

McLaughlin (1986) conducted a numerical chorological study of the Basin and Range and Colorado Plateau physiographic provinces of the southwestern United States based on the distributions of native species in 50 local floras. For a later paper (McLaughlin 1989), the database was expanded to 96 floras from the entire western United States west of the Great Plains. In a third study, the database was increased to 101 local floras and subdivided into three components—widespread, regional ("intermediate"), and narrow species (McLaughlin 1992). Widespread species were used to delineate five floristic provinces, regional species were used to delineate nine floristic subprovinces, and narrow species were used to delineate 20 floristic districts. These floristic areas, defined at different scales, approximated a nested hierarchy (McLaughlin 1992).

McLaughlin's (1986) analysis of local floras from the Basin and Range and Colorado Plateau physiographic provinces supported the traditional division of the desert into Mojavean and Sonoran floristic areas, with the Colorado Desert region included in the latter. The analysis using all species from the western United States combined the Sonoran and Mojave desert areas into a single floristic area (McLaughlin 1989). The third study recognized a single Sonoran Floristic Province composed of three floristic districts: a California ("Colorado-Mojave") Desert District, combining the traditional Colorado Desert area and the southern part of the Mojave Desert; an Arizona Upland District in southwestern Arizona; and a Northern Mojave Desert District (McLaughlin 1992). The latter result agrees with Shreve's (1924, 1925) earliest impressions of the chorological relationships in the southwestern deserts.

The "Sonoran floristic area" of McLaughlin (1986) actually corresponded quite closely to Shreve's Lower Colorado Valley subdivision of the Sonoran Desert (Shreve 1951), which extends far east of the Colorado River into Arizona at lower elevations. However, none of the floras used by McLaughlin (1986, 1992) lie wholly within the Arizona Upland subdivision as narrowly defined by Shreve (1951). In none of the analyses (McLaughlin 1986, 1989, 1992) is the Colorado Desert of southeastern California recognized as a distinct entity. In McLaughlin (1986), the two local floras from the Colorado Desert area are united with the floras of southwestern Arizona, and in McLaughlin (1992), they cluster with the floras of the Mojave Desert in the analysis of narrow species.

The Northern Mojave Desert District of McLaughlin (1992) corresponds reasonably closely to the "Inyo subdivision" used by Stebbins and Major (1965) in their study of California endemics and with the Northern Mojave Desert as defined by DeDecker (1984). Some local floras from southern Nevada, southeastern Utah, and northwestern Arizona are also included in the Northern Mojave Desert District of McLaughlin (1992). Stebbins and Major (1965) included some of the higher ranges of the eastern Mojave Desert (Thorne et al. 1981) in the Inyo area; these mountains fall within the California Desert District of McLaughlin (1992).

Species can be numerically grouped into floristic elements based on the principal-components analysis of local floras (McLaughlin 1994). The affinities of both local and regional floras can then be characterized in terms of the percentage composition of their various floristic elements; we have done this for 17 local floras from the Sonoran Floristic Province (see table 1.2). We selected these 17 to provide representative geographic coverage across the entire province. Based on the narrow-species elements that are most abundant in each flora, the first five local floras are placed in the Northern Mojave Desert District, the next five in the California Desert District, and the last seven in the Arizona Upland District.

The percentage of all Sonoran elements in these floras ranges from 44% to 79%. Widespread Sonoran species are those occurring throughout the Sonoran Floristic Province and extending well into adjacent floristic provinces; this element averages 11% for the 17 local floras in table 1.2. Regional Sonoran species, those largely restricted to but found throughout the Sonoran Floristic Province, average 25% for this sample of 17 local floras. The combined contribution of narrow species in the Northern Mojave Desert, California Desert, and Arizona Upland elements averages 27% and ranges from 9% for Tonto National Monument to 38% for the Black Mountains of Death Valley.

The Spring Range, which rises to an elevation of 3600 m, has a correspondingly large Cordilleran element. The Intermountain element constitutes a large fraction of the floras in the Northern Mojave Desert and California Desert Districts. The Californian element is predictably high in the one flora from the western edge of the Sonoran Floristic Province (Joshua Tree National Park); it is also high in the Eastern Imperial County flora, where aquatic habitats are important, and at Tonto National Monument in central Arizona. In general, the Californian element is best represented in Arizona in local floras from the central part of the state below the Mogollon Rim. Finally, the percentage of Madrean species is high in all floras from the Arizona Upland District.

The analyses of table 1.2 help to further define the floristic boundaries between the subdivisions of the Sonoran Floristic Province, at least in the United States. Based on these studies, the boundary between the California Desert and the Northern Mojave Desert Districts in California can be placed somewhere between the Black Mountains and Grapevine Mountains in California. In southern Nevada the dividing line occurs between the Newberry Mountains and the Muddy Mountains (closer to the latter),

Table 1.2 Floristic analyses of 17 local floras from within the Sonoran Floristic Province of western North America.

Flora[a]	Regional Elements (% of Total Flora)					Sonoran Elements (% of Total Flora)				
	CO	IN	CA	MA	SO	WIDE	REG	NMD	CD	AUP[b]
Northern Mojave Desert District										
Spring Range	21	19	4	11	45	7.6	14.5	17.9	5.1	0.3
Beaver Dam Mountains	7	27	4	16	46	9.0	19.9	11.9	4.4	0.8
Grapevine Mountains	9	18	5	4	64	8.2	19.8	21.6	14.1	0.4
Lower Grand Canyon	2	11	4	17	66	13.8	32.8	9.1	7.9	2.4
Muddy Mountains	2	15	7	12	64	10.9	26.5	12.8	12.3	1.4
California Desert District										
Black Mountains, Death Valley NP	2	11	5	3	79	8.8	30.0	14.3	23.5	0.3
Kingston Range	4	12	5	9	70	10.7	25.2	12.5	20.8	0.4
Newberry Mountains	1	9	6	10	74	13.8	34.5	4.1	21.0	1.0
Joshua Tree NP	3	8	12	9	68	9.0	22.5	1.8	32.5	2.5
Eastern Imperial County	1	5	10	9	75	11.9	30.2	0.0	24.8	7.6
Arizona Upland District										
Castle Dome Mountains	2	2	2	20	75	12.1	32.7	2.3	12.8	15.7
Cabeza Prieta NWR	1	4	5	22	68	10.7	26.5	0.6	14.3	16.4
South Mountains	1	2	6	20	71	14.3	30.6	0.8	6.1	18.8
Organ Pipe Cactus NM	1	3	7	29	60	9.2	18.8	0.9	5.0	25.7
McDowell Mountain RP	2	3	8	25	62	14.6	29.2	0.0	2.7	15.4
Tucson Mountains	1	3	6	43	47	9.0	16.1	0.4	2.7	19.1
Tonto NM	4	5	11	36	44	15.3	19.4	0.8	0.8	7.3

[a] NM = National Monument; NP = National Park; NWR = National Wildlife Refuge; RP = Regional Park.
[b] AUP = Arizona Upland; CA = Californian; CD = California Desert; CO = Cordilleran; IN = Intermountain; MA = Madrean; NMD = Northern Mojave Desert; REG = Regional Sonoran; SO = Sonoran; WIDE = Widespread Sonoran.

and between the Spring Range of Nevada and the Kingston Range of California. There is clearly a gradual transition between the floras of these two elements in California.

In contrast, the boundary between the California Desert and the Arizona Upland Districts is clearly at the Colorado River, much further west than Shreve's (1951) placement based on vegetation. Floras on either side of the Colorado River have similar percentages of widespread and regional Sonoran elements, but the percentage for the Arizona Upland element is much higher east of the Colorado River while that of the California Desert element is much higher west of the river. The Madrean element also increases markedly on the east side of the Colorado River. Russo

(1987) noted that Arizona Upland vegetation and Lower Colorado Valley vegetation showed little difference in species composition at her study site in the Castle Dome Mountains in western Yuma County. She also noted that the Castle Dome Mountains shared more species with the White Tank Mountains far to the east than with the Eastern Imperial County area just across the Colorado River.

The northern boundary of the Northern Mojave Desert District cannot be well defined since there are relatively few local floras available from the adjacent western Great Basin region to the north.

Species Diversity

The total (epsilon) species diversity of the Sonoran Floristic Province is low compared to more mesic areas (Shreve 1925). Parish (1930) conducted a census of the "xerophilous" flora of the Mojave Desert, estimating its size at 545 species. Takhtajan (1986) estimated that the floras of his Baja California and Sonoran Desert subprovinces consisted of 2500 to 3000 species each, with considerable overlap in species composition.

We have estimated the total number of species occurring in the Sonoran Floristic Province (table 1.3). Our upper estimate of the flora is based on native species listed either in local floras or in regional floras from within the Sonoran Floristic Province, especially Wiggins (1964), and also includes DeDecker's (1984) flora of the Northern Mojave Desert, Beatley's (1976) flora of central-southern Nevada, and Munz's (1974) flora of southern California. Wiggins (1964) included many species which occur primarily in the Madrean Floristic Province and in the Cape Region. About 250 of these probably do not occur in the Sonoran Floristic Province; excluding them produces our lower estimate of the flora. We have also totaled the number of species for the Sonoran Floristic Province within the United States in table 1.3.

Our classification of floristic elements produced using numerical analyses of local floras can only be applied to plant species occurring north of the international border, that is, within the study area of McLaughlin (1992). For those species listed by Wiggins (1964) that are not found within the United States, we subjectively assigned species to elements based on Wiggins's descriptions of ranges. Many of these species could be readily assigned to the Californian and Madrean elements. Species that range far south of the Sonoran Floristic Province are classified in table 1.3 as Neotropical-Widespread; many species found within the United States

Table 1.3 Numerical summary of the flora of the Sonoran Floristic Province.

| | United States and Mexico | | | | United States Only | |
| | Upper Estimate | | Lower Estimate | | | |
Floristic Element	N	(%)	N	(%)	N	(%)
Cordilleran	263	(7.7)	254	(8.0)	245	(11.6)
Intermountain	256	(7.5)	256	(8.0)	249	(11.8)
Californian	414	(12.1)	414	(13.0)	264	(12.5)
Madrean	629	(18.4)	392	(12.3)	381	(18.1)
Sonoran						
Widespread	54	(1.6)	54	(1.7)	54	(2.6)
Regional	157	(4.6)	157	(4.9)	157	(7.4)
Northern Mojave Desert	280	(8.2)	280	(8.8)	280	(13.3)
California Desert	300	(8.8)	300	(9.4)	284	(14.2)
Arizona Upland	179	(5.2)	179	(5.6)	179	(8.5)
Mexican-Sonoran Desert	131	(3.8)	131	(4.1)	—	—
Sanlucan	298	(8.7)	298	(9.4)	—	—
Total Sonoran	1399	(40.8)	1399	(44.0)	970	(46.0)
Neotropical						
Widespread	383	(11.2)	383	(12.0)	—	—
Cape Region	83	(2.4)	83	(2.6)	—	—
Total Species	3427		3181		2109	

portion of the Sonoran Desert probably belong in this category as well. Baja California endemics that occur primarily in the Cape Region but extend into desert habitats are classified in table 1.3 as a Neotropical-Cape Region element; those occurring primarily in the desert areas are classified as a Sanlucan element of the Sonoran Floristic Province. Finally, species occurring in the Sonoran Floristic Province and restricted to Sonora, Mexico, or found both in Sonora and Baja California, are classified in table 1.3 as a Mexican-Sonoran Desert element. An objective treatment of the Mexican portion of the Sonoran Floristic Province will be possible only when more local floras become available from the region.

The total flora of the Sonoran Floristic Province is 3200 to 3400 species; there are about 2100 species in the Sonoran Floristic Province of the United States. The total size of the flora, and the contribution of species from adjacent elements, depends on precisely where the boundaries between adjacent provinces are placed. The size of the Sonoran element, those species occurring primarily in the Sonoran Floristic Province, can be determined more precisely. The Sonoran element comprises approximately 1000 species north of the international border and about 1400 species from the entire Sonoran Floristic Province. Of these, about 50 are widespread Sonoran species and about 150 are regional Sonoran species. The Sanlucan, California Desert, and Northern Mojave Desert narrow-species elements each consist of about 300 species, as does the combined Arizona Upland and Mexican-Sonoran elements. The Arizona Upland District and most of the Sonoran Desert of northwestern Sonora likely are parts of a single floristic district.

Shreve (1925) estimated that only 3% of the flora of the southern Arizona deserts extended into the Mojave Desert. This estimate is clearly low, since the widespread and regional Sonoran elements, which occur in both areas, together constitute about 10% of the total flora of the Sonoran Floristic Province north of the international border (table 1.3), and account for 20% to 50% of the flora in any local area within the Sonoran Floristic Province (table 1.2).

We can further characterize the flora of the Sonoran Floristic Province by the largest families and genera found there. The Asteraceae (497 spp.), Fabaceae (335 spp.), and Poaceae (200 spp.) are the largest families, as is the case throughout most of North America. The next two largest families, the Cactaceae (133 spp.) and the Euphorbiaceae (119 spp.), are primarily Neotropical. The five next largest families are the Scrophulariaceae (113 spp.), Polygonaceae (108 spp.), Polemoniaceae (92 spp.), Brassicaceae (90 spp.), and Boraginaceae (86 spp.), all large families throughout the western United States. Most of the species in the largest genera, such as *Eriogonum* (69 spp.), *Astragalus* (60 spp.), *Phacelia* (40 spp.), *Cryptantha* (39 spp.), *Gilia s.s.* (37 spp.), *Chamaesyce* (36 spp.), *Mentzelia* (36 spp.), *Penstemon* (33 spp.), *Lupinus* (33 spp.), and *Camissonia* (28 spp.) are narrowly distributed annuals and herbaceous perennials. Large genera in the flora that contain many succulent or woody species are *Opuntia* (40 spp.), *Mammillaria* (38 spp.), *Dalea s.s.* (33 spp.), *Salvia* (31 spp.), *Brickellia* (26 spp.), *Atriplex* (21 spp.), and *Agave* (20 spp.).

Variation in landscape (gamma) diversity within the Sonoran Floristic Province has been discussed by many authors. Parish (1903) stated that the desert flora of southern California, in comparison with the floras of the Sierra Nevada and cismontane California, is notably depauperate in the number of herbaceous perennials. This is clearly the case for the flora of Eastern Imperial County (McLaughlin et al. 1987). Shreve (1925) pointed out that the Mojave Desert area of California is much more diverse than the Colorado Desert area. MacMahon and Wagner (1985) reported that the diversity of perennial species in the Sonoran Desert is highest in the Mexican portions, both in Sonora and Baja California, and that the diversity of annuals is highest in the Mojave Desert.

Stebbins and Major (1965) found a decrease in the number of endemic species in large genera in going from the Inyo area to the Mojave Desert to the Colorado Desert. Richerson and Lum (1980) also found a north-to-south gradient of decreasing regional species diversity in the California deserts. Shreve (1942) misidentified the southwestern portion of the Mojave Desert as the portion with the highest species diversity.

We provide here an analysis of species diversity using a sample of 25 local floras from the Sonoran Floristic Province; all except one (Sierra Rosario, Felger 1980) are from the United States. The sample represents recent (since 1950) mainland floras for which the number of native species, area, minimum and maximum elevations, and latitude and longitude could be readily determined. Areas and elevation ranges were usually provided by the authors of the local floras; where these were not given, we estimated them from topographic maps. Minimum elevation should provide a measure of aridity while maximum elevation should indicate the potential for moister habitats. Relief (maximum to minimum elevation) should be correlated with total habitat diversity (Bowers and McLaughlin 1982). We also calculated the natural log of area (Ln) and the product of the latitude and longitude (NWSE). The latter variable provides a measure of the local flora's position on a northwest to southeast gradient, which should parallel the northwest to southeast gradient in the ratio of winter to summer precipitation across the Sonoran Floristic Province in the United States.

Correlations between landscape diversity and the environmental variables are presented in table 1.4. These correlations provide a preliminary indication of the determinants of landscape diversity in the Sonoran Floristic Province. All variables except latitude are positively and significantly correlated with species number.

The correlations among independent variables help to better characterize the physical geography of the region. Ln area is significantly correlated with maximum elevation ($r = 0.43$, $p < .05$), relief ($r = 0.54$, $p < .05$), longitude ($r = 0.62$, $p < .001$), and NWSE ($r = 0.49$, $p < .05$); maximum elevation is correlated with relief ($r = 0.91$, $p < .001$), latitude ($r = 0.79$, $p < .001$), longitude ($r = 0.67$, $p < .001$), and NWSE ($r = 0.81$, $p < .001$);

Table 1.4 Correlations (r) between local gamma diversity (number of native species) and several environmental variables in the Sonoran Floristic Province.

Variable	r
Area	0.43*
Natural log of area	0.63**
Maximum elevation	0.71**
Minimum elevation	0.40*
Relief	0.63**
Latitude	0.33
Longitude	0.40*
Latitude × longitude	0.44*

*$p < .05$.   **$p < .001$.

and relief is correlated with latitude ($r = 0.78$, $p < .001$), longitude ($r = 0.76$, $p < .001$), and NWSE ($r = 0.87$, $p < .001$). Within this sample of 25 local floras, those areas with the highest maximum elevations are also those with the greatest relief, and they occur within the northwestern part of the region; that is, within the Northern Mojave Desert and the northern parts of the California Desert Districts. This greater habitat diversity in the northwestern sections accounts for the north to south decline in species diversity noted by Stebbins and Major (1965) and Richerson and Lum (1980), a gradient running counter to the global latitudinal gradient in species diversity.

We calculated a stepwise multiple regression to investigate the independent contributions of area, elevation, and geographic location to landscape diversity. The best equation contains three variables—maximum elevation (MAXELEV), latitude (LAT), and ln area (LNAREA)—and accounts for 74% of the variation in the number of species (SPP):

$$SPP = 1454 + 0.167(MAXELEV) - 44.1(LAT) + 25.0(LNAREA)$$

where maximum elevation is in meters, latitude is in degrees, and area is in square kilometers. R^2 changes for this model are 0.508 for maximum elevation ($p < .001$), 0.145 for latitude ($p < .01$), and 0.089 for ln area ($p < .05$). Since maximum elevation is highly correlated with relief in this region, it probably combines the influences of both habitat diversity and moisture availability, both of which should promote increased species diversity. It is interesting that the slope for latitude is negative. Once we correct for the fact that the more northern local floras are for areas that generally have more relief and reach higher elevations, we find a significant latitudinal gradient in species diversity that follows the expected trend, i.e., diversity increasing with decreasing latitude. Finally, area contributes significantly to species diversity in the Sonoran Floristic Province but is less important than maximum elevation and latitude.

We define "relative richness" (RR) in the same way as Bowers and McLaughlin (1982) — as the percentage difference between the observed (O) and expected (E) number of species: $[100 \times (O - E)/E]$. The expected number of species, determined using the multiple regression given above, was used to calculate the relative richness values for each of the 25 floras used in the analysis. The high relative richness of Joshua Tree National Park flora (RR = 37%) reflects both its proximity to the Californian Floristic Province and the large size of the California Desert element in its flora (table 1.2). In general, floras from the eastern part of the Arizona Upland District, such as McDowell Mountain Regional Park (RR = 26%), Tucson Mountains (RR = 21%), and Tonto National Monument (RR = 21%), have the highest relative richness, while floras from the Colorado Desert and the western part of the Arizona Upland District, such as Sierra Rosario (RR = −56%), Castle Dome Mountains (RR = −27%), and Cabeza Prieta National Wildlife Refuge (RR = −18%) have the lowest relative richness. Mojave Desert floras span the range of relative richness values.

A large portion of the variation in relative richness is undoubtedly attributable to collecting effort (Bowers and McLaughlin 1982). Unfortunately, compilers of local floras rarely provide the information necessary to meaningfully compare different floras in the amount of effort that has gone into compiling them. The number of collections would probably be suitable; number of person-days in the field would be even better. Even in relatively well-studied areas, such as Organ Pipe Cactus National Monument, additional effort inevitably results in finding species new to the flora (Pinkava et al. 1992).

Conclusions

The Sonoran Floristic Province probably has one of the best-known floras, especially for an arid area, of all of the earth's floristic provinces. Further exploration is unlikely to significantly change our understanding of the diversity and affinities of this region. Additional work on local floras, however, will help to further define the northern and southern boundaries of the province. The greatest need is for local floras from Mexico — in this case, from all parts of Sonora, Baja California, and Sinaloa. Additional local floras from the Colorado Desert area, the southwestern Mojave Desert, and the southwestern Great Basin, along with more floras from Mexico, are needed to better define the subdivisions of the Sonoran Floristic Province.

Literature Cited

Adams C. F. (1957) *Checklist of Plants of Joshua Tree National Monument.* Southwestern Monuments Association, Globe, Ariz.

Annable C. R. (1985) Vegetation and flora of the Funeral Mountains, Death Valley National Monument, California-Nevada. Master's thesis, University of Nevada, Las Vegas.

Beatley J. C. (1976) *Vascular Plants of the Nevada Test Site and Central-Southern Nevada: Ecologic and Geographic Distributions.* Energy Research and Development Administration, Washington, D.C.

Beauchamp R. M. (1986) *A Flora of San Diego County, California.* Sweetwater River Press, National City, Calif.

Benson L. (1982) *The Cacti of the United States and Canada.* Stanford University Press, Stanford.

Blake W. P. (1914) The Cahuilla Basin and desert of the Colorado. In: MacDougal D. T. (ed) *The Salton Sink: A Study of the Geography, the Geology, the Floristics, and the Ecology of a Desert Basin,* pp 1–12. Carnegie Institution of Washington Publication no. 193, Washington, D.C.

Bostick V. B. (1973) Vegetation of the McCullough Mountains, Clark County, Nevada. Master's thesis, University of Nevada, Las Vegas.

Bowers J. E. (1980) Flora of Organ Pipe Cactus National Monument. *Journal of the Arizona-Nevada Academy of Science* 15: 1–11, 33–37.

Bowers J. E. (1982) Local floras of the Southwest, 1920–1980: an annotated bibliography. *Great Basin Naturalist* 42: 105–112.

Bowers J. E. (1988) *A Sense of Place: The Life and Work of Forrest Shreve.* University of Arizona Press, Tucson.

Bowers J. E. (1990) A debt to the future: scientific achievements of the Desert Laboratory, Tumamoc Hill, Tucson, Arizona. *Desert Plants* 10: 9–12, 35–47.

Bowers J. E., McLaughlin S. P. (1982) Plant species diversity in Arizona. *Madroño* 29: 227–233.

Bowers J. E., McLaughlin S. P. (1987) Flora and vegetation of the Rincon Mountains, Pima County, Arizona. *Desert Plants* 8: 49–96.

Bowers J. E., McLaughlin S. P. (1996) Flora of the Huachuca Mountains, a botanically rich and historically significant sky island in Cochise County, Arizona. *Journal of the Arizona-Nevada Academy of Science* 29: 66–107.

Bowers J. E., Turner R. M. (1985) A revised vascular flora of Tumamoc Hill, Tucson, Arizona. *Madroño* 32: 225–252.

Burgess R. L. (1965) A checklist of the vascular flora of Tonto National Monument. *Journal of the Arizona-Nevada Academy of Science* 3: 213–223.

Clokey I. W. (1951) *Flora of the Charleston Mountains, Clark County, Nevada.* University of California Press, Berkeley.

Coville F. V. (1893) Botany of the Death Valley expedition. *Contributions from the U.S. National Herbarium* 4: 1–363.

Daniel T. F., Butterwick M. L. (1992) Flora of the South Mountains of south-central Arizona. *Desert Plants* 10: 99–119.

DeDecker M. (1984) *Flora of the Northern Mojave Desert, California.* California Native Plant Society Special Publication no. 7, Berkeley.

Dice L. R. (1939) The Sonoran Biotic Province. *Ecology* 23: 199–208.

Dice L. R. (1943) *The Biotic Provinces of North America.* University of Michigan Press, Ann Arbor.

Felger R. S. (1980) Vegetation and flora of the Gran Desierto, Sonora, Mexico. *Desert Plants* 2: 87–114.

Felger R. S. (1992) Synopsis of the vascular plants of northwestern Sonora, Mexico. *Ecologia* 2: 11–44.

Felger R. S., Lowe C. H. (1976) *The Island and Coastal Vegetation and Flora of the Northern Part of the Gulf of California.* Los Angeles County Natural History Museum Contributions in Science no. 285, Los Angeles.

Gentry H. S. (1942) *Rio Mayo Plants: A Study of the Flora and Vegetation of the Valley of the Rio Mayo, Sonora.* Carnegie Institution of Washington Publication no. 527, Washington, D.C.

Hickman J. C. (ed) (1993) *The Jepson Manual: Higher Plants of California.* University of California Press, Berkeley.

Higgins L. C. (1967) A flora of the Beaver Dam Mountains. Master's thesis, Brigham Young University, Provo, Utah.

Holland J. S. (1982) A floristic and vegetation analysis of the Newberry Mountains, Clark County, Nevada. Master's thesis, University of Nevada, Las Vegas.

Kearney T. H., Peebles R. H. (1960) *Arizona Flora.* University of California Press, Berkeley.

Keil D. J. (1973) Vegetation and flora of the White Tank Mountains Regional Park, Maricopa County, Arizona. *Journal of the Arizona-Nevada Academy of Science* 8: 35–48.

Kurzius M. A. (1981) Vegetation and flora of the Grapevine Mountains, Death

Valley National Monument, California-Nevada. Master's thesis, University of Nevada, Las Vegas.

Lane M. A. (1981) Vegetation and flora of McDowell Mountain Regional Park, Maricopa County, Arizona. Master's thesis, Arizona State University, Tempe.

Lehto E. (1970) A floristic study of Lake Pleasant Regional Park, Maricopa County, Arizona. Master's thesis, Arizona State University, Tempe.

MacMahon J. A., Wagner F. H. (1985) The Mojave, Sonoran and Chihuahuan Deserts of North America. In: Evenari M., Noy-Meir I., Goodall D. W. (eds) *Ecosystems of the World 12A: Hot Deserts and Arid Shrublands,* pp 105–202. Elsevier, Amsterdam.

McLaughlin S. P. (1986) Floristic analysis of the southwestern United States. *Great Basin Naturalist* 46: 46–65.

McLaughlin S. P. (1989) Natural floristic areas of the western United States. *Journal of Biogeography* 16: 239–248.

McLaughlin S. P. (1992) Are floristic areas hierarchically arranged? *Journal of Biogeography* 19: 21–32.

McLaughlin S. P. (1994) Floristic plant geography: the classification of floristic elements and floristic areas. *Progress in Physical Geography* 18: 185–208.

McLaughlin S. P., Bowers J. E., Hall K. R. F. (1987) Vascular plants of eastern Imperial County, California. *Madroño* 34: 359–378.

Moran R. (1983a) The vascular flora of Isla Angel de la Guarda. In: Case T. J., Cody M. L. (eds) *Island Biogeography in the Sea of Cortez,* pp 382–403. University of California Press, Berkeley.

Moran R. (1983b) Vascular plants of the gulf islands. In: Case T. J., Cody M. L. (eds) *Island Biogeography in the Sea of Cortez,* pp 348–381. University of California Press, Berkeley.

Munz P. A. (1935) *A Manual of Southern California Botany.* J. W. Stacey, San Francisco.

Munz P. A. (1974) *A Flora of Southern California.* University of California Press, Berkeley.

Munz P. A., Keck D. D. (1959) *A California Flora.* University of California Press, Berkeley.

Norris, L. L. (1982) *A Checklist of the Vascular Plants of Death Valley National Monument.* Death Valley Natural History Association, Death Valley.

Parish S. B. (1903) A sketch of the flora of southern California. *Botanical Gazette* 36: 203–222, 259–279.

Parish S. B. (1914) Plant ecology and floristics of Salton Sink. In: MacDougal D. T. (ed) *The Salton Sink: A Study of the Geography, the Geology, the Floristics, and the Ecology of a Desert Basin,* pp 85–114. Carnegie Institution of Washington Publication no. 193, Washington, D.C.

Parish S. B. (1930) Vegetation of the Mohave and Colorado Deserts of Southern California. *Ecology* 11: 481–499.

Peterson P. M. (1984) Flora and physiognomy of the Cottonwood Mountains, Death Valley National Monument, California. Master's thesis, University of Nevada, Las Vegas.

Phillips A. M. III (1975) Flora of the Rampart Cave area, lower Grand Canyon, Arizona. *Journal of the Arizona Academy of Science* 10: 148–159.

Pierce A. L. (1979) Vegetation and flora of Buckeye Hills Recreation Area, Maricopa County, Arizona. Master's thesis, Arizona State University, Tempe.

Pinkava D. J., Baker M. C., Johnson R. A., Trushell N., Ruffner G. A., Felger R. S., Van Devender R. K. (1992) Additions, notes and chromosome numbers for the flora of vascular plants at Organ Pipe Cactus National Monument, Arizona. *Journal of the Arizona-Nevada Academy of Science* 24–25: 13–18.

Richerson P. J., Lum K. (1980) Patterns of plant species diversity in California: relation to weather and topography. *American Naturalist* 116: 504–536.

Rondeau R. J. (1991) Flora and vegetation of the Tucson Mountains, Pima County, Arizona. Master's thesis, University of Arizona, Tucson.

Russo M. J. (1987) Flora and vegetation of the Castle Dome Mountains, Kofa National Wildlife Refuge, Yuma County, Arizona. Master's thesis, Arizona State University, Tempe.

Schramm D. R. (1982) Floristics and vegetation of the Black Mountains, Death Valley National Monument, California. Master's thesis, University of Nevada, Las Vegas.

Shreve F. (1915) *The Vegetation of a Desert Mountain Range as Conditioned by Climatic Factors.* Carnegie Institution of Washington Publication no. 217, Washington, D.C.

Shreve F. (1924) Across the Sonoran Desert. *Bulletin of the Torrey Botanical Club* 51: 283–293.

Shreve F. (1925) Ecological aspects of the deserts of California. *Ecology* 6: 93–103.

Shreve F. (1942) The desert vegetation of North America. *Botanical Review* 8: 195–246.

Shreve F. (1951) *Vegetation of the Sonoran Desert.* Carnegie Institution of Washington Publication no. 591, Washington, D.C.

Simmons N. M. (1966) Flora of the Cabeza Prieta Range. *Journal of the Arizona Academy of Science* 4: 93–104.

Stebbins G. L., Major J. (1965) Endemism and speciation in the California flora. *Ecological Monographs* 35: 1–35.

Sundell E. G. (1974) Vegetation and flora of the Sierra Estrella Regional Park, Maricopa County, Arizona. Master's thesis, Arizona State University, Tempe.

Swearingen T. A. (1981) The vascular flora of the Muddy Mountains, Clark County, Nevada. Master's thesis, University of Nevada, Las Vegas.

Takhtajan A. (1986) *Floristic Regions of the World.* University of California Press, Berkeley.

Thornber J. J. (1909) Vegetation groups of the Desert Laboratory domain. In: Spalding V. M. (ed) *Distribution and Movements of Desert Plants,* pp 103–113. Carnegie Institution of Washington Publication no. 113, Washington, D.C.

Thorne R. F., Prigge B. A., Henrickson J. (1981) A flora of the higher ranges and the Kelso Dunes of the eastern Mojave Desert in California. *Aliso* 10: 71–186.

Twisselmann E. D. (1967) A flora of Kern County, California. *Wasmann Journal of Biology* 25: 1–395.

Whittaker, R. H. (1977) Evolution of species diversity in land communities. *Evolutionary Biology* 10: 1–67.

Wiggins I. L. (1964) Flora of the Sonoran Desert. In: Shreve F., Wiggins I. L. *Vegetation and Flora of the Sonoran Desert,* pp 187–1740. Stanford University Press, Stanford.

Wiggins I. L. (1980) *Flora of Baja California.* Stanford University Press, Stanford.

2 Vegetation and Habitat Diversity at the Southern Edge of the Sonoran Desert

Alberto Búrquez, Angelina Martínez-Yrízar,
Richard S. Felger, and David Yetman

More than one third of Forrest Shreve's (1951) Sonoran Desert lies within the boundaries of Sonora, the second largest state in Mexico. Five of the seven major vegetational subdivisions of the Sonoran Desert are found within Sonora as well, more than in any other state. It is safe to say that the specific biological diversity found in the Sonoran portion of the Sonoran Desert is greater than in any other desert in the world.

This abundance and diversity of species are due in large part to the extensive variation in topography and the degree of continentality that occur within the desert's boundaries; these combine to create a complex mosaic of plant associations influenced by different climatic and edaphic conditions. At its northeastern and eastern limits, the Sonoran Desert is variously replaced by Chihuahuan desertscrub, grasslands, and oak woodlands. To the northwest lies the drier and less varied Mojave Desert. Along the northern boundary extends a broad band of chaparral. To the east lies the Sierra Madre, to the west the Sea of Cortez and the Pacific Ocean. In the south the desert merges with thornscrub and tropical deciduous forest, which make the southern region of the Sonoran Desert rich in tropical elements.

The major features of the southern portion of the Sonoran Desert are (1) the islands of the Gulf of California, where maritime influence and isolation have fostered biological endemism; (2) the extensive xeric plains along the coast of the Gulf of California; (3) the once-forested river deltas and river basins, both now nearly devoid of vegetation due to reductions in stream flow; (4) the more mesic hills, canyons, and small sierras within the desert; and (5) the low-elevation escarpments and foothills of the western slope of the Sierra Madre.

In this chapter we review the relationships between physical environment, habitat diversity, vegetation structure, and ecosystem dynamics in the central and southern desert regions of Sonora, which produce a remarkable richness of biological diversity.

Physical Environment
Climate

The Sonoran Desert has the highest temperatures and the lowest precipitation in North America (Schmidt 1989; Turner et al. 1995). The climate is characterized by very hot summers, mild winters, large day/night temperature variations, high levels of sunshine, and highly variable annual rainfall of bimodal distribution with peaks in summer and winter. Temperature and precipitation also vary markedly depending on elevation and the relative importance of continental or coastal influence.

Because of this climatic variability, rainfall is better described by fitting the data to gamma distributions than by calculating the arithmetic means (unfortunately, most of the available data are averaged; Mosiño and García 1981; Ezcurra and Rodríguez 1986). Extended rainless periods occur on the plains along the Gulf Coast northward from the Río Sonora delta. May (1973) recorded 34 consecutive months with no precipitation in the Gran Desierto. The reported mean of 9.6 rainless months/year for Bahía Magdalena in Baja California Sur (Schmidt 1989) does not show the large interannual variability.

Much of the rain in the Sonoran Desert is produced by thunderstorms during the "monsoon" season of July to early September. Generally localized, these storms are often accompanied by strong winds and flash flooding. Most of the remaining precipitation occurs during the winter and early spring in the form of gentle and more widespread rains. Rainfall totals generally increase from west to east. Summer rains contribute more than half the annual precipitation in the eastern portion of the desert, decreasing in percentage as one moves west (fig. 2.1). Mean annual rainfall in the Sonoran Desert ranges from less than 30 mm near the Colorado River delta to more than 350 mm at the eastern desert margin (Hastings and Turner 1965; Hastings and Humphrey 1969; García 1973; Schmidt 1989). Rainfall normally begins in the southeastern portion of the desert in late

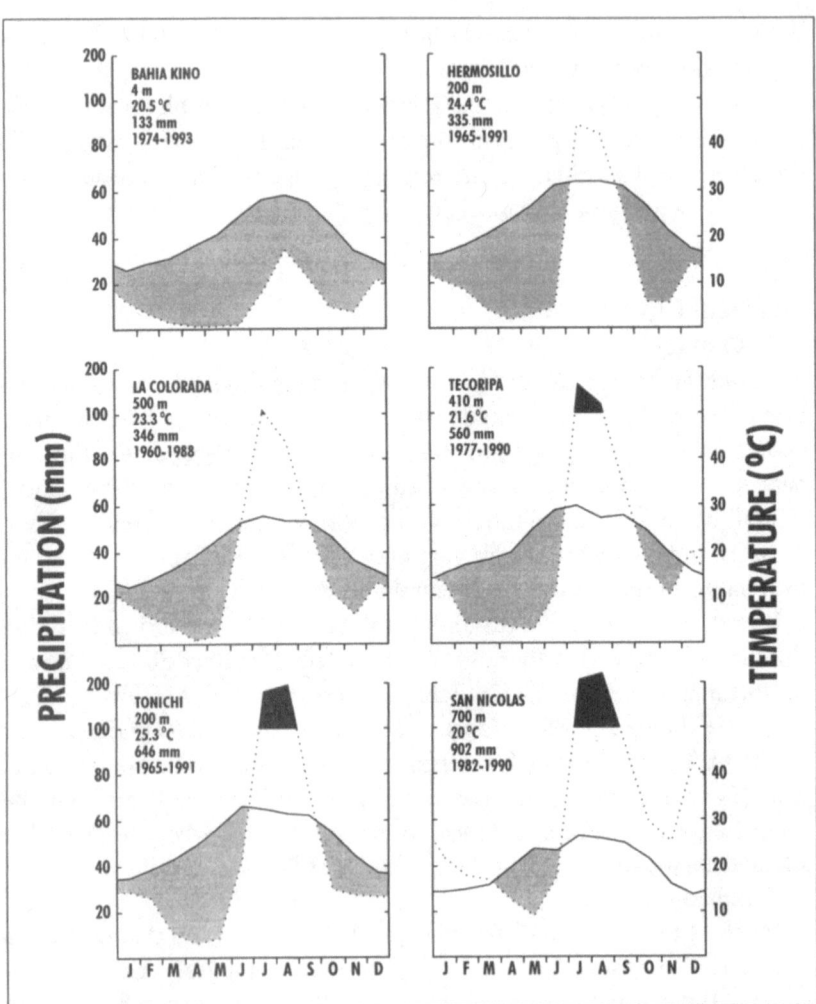

Figure 2.1 Ombrothermal climograms for selected localities along Highway 16, Sonora, Mexico. Solid line = mean monthly temperature, dotted line = monthly precipitation. Values were recalculated from Comisión Nacional del Agua data, accounting for missing values of precipitation and temperature. The values on the upper left corner on each diagram indicate locality, elevation, mean annual temperature, mean annual precipation, and period of observation. The gray areas indicate periods of water deficit and the black areas, water surplus. Localities are arranged from west to east.

June (traditionally on June 24, "el día de San Juan"), while in the western portion, rains seldom begin until well into July. Rainfall is more reliable in the southeastern Sonoran Desert than in the northwestern part. In general, then, rainfall totals and seasonality form a southeast-to-northwest gradient in the region.

Temperatures exhibit a similar trend, with the greatest extremes occurring in northwestern Sonora, where summer highs frequently exceed 49°C and winter lows occasionally fall below freezing. The southern limit of periodic, damaging freezes establishes the northern limits of many Sonoran Desert plant species (Shreve 1914, 1951; Hastings and Turner 1965; Steenbergh and Lowe 1977; Bowers 1980; Felger and Moser 1985; Turner et al. 1995). Perhaps once a decade, cold air masses blast down through the Great Basin into the Sonoran Desert, with catastrophic results for tropical outliers and with important consequences for vegetation structure as well as plant and animal population dynamics. Subfreezing temperatures have been recorded as far south as the Guaymas region, damaging mainly species with distributions that extend well south of the Sonoran Desert (Gentry 1942; Krizman 1972). For example, mangroves at Estero Sargento, north of Bahía Kino, were severely damaged in 1971 and again in 1978. Catastrophic freezes have been reviewed by Bowers (1980) and Felger and Moser (1985).

The climatic extremes along the region's northern border gradually give way to a more moderate climate along the southern coastal plains and foothills in Sonora. For example, the 20°C isotherm roughly follows a northwest-southeast line abruptly veering when approaching the 22°C and 24°C isotherms northeast of Hermosillo (INEGI 1988).

South of the plains of the Río Sonora, frosts seldom occur, temperatures are milder, humidity increases, and summer rainfall provides the bulk of annual precipitation. Similarly, maritime influences ameliorate the aridity on the Baja California peninsula, the Gulf islands, and a narrow band along the coast of Sonora. In these areas, temperatures are moderated by the ocean and maritime fog condenses as dew, which is an important source of moisture.

Geomorphology and Hydrology

The most prominent features of the southern portion of the Sonoran Desert are the tectonics that produced the Basin and Range Province (geomorphologically the Buried Ranges, *sensu* Raisz 1964): a marked northwest-

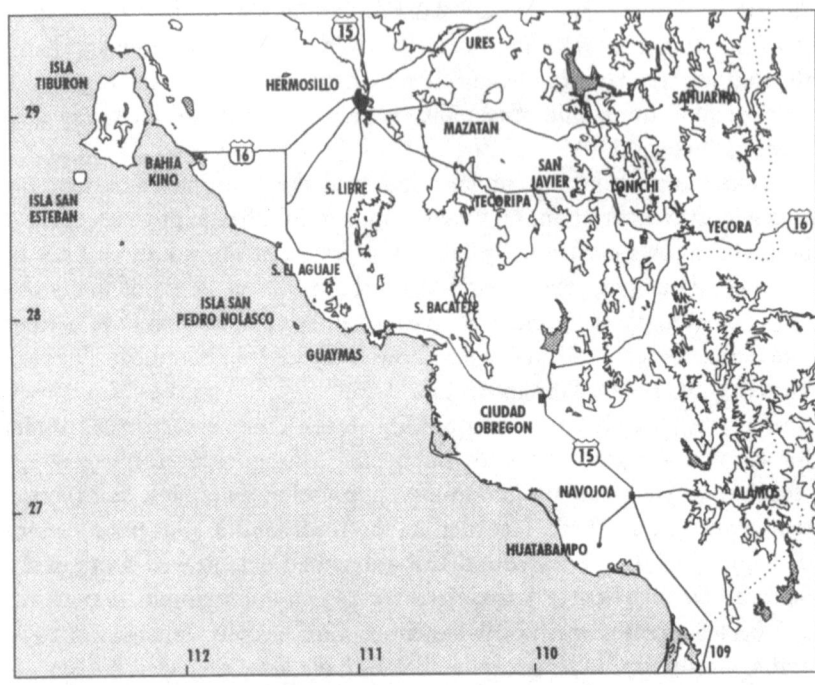

Figure 2.2 Map of southern Sonora, Mexico, showing main roads, localities, and topography at 0, 400, and 1000 m altitudes. Based on INEGI (1988).

southeast alignment of mountain ranges with broad valleys in between. Toward the west are broad plains broken by isolated granitic sierras of the Sonoran batholith and, on top of it, volcanic ranges. Toward the east, ash flows and basalts mark the start of the extensive volcanism of the Sierra Madre Occidental (fig. 2.2).

The region is geologically complex, including an extensive stratigraphic column (Crosswhite and Crosswhite 1982; INEGI 1988); Quaternary regolith and alluvial sediments predominate in the plains. The plains are also dotted by inselbergs and small mountains of Paleozoic shales, Cretaceous granites, Tertiary conglomerates, and Tertiary-Quaternary extrusive rocks (basalts and ignimbrites). Conglomerates of the Baucarit Formation (Miocene) show striking erosive patterns along the Río Yaqui and elsewhere in the sierra foothills. The coastal plain narrows near Guaymas

Vegetation and Habitat Diversity   41

because of three moderately high (about 1000 m elevation) Tertiary volcanic ranges: the Sierras El Aguaje, Libre, and El Bacatete.

Three major river systems run through the Sonoran Desert in Sonora: the Río Magdalena (comprising the Ríos Magdalena, Asunción, and Concepción), the Río Sonora, and the Río Yaqui. The three river systems have extensive deltas with fine sediments and high water tables. However, on the Magdalena and Sonora river deltas, water table levels have been drastically lowered by agricultural pumping. The Río Yaqui, with headwaters in the Sierra Madre in Chihuahua, has a flow at times rivalling that of the Colorado. In recent times, only the Yaqui reached the sea with a continuous flow. All three rivers owe their flow to montane watersheds to the east, and all have been dammed to such an extent that flows are greatly reduced and intermittent. Their upper basins run along Sierra Madre valleys with a gross north-south orientation, while on the coastal plains they flow east to west.

Four small rivers of purely Sonoran Desert origin are worth noting: the Río Sonoyta, which flows along the east flank of the Sierra El Pinacate and disappears into the dunes of the Gran Desierto a few kilometers east of Puerto Peñasco; the San Ignacio/Arivaipa and the Bacoachi, whose small basins are located north of Bahía Kino; and the Río Mátape, which originates in the isolated granitic dome of the Sierra Mazatán and drains into the Estero El Rancho near Guaymas.

The prevailing soils in the plains and undulating hills of the coast are Xerosols and Yermosols (INEGI 1988). Regosols are the most widespread soil unit in the Sonoran Desert. These soils are without developed horizons and resemble their parent rock. Substantial areas of Lithosols occur in the Sierra Madre foothills and isolated coastal sierras. More restricted are the Feozem and Vertisols, which occur as long, narrow strips along the major waterways and large arroyos.

Soils determine the vegetation character by changes in their nutrient content, in the particle sizes, and in the degree of infiltration of rainfall. Soils with a high proportion of sand and rock are more favorable to plant development than soils with clay, caliche layers, or pavement. The former allow for rapid water infiltration, while the latter—with a hard, impermeable crust—impede water penetration and favor what Shreve (1951) called sheet flood erosion. Extreme examples of this relationship are the Lithosols found in rocky slopes that act as rain catchment areas. Water percolates through fractures and crevices, sustaining vegetation with a much larger biomass than that found in neighboring level areas where most of the water runs off the surface with relatively little infiltration (Shreve 1951; Rzedowski 1978). Along a precipitation gradient, thornscrub appears first on rocky slopes with Lithosols. The vegetation of the Plains of Sonora owes most of its character to the deep, relatively fine, Pleistocene alluvial soils that produce a savannoid vegetation with scattered large trees within a matrix of grasses and subshrubs (see Búrquez et al. 1998). In more northern desert latitudes, these soils harbor arid grasslands (McAuliffe 1995). Perhaps the essence of the desert is embodied in the Solonchaks — highly saline soils near the coast and in closed basins or playas that in wetter times had permanent standing water. The largest of these is Playa San Bartolo, north of Bahía Kino (Petit-Maire and Casta 1977). These areas have fine, highly saline alluvial soils almost devoid of vegetation.

Vegetation
Vegetation Units and Structure

Shreve's (1951) seven major subdivisions of the Sonoran Desert featured distinct vegetational units. He associated these with specific geographic regions that exhibit increasing geomorphological complexity from west to east and increasing importance of tropical taxa toward the south and east.

In the Lower Colorado Valley, perennials are sparse due to unpredictable rainfall and long periods of drought. Still, the region is well endowed with ephemerals and cryptophytes (Shreve 1951; Felger 1980, 1992). Many seeds lie quiescent in the soil for years, producing prodigious growth after exceptional rainfall — as during 1909–1910 (Lumholtz 1912) and a more recent mass flowering of winter annuals in 1992. Ephemeral species constitute the major proportion of the flora for the northwestern edge of the desert — 55% to 59% for the Gran Desierto (Felger 1980, 1992). Their proportional contribution decreases toward the southeast. Desert vegetation becomes increasingly arborescent and more tropical in composition and structure as the desert merges first into thornscrub, then into the tropical deciduous forests of southern Sonora. Likewise, a similar gradient is apparent as vegetation structure and composition changes with increasing altitude to the east (Gentry 1942; White 1948; Felger and Lowe 1967; Turner and Brown 1982; Van Devender 1990).

Tropical deciduous forest and thornscrub predominate at medium and low elevations along the Pacific coast of Mexico (Rzedowski 1978). In Sonora and northern Sinaloa, however, increasing aridity northward is responsible for increasing sparseness and the smaller stature of the forest. The more complex and better developed vegetation is restricted to the wetter foothills of the Sierra Madre while desertscrub occurs on the drier coast. These patterns of distribution give the impression that tropical deciduous forest and thornscrub form a wedge between the desert and the oak woodlands. In fact, tropical deciduous forests occur in an ever-narrowing band of moist tropical climates (type A of Köppen [García 1973]) on the upper Mayo and Yaqui river valleys (and some small stretches of the Río Sonora), confined between the desert to the west and the temperate woodlands and forests to the east (for insect taxa see, for example, Búrquez 1997).

Thornscrub and tropical deciduous forests were first described for the Río Mayo and the Guaymas region by Gentry (1942, 1949). The northern and western limits of the tropical vegetation and transitional ecotones leading into the Sonoran Desert were more precisely delineated by Wiseman (1980) and mapped by Brown and Lowe (1980). This tropical vegetation contributes a great number of perennials to the flora of the desert. The Foothills of Sonora subdivision was originally classified as Sonoran Desert by Shreve (1951) but has more recently been referred to as thornscrub (Felger and Lowe 1976; Rzedowski 1978; Felger and Moser 1985; Búrquez et al. 1992a). Continuing disagreements over the appropriate southern boundary of the desert demonstrate the elusiveness of a clear limit for either vegetational regime.

Habitat Diversity: Major Ecoclines
Large-Scale Habitat Variation

Across the Sonoran Desert from northwest to southeast, temperature, precipitation, and then light replace each other as main axes for niche differentiation. While in the northern reaches freezing temperatures and soil moisture determine the distribution of plant species and communities (see Turner and Brown 1982), the distribution of the major vegetation units in the southern Sonoran Desert are established mainly by soil moisture. The west-east rainfall gradient and the increasing proportion of "monsoon" rains from northwest to southeast create major differences in species occurrence, life-form spectra, and vegetation profiles. As perennials attain larger statures and greater canopy cover in the southern and eastern boundaries and desert gives way to thornscrub and tropical deciduous forest, light competition heralds a major community change.

Major trends along these ecoclines are (1) the proportional decrease of ephemerals toward the east and south with an attendant increase in perennials and overall diversity, (2) the loss of cold-hardiness in southern populations of species that are frost-resistant in the north, and (3) the change in morphology and function associated with heat loss (Felger and Lowe 1967; Gibson and Nobel 1986). The differences in cold-hardiness between southern and northern populations of many Sonoran Desert species are well known among horticulturists but remain poorly documented and offer intriguing areas for research.

Intermediate-Scale Habitat Variation

On a local scale, species of mesic affinities occur as patches within desert environments. Topography, soils, or slope orientation are major factors affecting local vegetation. Species growing in a particular environment in one phytogeographic desert subdivision are often present in a different habitat in another subdivision.

The most striking feature of desert vegetation when viewed from the air is its patchiness. Flat areas exhibit an almost regular pattern of distribution of trees and shrubs with large gaps between them. Small arroyos have a linear canopy, while large arroyos are wide corridors with a closed canopy of trees along the margins. These distributional patterns have important consequences for the population biology of single species. For example, near Hermosillo, *Ipomoea arborescens* (tree morning-glory) is present in dense stands on steep north-facing slopes within localized thornscrub, but it is confined to linear populations in the small arroyos flowing from the mountains into more desertlike vegetation. This distribution suggests that the main source of *Ipomoea* propagules are the mountain populations and that individuals in the desert are stragglers (A. Búrquez and F. Molina, unpublished data).

Such linear patterns of distribution along ephemeral watercourses seem to be the rule for many desert species. On the Plains of Sonora, the vine *Merremia palmeri* is restricted to small desert arroyos, a pattern that imposes a very special genetic structure on its population—pollen flow and seed dispersal occur primarily along a linear vector (Willmott and Búrquez 1996). True desert dwellers, such as *Olneya tesota* (ironwood),

Larrea tridentata (creosote bush), *Cercidium microphyllum* (foothill paloverde), different species of *Fouquieria* (ocotillo), columnar cacti, and many other species, are usually not restricted to arroyos. Yet even these can be confined to dry watercourses in the driest habitats, such as in Reserva de la Biosfera El Pinacate y El Gran Desierto de Altar (El Pinacate and Gran Desierto de Altar Biosphere Reserve) in northwestern Sonora (Felger 1980; Turner 1990).

The degree of rockiness, the geological origin of the substrate, and slope aspect and steepness also restrict the distribution of many species. For example, in El Pinacate and Gran Desierto de Altar Biosphere Reserve, *Fouquieria splendens* (ocotillo) largely determines the structure of vegetation on rocky soils (Ezcurra et al. 1987). Similarly, throughout its range, *Acacia willardiana* (palo blanco) is the most prominent species on rocky soils, usually on steep slopes.

Even on modest hills, slopes with coarse rocky soil, particularly those oriented north and east, harbor a very different species assemblage from the fine-textured soils on the plains (and from those slopes facing south and west). For example, in transects of less than 1000 m near Hermosillo, the typical Plains of Sonora dominants — *Encelia farinosa* (brittlebush), *Jatropha cardiophylla* (sangrengado), and *Olneya tesota* — give way on gentle slopes to *Cercidium microphyllum*. On steeper, rocky slopes, this association is replaced by thornscrub of *Acacia willardiana*, *Agave angustifolia*, *Ambrosia cordifolia* (chicurilla), *Bursera fagaroides* (torote blanco), *Croton sonorae*, *Hechtia montana*, *Ipomoea arborescens*, and *J. cordata* (torote papelillo). *Olneya* is rarely found on slopes on the Plains of Sonora, yet to the west, on small extrusive volcanic hills, it grows with *J. cuneata* (matacora) and *Carnegiea gigantea* (saguaro) in vegetation more closely resembling the Central Gulf Coast subdivision of the Sonoran Desert.

Certain mesic habitats within the desert, such as mountain canyons, ravines, and north-facing slopes, support a wealth of species with more tropical affinities. The Sierras Libre, El Aguaje, and El Bacatete share many species with tropical deciduous forests to the south. Within the desert, tropical "islands" develop where areas are protected from extreme temperature oscillations, furnished with shade and humidity, and provided with more available water. These mountain oases are like closed microcosms that maintain relict and disjunct populations of organisms. Their volcanic origin allows greater biological diversity than if they were granitic or metamorphic: the welded tuffs and ignimbrites are porous, full of contact zones where seeps develop, and sufficiently cracked and eroded to produce deep canyons with numerous *aguajes* (waterholes). An excellent example of a special mountain/canyon habitat is Cañón del Nacapule in the Sierra El Aguaje, which hosts numerous species with restricted distributions (Felger, in press). Wetland plants, mostly of tropical origin, grow along the permanent streams. Tropical vines climb through the shrubs and small trees, lacing into gallery groves of tall palms. Several plant species, including *Vallesia baileyana* and *Verbesina felgeri*, are known only from here and certain other nearby canyons. The rich flora includes three species of fig (*Ficus insipida, F. palmeri,* and *F. pertusa*) and three native palms (*Brahea elegans, Sabal uresana,* and *Washingtonia robusta*). One fig, *F. insipida*, forms short buttresses from a massive trunk. Despite its small population size and extreme isolation, this fig maintains healthy populations of its pollinators by flowering throughout the year (Smith 1994). Nacapule and nearby canyons are the only places this species occurs within the Sonoran Desert, as is the case with the large shrubs *Coccoloba goldmanii* and *Zanthoxylum mazatlanum*. Likewise, other riparian canyons in the region support disjunct populations of plants from more tropical regions in southern Sonora or Baja California Sur (Felger 1966; Yetman and Búrquez 1995).

Microhabitat Variation
Long-lived plants, the desert old growth, create islands of much greater biological diversity than the surrounding desert (Búrquez and Quintana 1994; Tewksbury and Petrovich 1994). The role of paloverde canopies as a major element in the recruitment of *Carnegiea gigantea*, the "nurseplant syndrome," was recognized by Turner et al. (1966) and Steenbergh and Lowe (1977). Although the causes of increased survival of cacti seedlings is still debated, the role of shrubs and trees in the regeneration niche (*sensu* Grubb 1977) of columnar cacti is well-established (Valiente-Banuet and Ezcurra 1991; Nabhan and Suzán 1994). *Olneya tesota* is associated not only with cacti but also with dozens of other desert species (Felger 1966; Búrquez and Quintana 1994; Tewksbury and Petrovich 1994). Larger trees have more species per unit area under them than smaller trees. Graphical data representing species diversity as a function of area clearly demonstrate that habitat beneath *O. tesota* has a richer flora than areas outside the canopy shadow (fig. 2.3, table 2.1).

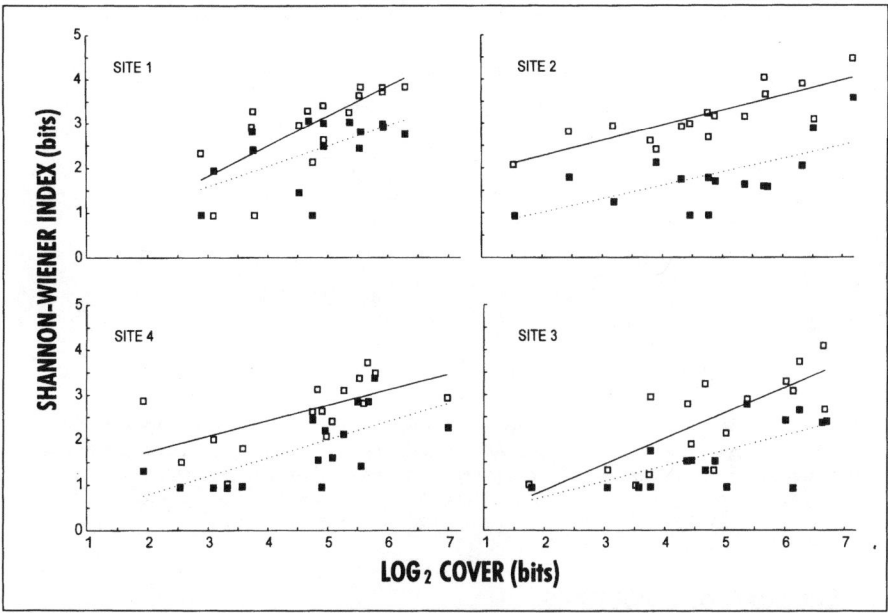

Figure 2.3 Shannon-Wiener index of heterogeneity used as a measure of species diversity for plots of different size located under (solid line) and outside (dotted line) ironwood trees at four sites near Hermosillo, Sonora. Sites 1 and 3 are on dry level terrain; sites 2 and 4 are along temporary desert arroyos. Curves are significantly different either in slope or intercept (see statistics in table 3 in Búrquez and Quintana 1994). Axes units in bits as indicated by Krebs (1985, pp 523).

Greater diversity associated with nurse plants is related to several factors. The environment under the canopy of desert trees has a lowered irradiance (Franco and Nobel 1989; Valiente-Banuet and Ezcurra 1991) and an increased water infiltration and nutrient availability (Paulsen 1953; García-Moya and McKell 1970). Plants that have animal-dispersed seeds are better represented under the canopy of desert trees, probably because trees provide shelter, nesting, or roosting sites for various organisms (Búrquez and Quintana 1994). Plants growing under trees are also protected from herbivory (McAuliffe 1984). The most important nurse trees in the southern Sonoran Desert include spreading desert legumes such as *Cercidium*, *Olneya*, and *Prosopis*. These trees promote a further partitioning of habitat at a smaller scale.

Table 2.1 Similarity matrix comprising all the species found in four sites where ironwood grows.

		Under			Outside		
		3	2	1	3	2	1
	3	*52*	.77	.72	.63	.56	.42
Under	2	35	*39*	.71	.63	.55	.52
	1	30	25	*31*	.62	.67	.57
	3	27	23	20	*34*	.74	.61
Outside	2	21	17	18	21	*23*	.68
	1	14	14	13	15	13	*15*

Source: Búrquez and Quintana (1994).
Note: Values in italics along the main diagonal indicate the number of species in each plot size (1 = small; 2 = medium; 3 = large), either under or outside canopies of ironwood. Values below the main diagonal are the numbers of species in common between samples, whereas those above it are the similarity ratios using Sorensens index. $N = 20$ for each plot size, except for plot size 3, where $N = 24$.

Ecosystem Dynamics

The vegetation, flora, fauna, and physical environment establish complex functional relationships. Energy flow and nutrient cycling are important features of the community metabolism. These include the processes of above- and belowground accumulation of organic matter, and the production and decomposition of litter.

Variation in standing crop biomass and primary productivity of terrestrial ecosystems is correlated to rainfall (see Lieth and Whittaker 1975; Noy-Meir 1985). In the Sonoran Desert, primary productivity increases from west to east with increasing rainfall. Productivity in the Lower Colorado Valley and Central Gulf Coast is much lower than that of the other Sonoran Desert subdivisions. However, much temporal and spatial heterogeneity occurs because site-specific water catchment and between-year rainfall are not uniform. Furthermore, nutrients and their variable distribution in the soil are critical to primary productivity, standing crop biomass, and floristic composition (Ehleringer and Mooney 1983; Webb et al. 1983; Ludwig 1987; Polis 1991).

Few studies have documented the dynamics of the Sonoran Desert at the ecosystem level, and most of them have focused on the northern region, emphasizing single species measurements. Annual aboveground net primary productivity varies from a low of 55 g m^{-2} in a desertscrub community in the Tucson Basin, Arizona (Szarek 1979), to a high of nearly 130 g m^{-2} in Arizona Upland desertscrub in the Santa Catalina Mountains in Tucson (Whittaker and Niering 1975). Dramatic changes in primary productivity also occur at local scales. Plant communities in the plains near Hermosillo have a litter production of about 90 g m^{-2} y^{-1}. In contrast, adjacent sites in the densely vegetated xeroriparian habitats produce around four times more litter (370 g m^{-2} y^{-1}). Thornscrub on adjacent hillsides produces an intermediate value of 178 g m^{-2} y^{-1} (Martínez-Yrízar et al. 1993). This extreme variability in productivity between sites has also been observed in other desert ecosystems (Ludwig 1986; Polis 1991; Martínez-Carretero and Dalmasso 1992).

The shedding of plant parts and the fate of the litter are important elements of ecosystem nutrient cycling and energy flow. Surface litter on desert substrates is usually sparse and patchy in distribution due to its accumulation around the base of shrubs or in wind-protected areas (West 1979). On the Plains of Sonora near Hermosillo, standing crop litter varies spatially from 21 g m^{-2} in open areas to 210 g m^{-2} beneath trees and shrubs (A. Martínez-Yrízar, A. Búrquez, and S. Núñez, unpublished data).

Ephemerals, despite their brief presence in the desert, can make an appreciable contribution to productivity (Inouye 1991). Near Cave Creek, Arizona, Halvorson and Patten (1975) found that annuals produced a mean biomass of 45 g m^{-2}. Annuals growing under shrubs contributed more than twice as much biomass as those growing outside plant canopies (Halvorson and Patten 1975; Patten 1978). The magnitude of temporal variation in productivity is illustrated by the study of Patten (1978), who found a tenfold increase in aboveground net primary productivity of winter annuals during a wet versus a dry year.

Studies of standing crop biomass for the Sonoran Desert are very limited. *Simmondsia chinensis* (jojoba) at Punta Chueca, Sonora, has an aboveground biomass of 1.57 Mg ha^{-1} (Braun and Espericueta 1979). A lower value, 1.16 Mg ha^{-1}, was reported for an *Ambrosia deltoidea* (triangle-leaf bursage)–*Larrea tridentata*–*Olneya tesota* desertscrub community at Silverbell, Arizona (Thames 1973). In contrast to these low biomass values, communities in the southern Sonoran Desert yield considerably larger amounts (5 to 20 Mg ha^{-1}; Búrquez et al. 1992b). These higher values are related to the less harsh climate that supports a more complex structure of vegetation.

Mesquite forests have a much higher standing crop biomass and productivity than those of desertscrub. In southeastern California, close to the Salton Sea, a stand of *Prosopis glandulosa* var. *torreyana* (mesquite) had 13 Mg ha^{-1} of aboveground biomass (Rundel et al. 1982). Similar or higher values are likely to be found in southern localities. Large mesquites have restricted distributions along major drainage channels and fine silty plains on river deltas where the water table is shallow and most nutrients are not a limiting factor. Other exceptions to the relationship between precipitation and productivity are coastal swamps and lagoons, and wetlands with underground water sources or remote sources. These oases have been studied mainly in northwestern Sonora, including the Ciénaga de Santa Clara and El Doctor wetlands (Glenn et al. 1992) and the *pozos* ("springs") at Bahía Adair (Ezcurra et al. 1988), Quitovac (Nabhan et al. 1982), and Quitobaquito (Felger et al. 1992). These wetland habitats exhibit high productivity and standing crop biomass. However, there are no formal studies on their productivity. Desert oases also occur as seeps in the foothills of most large ranges (e.g., Felger and Moser 1985) but have not been studied for localities in the southern Sonoran Desert. Despite their relevance in ecosystems in which water is the main limiting factor, no studies of ecosystem dynamics have been carried out in these communities.

Geographic Limits

The delimitation of the Sonoran Desert has posed problems since its formalization by Shreve (1951). As Schmidt (1989) has shown, the Mexican portion of the Sonoran Desert has been broadly interpreted to include northern Sinaloa, most of Sonora, and the peninsula of Baja California, or narrowly interpreted to include only a coastal strip from west-central Sonora to the Colorado River delta (fig. 2.4).

There is general agreement about the northern boundaries of the Sonoran Desert, which are largely determined by winter freezing (e.g. White 1948; Shreve 1951; Steenbergh and Lowe 1977; Crosswhite and Crosswhite 1982; Turner and Brown 1982). However, the limits in the south and east have been much debated. The lack of agreement illustrates well the Gleasonian paradigm that in the absence of major environmental discontinuities, populations distribute independently of each other, and communities merge seamlessly. The ecosystem dynamics at the southern desert edge are remarkably similar to those of tropical deciduous forest (Martínez-Yrízar et al., in press). In fact, the southern reaches of the Sonoran

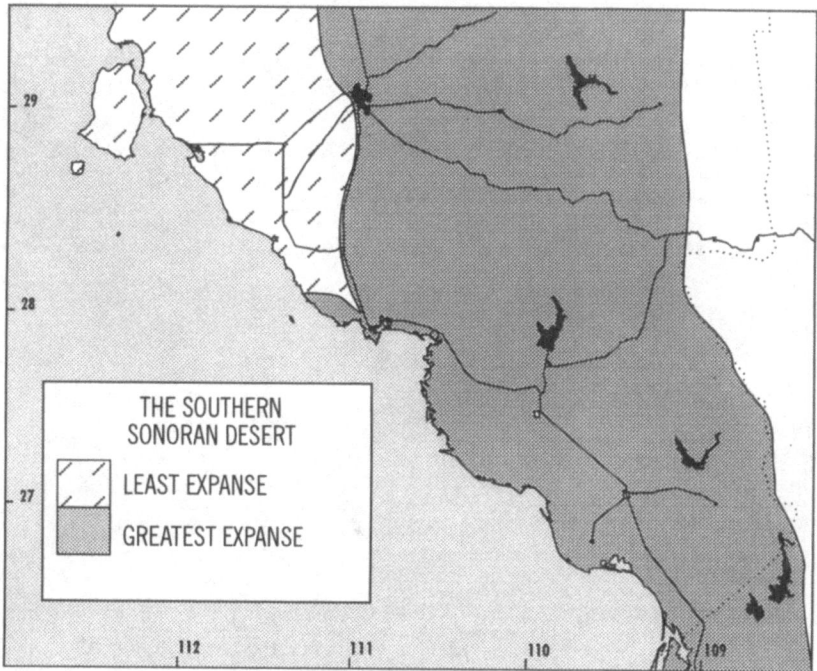

Figure 2.4 Least and greatest expanse of the Sonoran Desert in southern Sonora, Mexico, according to Schmidt (1989).

Desert resemble dwarfed tropical deciduous forest. Many species are common to both these vegetation types, but individual species of trees and shrubs are often much smaller and different in growth form in the desert. The proportion of climbers is comparable in both habitats (Rundel and Franklin 1991), as are the complex structural and diversity patterns.

Several authors have defined the southern and eastern boundaries of the Sonoran Desert using mainly temperature and rainfall data (see Arbingast et al. 1975; Schmidt 1989). This scheme includes most of the Baja California peninsula and a narrow coastal strip continuing well into Sinaloa, and excludes the relatively narrow valleys in the foothills of the Sierra Madre in Sonora. However, a purely climatic delimitation of the desert suffers from oversimplification. Climatic boundaries do not take into consideration differences in soil permeability, water table height, local maritime influence, or slope aspect and steepness. The use of mean annual values

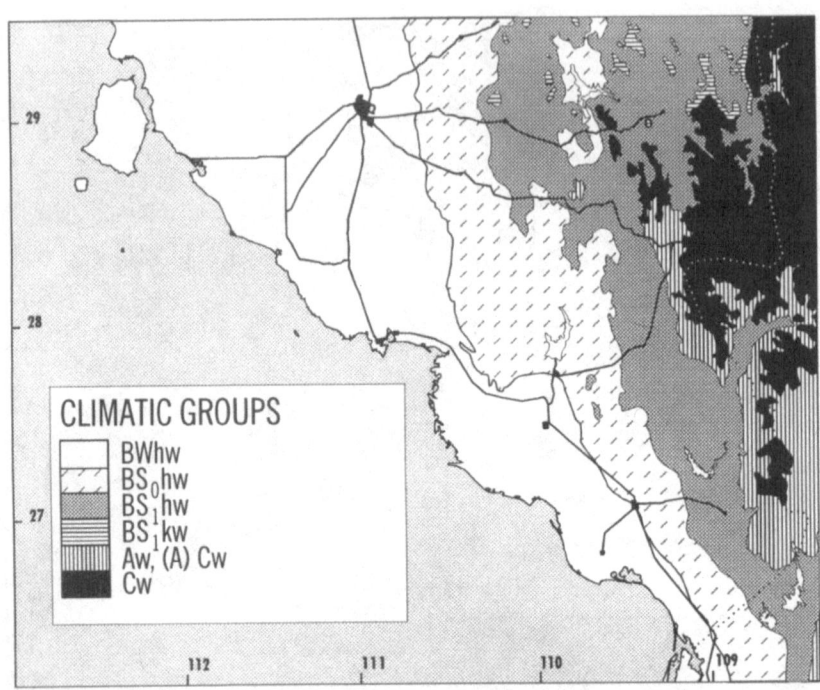

Figure 2.5 Climates of southern Sonora, Mexico, based upon INEGI (1988).

ignores the temporal distribution of temperature and rainfall. These factors are of paramount importance in understanding vegetation structure and ecosystem dynamics. For example, the large deltas of the Sonoran rivers (about 40 km wide for the Río Sonora and more than 150 km wide for the Yaqui-Mayo Rivers) have formed expansive plains of fine-textured, level soils with a historically shallow water table. These deltas harbor a vegetation unlike that which would be predicted by considering major climatic parameters alone, namely desertscrub, grasslands, or open shrubland (Whittaker 1975). The presence of dense mesquite forests and impenetrable thornscrub at the coastal southern boundary of the Sonoran Desert seems similar to the occurrence of some Australian *Eucalyptus* associations outside forest climatic zones.

The modification to the climate scheme of Köppen by García (1973) gives more resolution to the correlation between climate and vegetation (see fig. 2.5). The very dry–very warm climates (BWhw group) represent the more xeric Sonoran Desert communities, while the dry-warm climates (BShw group) indicate different stages in the transition to tropical communities. The presence of tropical climates (Aw group) indicates unequivocally tropical forests, while desert-steppe and temperate climates (BS_1 and Cw groups) are well correlated with dry woodlands and pine-oak forests.

Other authors have relied on the use of vegetation and plant and animal distributions to define the boundaries of the Sonoran Desert (Shreve 1951; Leopold 1950; Felger and Lowe 1976; MacMahon 1979; Brown 1982). By these criteria, of multivariate nature, the southern limits of the desert in Sonora are set around the latitude of Guaymas and exclude the long coastal strip southward. To the east, the limits are determined by the effects of elevation. Mountain ranges, having small areas of nondesert vegetation, appear as isolated islands within the desert. Examples of these nondesert islands are Sierras La Giganta and La Laguna in Baja California Sur, Sierra Kunkaak on Isla Tiburón, and Sierras Mazatán, San Javier, Libre, El Bacatete, and El Aguaje in Sonora.

The west-to-east increase in summer precipitation and the north-to-south decrease in frost damage induce changes in the stature and structure of vegetation—to the south and east vegetation is generally denser and taller. At the southern and eastern limits of the Sonoran Desert, unlike the northern and western portions, the vegetation lacks the winter-spring flurry of biological activity brought about by winter-spring Pacific frontal storms. At the southern and eastern boundaries, the desert merges with a type of vegetation that has been called thorn forest (Gentry 1942), thornscrub (Felger and Lowe 1976), and Sinaloan thornscrub (Gentry 1949; Brown 1982). Thornscrub varies greatly in complexity, from relative simplicity at the desert margin to complex communities that intergrade into tropical deciduous forest. Rzedowski (1978) relates thornscrub to his "Bosque Espinoso" type, but indicates considerable overlap with both desert and tropical deciduous forest.

Two major types of thornscrub can be recognized for Sonora (*sensu* Felger and Lowe 1976): Foothills and Coastal Thornscrub. The former is broadly equivalent to the Foothills of Sonora, as originally proposed by Shreve (1951) as a subdivision of the Sonoran Desert. This, however, has been excluded as part of the desert in most modern accounts. On the basis of structural criteria, some areas of the Arizona Upland subdivision could be considered thornscrub (Felger and Lowe 1976; Turner and Brown

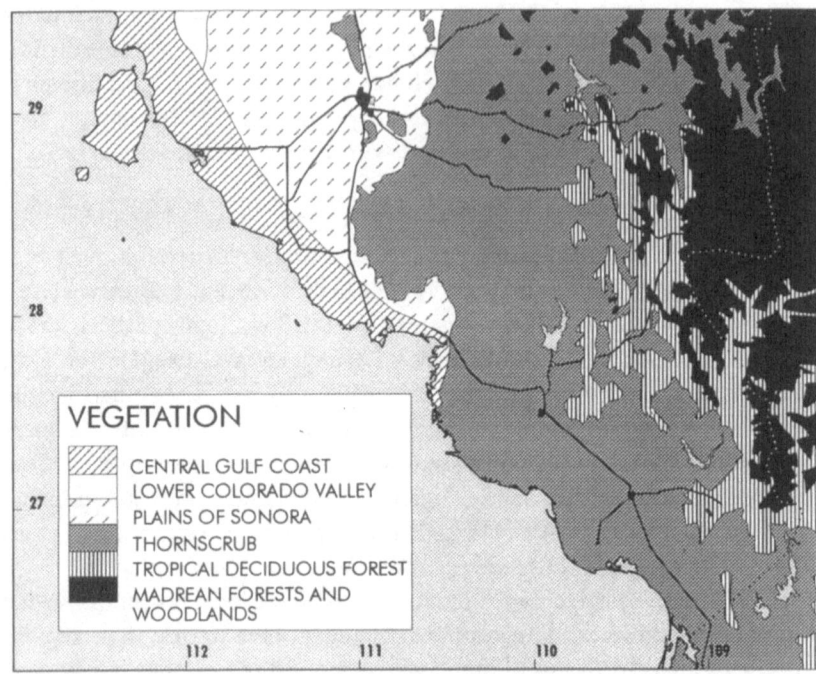

Figure 2.6 Delineation of the vegetation of southern Sonora, Mexico, based upon Shreve (1951), Brown and Lowe (1980), Wiseman (1980), INEGI (1988), Búrquez et al. (1992a), and personal observations.

1982). Also, thornscrub communities are found in coastal southern Sonora and northern Sinaloa. Other vegetation types related to thornscrub include the riparian woodlands and thornscrub that occur near large desert arroyos, canyons, and springs.

Coastal thornscrub shows a gradient of structural change from the coast to the interior. Near the coast it is low and poorly structured, while farther inland a more intricate vertical structure develops with the addition of more arborescent species. Probably the most striking features of these communities are the remarkably dense growth of columnar cacti, mainly *Stenocereus alamosensis, S. thurberi* (organ pipe cactus) and *Pachycereus pecten-aboriginum* (etcho), and the occurrence of several species of epiphytic *Tillandsia*. Gross ecosystem changes occurring along the Pacific coast of Mexico from Jalisco and Nayarit northward are mirrored in short transects (less than 100 km) from the desert to the Sierra Madre at latitudes 27° to 30° N (fig. 2.6 and fig. 2.7).

Foothills thornscrub represents the transitional vegetation between the Sonoran Desert proper in western Sonora and the most northern tropical deciduous forests on the western flank of the Sierra Madre. It is also prevalent on desert mountain slopes (mainly northern slopes). This vegetation can be broken into several units or associations. Examples of this are the *Acacia willardiana–Jatropha cordata–Mimosa laxiflora–Croton sonorae* mixed scrub on hillsides (mostly north-facing) within the Sonoran Desert proper, and the *Pachycereus pecten-aboriginum–Ipomoea arborescens–Fouquieria macdougalii–Ceiba acuminata* associations on the plains near Tecoripa.

Structural criteria such as vegetation cover have been used to distinguish desert from thornscrub (Gentry 1949; Felger and Lowe 1976; Brown 1982: wide open spaces = desert; dense cover with few open spaces = thornscrub). Gentry (1942) used vegetation structure (such as the height of the columnar cactus *Pachycereus pecten-aboriginum* taller than the canopy = thornscrub) as well as floristics (e.g., *Acacia cochliacantha* (boat-thorn acacia) = thornscrub; *Lysiloma divaricatum* = tropical deciduous forest) to separate thornscrub from tropical deciduous forest. The difficulties of defining thornscrub have been elegantly summarized by Rzedowski (1978) as translated here "thornscrub communities . . . often are not well delimited because they imperceptibly change into other vegetation types such as tropical deciduous forest, desertscrub, and grasslands."

Delimiting the southern and eastern Sonoran Desert boundaries, as happens with ecotones, seems to depend on researchers' biases. Researchers with a northern hemisphere bias, accustomed to the more discrete nature of temperate communities, seem to include thornscrub as allied to the Sonoran Desert. Researchers with a southern bias, who place greater emphasis on the tropics, include thornscrub as a subset of the more developed tropical deciduous forests.

A Transect along Highway 16

Mexico Highway 16 running between Bahía Kino and Maycoba in Sonora (28°30'–29°15'N, 109°30'–113°W) provides a transect of plant communities from marine to montane forests and reveals their extensive intergradation with desert associations (Búrquez et al. 1992a; fig. 2.7).

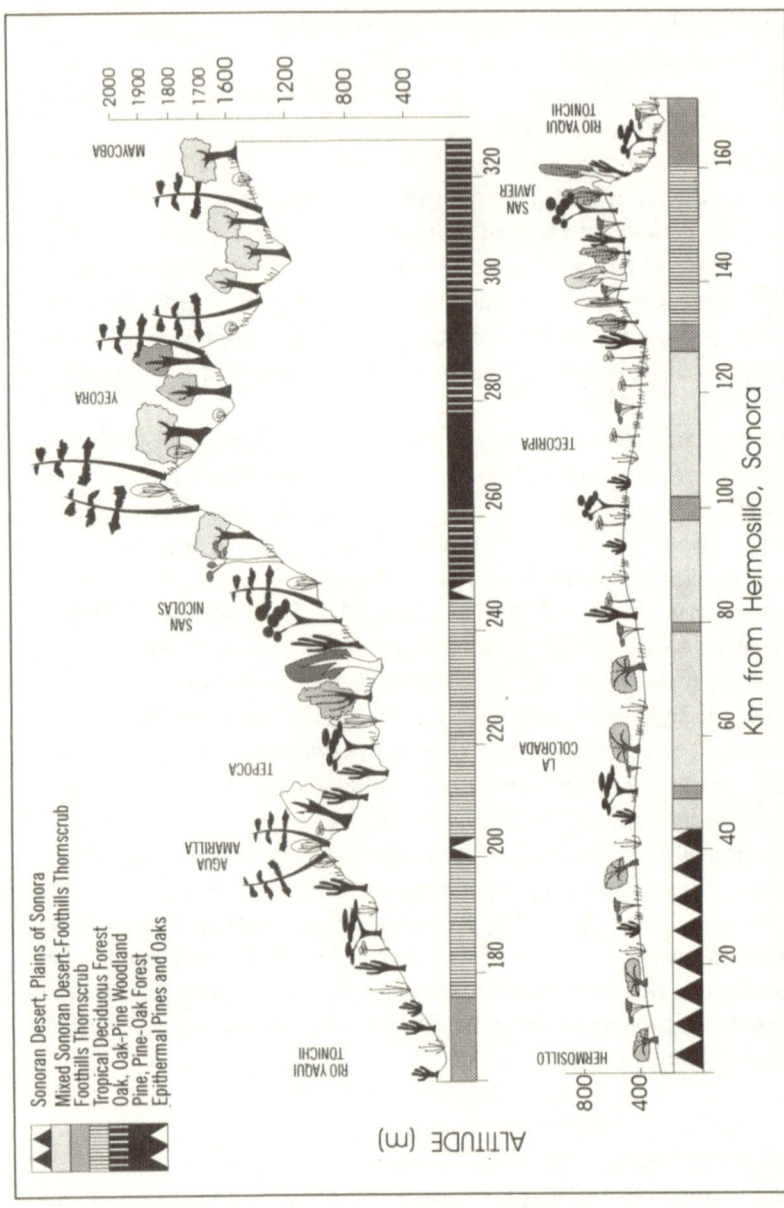

Figure 2.7 Diagramatic view of changes in vegetation along the road from Hermosillo to Maycoba (Highway 16, Sonora, Mexico). Distances are as marked on the road signs. The section from Bahía Kino to Hermosillo, although not illustrated, is discussed in the text. Modified from Búrquez et al. (1992a).

Altitudinal changes along this route are typical of the transition from the southern Sonoran Desert to tropical deciduous forest to temperate forests. In this section we discuss these changes from sea level at the Gulf Coast eastward to the Sierra Madre foothills.

Marine macroalgae communities are well developed in shallow coastal waters (Norris 1978). In addition, there are extensive sea grass meadows. These consist of submerged pastures of *Zostera marina* and *Ruppia maritima* growing along protected, subtidal waters in sandy-muddy substrates (Felger et al. 1980). These are best developed in the Canal del Infiernillo and also occur at Estero de la Cruz at Bahía Kino and from the Guaymas region southward to central Sinaloa.

Littoral scrub, including mangroves and saltscrub, occurs at desert river deltas, lagoons, and esteros. Mangroves (*Avicennia germinans, Laguncularia racemosa,* and *Rhizophora mangle*) reach their distribution limits in North America at about 29°20'N along the shores of the Canal del Infiernillo, with a small outlier population of *Avicennia* at Puerto Lobos (30°15'N). Mangroves bordering the desert are relatively dwarfed and usually shrubs less than 5 m tall. Southward, mangroves gradually increase in extent and height.

Coastal vegetation is varied and has sharp boundaries imposed mainly by environmental factors such as exposure to prevailing winds, degree of rockiness, drainage, salinity, and dew deposition. These factors are usually represented at the extremes, resulting in a few dominant species. Coastal dunes are common at the mouths of large arroyos or major rivers, forming narrow bars along the coast. These are different from continental dunes like the ones in the Gran Desierto region that result from long-distance aeolian transport (Felger 1980) and do not have a strong coastal influence. However, these dune systems share a large proportion of their specialized flora, having a remarkable uniformity of species. Aside from Felger and Lowe (1976) and Felger (1980, 1992), few studies have been conducted on the distribution of the vegetation and flora of these dunes.

Along tidal channels, leeward sides of sand dunes, and depressions isolated from the sea, saltscrub appears next as a narrow band of salt grasses (*Jouvea pilosa, Monanthchloe littoralis,* and *Sporobolus virginicus*), halophytic shrubs (*Allenrolfea occidentalis, Atriplex barclayana, Maytenus phyllanthoides,* and *Suaeda moquinii*) and succulent forbs (*Batis maritima, Salicornia* spp., and *Sesuvium* cf. *verrucosum*) as the major elements (Felger and Lowe 1976).

Desertscrub is well represented in the vicinity of Bahía Kino. This vegetation is readily ascribed to the Central Gulf Coast subdivision of the Sonoran Desert. However, further inland (eastward) the vegetation is a complex patchwork without clear boundaries unless obvious environmental changes occur. Common Central Gulf Coast perennials include *Bursera microphylla* (elephant tree), *Fouquieria splendens, Jacquinia macrocarpa* subsp. *pungens* (san juanico), *Jatropha cuneata, Olneya tesota, Viscainoa geniculata* (guayacán), and many others. The columnar cacti *Carnegiea gigantea, Pachycereus pringlei,* and *Stenocereus thurberi* are dense and have high recruitments, while *F. splendens, Krameria grayi* (white ratany), *Lophocereus schottii* (senita), and *Prosopis glandulosa* occur in close proximity to halophyte associations but on higher, non-saline ground. These species form veritable islands on sandy mounds within salt flats near the coast.

Mesquites and columnar cacti are dominant on slightly higher terrain. *Pachycereus pringlei* attains truly gigantic sizes, being by far the largest columnar cactus in the Sonoran Desert and one of the largest succulents in the world. The mainland *P. pringlei* forests occur along a narrow band less than 30 km wide, from Caborca southward near the coast to the Guaymas region, where it is replaced by its close relative *P. pecten-aboriginum* (Turner et al. 1995). To the east the mesquite forests on the ancient Río Sonora delta have been replaced by large-scale agriculture. Derelict farmland in this region attests to the failure of groundwater irrigation in the desert.

At Siete Cerros, a small linear succession of rocky hills surrounded by deltaic silt, the Central Gulf Coast vegetation gives way to the Plains of Sonora (about 55 km east of Bahía Kino). The vegetation becomes denser, with *Bursera laxiflora* (torote prieto), *B. microphylla, Carnegiea gigantea,* and *Olneya tesota* on the slopes. These hills also show a dense cover of the invasive newcomer *Pennisetum ciliare* (buffelgrass). Prominent species in the plains include *Cercidium microphyllum, C. praecox* (palo brea), *Encelia farinosa, Larrea tridentata, O. tesota, Stenocereus alamosensis* (sina), and *S. thurberi. Prosopis velutina* (velvet mesquite) occurs in the more mesic areas next to the main road and along irrigation canals. At Siete Cerros, *S. alamosensis* reaches its westernmost limit (Felger and Moser 1985). This vegetation continues eastward with little variation until reaching Hermosillo.

In the vicinity of Hermosillo, increased precipitation and changes in

Vegetation and Habitat Diversity 59

topography and geological substrate produce discontinuities in the vegetation. Granitic and calcareous ranges (Espinazo Prieto, La Cementera, Agualurca, and Santa Teresa) support thornscrub primarily on the north- and east-facing slopes, and desertscrub on the plains and more arid slopes. Prominent species of these sierras include *Acacia willardiana, Agave angustifolia, Bursera fagaroides, B. microphylla, Cercidium floridum* (blue paloverde), *C. microphyllum, C. praecox, Croton sonorae, Fouquieria macdougallii, Hechtia montana, Ipomoea arborescens, Jatropha cordata,* and *Stenocereus thurberi.* On the desert plains *B. laxiflora, B. microphylla, Encelia farinosa, Guaiacum coulteri* (guayacán), *Mimosa laxiflora, Olneya tesota,* and *S. alamosensis* are common. Large areas of the Plains of Sonora have been cleared of natural vegetation to create *Pennisetum ciliare* pasture. This grass is widely naturalized, forming dense, almost monospecific stands in many areas. Colonization by *P. ciliare* introduces fire as a major factor in the dynamics of desert vegetation. Desert and thornscrub communities are not fire-adapted, and the establishment of *P. ciliare* is leading to irreparable destruction of the Sonoran Desert (Búrquez and Quintana 1994; Yetman and Búrquez 1994; Búrquez et al. 1998; M. Miller and A. Búrquez, unpublished data).

The Plains of Sonora desertscrub and thornscrub alternate with elevational changes along the transect eastward from Hermosillo. Initially, desertscrub vegetation is dominant with thornscrub as isolated islands, especially on north-facing slopes. Moving eastward, there is an increased diversity, density, and size of individual plants, particularly of tropical species including *Ceiba acuminata, Erythrina flabelliformis* (coral bean), *Ipomoea arborescens, Pachycereus pecten-aboriginum,* and *Piscidia mollis* (palo blanco). Farther east, near Tecoripa, thornscrub almost completely replaces the desertscrub. Before reaching the Río Yaqui crossing at Tonichi, the first tropical deciduous forest is seen on Sierra San Javier. The more xerophytic thornscrub and desertscrub lose their identity a few kilometers after the river crossing. Here many species typical of tropical deciduous forests are present, including *Bursera lancifolia, Ficus petiolaris* and scattered *F. insipida, Helicteres baruensis, Heliocarpus attenuatus, Lysiloma divaricatum, L. watsoni* (tepeguaje), *Montanoa leucantha, Randia echinocarpa* (papache), *Senna atomaria* (palo zorrillo), *Solanum erianthum, Tabebuia chrysantha* (amapa amarilla), *T. impetiginosa, Lonchocarpus hermannii,* and many others. At the Río Yaqui crossing, Sonoran

Desert species reappear only to quickly disappear as the elevation increases toward the Sierra Madre. From here the ascent is rapid as one traverses tropical deciduous forest, oak woodlands, and pine-oak forests.

The vegetation of the Gulf islands represents a special case. Due to differences in isolation, size, and topography, each island harbors distinct assemblages of animal and plant species. The vegetation and flora of Isla Tiburón, Mexico's largest island, is remarkably similar to that of the mainland — no doubt due to its close proximity and relatively recent connection to Sonora as well as its large area. Its vegetation includes sea grass meadows, mangroves, saltscrub, and dune vegetation along its coast, desertscrub on the plains, and thornscrub on the north-facing slopes of the highest mountain range (Felger and Lowe 1976). In contrast, Isla Angel de la Guarda has not been recently connected to the mainland and, despite its large size and relief, has a rich but relatively arid vegetation lacking thornscrub. Evidence of elevational zonation is exemplified by the presence of a population of *Fouquieria columnaris* (boojum tree) near the summit of the island (Moran 1983). The other smaller islands have special physical features (such as differences in topography, elevation, and geology), an absence of larger herbivores, and probably haphazard colonization, which make them singular in terms of biodiversity. Angel de la Guarda and smaller islands all support endemic plant and animal species (see Case and Cody 1983).

Conclusions

Vegetation in northwestern Mexico forms a continuum from the dry shores of the Gulf of California to the foothills of the Sierra Madre Occidental. The climate changes from extremely xeric to tropical and temperate in the short distance of 200 km. The multitude of plant communities and ecotones are an indication of the exceeding richness of habitats. The gradual change in floristics, structure, and ecosystem dynamics makes vegetation difficult to classify and complicates the delimitation of the boundaries of the Sonoran Desert. Present land-use policies in Sonora, particularly the clearing of large tracts of land for introduced exotic grasslands, are forcing the Sonoran Desert toward a new, less complex, and less diverse ecological status. Given the paucity of data for this region, there is an urgent need to document the pace of change and to set aside significant areas for conservation.

Acknowledgments

We thank Adrían Quijada, Silvia Núñez, Trinidad Quintero, and Ma. de los Angeles Quintana for technical support. Alberto Búrquez and Angelina Martínez-Yrízar are deeply indebted to Prof. Paul S. Martin for sharing his enthusiasm and knowledge of the transition between the desert, the tropical deciduous forest, and the oak woodlands. We thank Dr. Thomas Van Devender for his valuable comments. This work was partly funded by grants to Alberto Búrquez from Consejo Nacional de Ciencia y Tecnología, México, 080N-9106, and Programa de Apoyos a Proyectos de Investigación e Innovación Tecnológica, UNAM IN-212894. Richard Felger acknowledges support from the Wallace Genetic Foundation and the Wallace Research Foundation and thanks Michael F. Wilson for assistance with the manuscript.

Note

Added in proof: Several important papers concerning the region have recently appeared. Among these are contributions in a special issue of the *Journal of the Southwest* 39 (3–4), 1997, devoted to northwestern Sonora, Baja California, and southwestern Arizona.

Literature Cited

Arbingast S. A., Blair C. P., Buchanan J. R., Gill C. C., Holz R. K., Martin R., Morris C. A., Ryan R. H., Bonero J. E., Weiler J. R. (1975) *Atlas of Mexico*. University of Texas Bureau of Business Research, Austin.

Bowers J. E. (1980) Catastrophic freezes in the Sonoran Desert. *Desert Plants* 2: 232–236.

Braun R. H., Espericueta M. (1979) Biomasa y producción ecológica de jojoba (*Simmondsia chinensis* Link) en el desierto costero de Sonora. *Deserta* 5: 57–72.

Brown D. E. (1982) Sinaloan thornscrub. *Desert Plants* 4: 101–105.

Brown D. E., Lowe C. H. (1980) *Biotic Communities of the Southwest*. USDA Forest Service General Technical Report RM-78. Rocky Mountain Forest and Range Experiment Station, Fort Collins, Colo.

Búrquez A. (1997) Distributional limits of Euglossine and Meliponine bees (*Hymenoptera: Apidae*) in northwestern Mexico. *Pan-Pacific Entomologist* 73: 137–140.

Búrquez A., Martínez-Yrízar A., Martin P. S. (1992a) From the High Sierra Madre to the coast: changes in vegetation along Highway 16, Maycoba to Hermosillo, Sonora. In: Clark K. F., Roldán-Quintana J., Schmidt R. H. (eds) *Geology and Mineral Resources of Northern Sierra Madre Occidental, Mexico*, pp 239–252. El Paso Geological Society, El Paso.

Búrquez A., Martínez-Yrízar A., Miller M. E., Rojas K., Quintana M. A., Yetman D. (1998) Mexican grasslands and the changing aridlands of Mexico: an overview and a case study in northwestern Mexico. In: Tellman B., Finch D., Hamre R., Edminster, C. (eds) *The Future of Arid Grasslands: Identifying Issues, Seeking Solutions*. USDA Forest Service Proceedings RMRS-P-3, Rocky Mountain Forest and Range Experiment Station, Fort Collins, Colo.

Búrquez A., Martínez-Yrízar A., Núñez S., Quintero T., Aparicio A. (1992b) Above-ground phytomass in a Sonoran Desert community. *American Journal of Botany* 79(Suppl 6): 186.

Búrquez A., Quintana M. A. (1994) Islands of diversity: ironwood ecology and the richness of perennials in a Sonoran Desert biological reserve. In: Nabhan G. P., Carr J. L. (eds) *Ironwood: An Ecological and Cultural Keystone of the Sonoran Desert*, pp 9–27. Conservation Biology Occasional Papers no. 1, Conservation International, Washington, D.C.

Case T. J., Cody M. L. (eds) (1983) *Island Biogeography in the Sea of Cortéz*. University of California Press, Berkeley.

Crosswhite R. K., Crosswhite C. D. (1982) The Sonoran Desert. In: Gordon G. L. (ed) *Reference Handbook on the Deserts of North America*, pp 163–319. Greenwood Press, Westport, Conn.

Ehleringer J., Mooney H. A. (1983) Productivity of desert and mediterranean-climate plants. In: Lange O. L., Nobel P. S., Osmond C. B., Ziegler H (eds) *Encyclopedia of Plant Physiology*. Physiological Plant Ecology IV, n.s., 12D: 205–231. Springer-Verlag, Berlin.

Ezcurra E., Equihua M., López-Portillo J. (1987) The desert vegetation of El Pinacate, Sonora, Mexico. *Vegetatio* 71: 49–60.

Ezcurra E., Felger R. S., Rusell A. D., Equihua M. (1988) Freshwater islands in a desert sand sea: the hydrology, flora, and phytogeography of the Gran Desierto oases of northwestern Mexico. *Desert Plants* 9: 35–44, 55–63.

Ezcurra E., Rodríguez V. (1986) Rainfall patterns in the Gran Desierto, Sonora, Mexico. *Journal of Arid Environments* 10: 13–28.

Felger R. S. (1966) Ecology of the Gulf Coast and islands of the Gulf of California. Ph.D. dissertation, University of Arizona, Tucson.

Felger R. S. (1980) Vegetation and flora of the Gran Desierto, Sonora, Mexico. *Desert Plants* 2: 87–114.

Felger R. S. (1992) Synopsis of the vascular plants of Northwestern Sonora, Mexico. *Ecologica* 2: 11–44.

Felger, R. S. (in press). Flora of Cañón del Nacapule: a desert-bounded tropical canyon near Guaymas, Sonora, Mexico. *Proceedings of the San Diego Society of Natural History*.

Felger R. S., Lowe C. H. (1967) Clinal variation in the surface-volume relationships of the columnar cactus *Lophocereus schottii* in northwestern Mexico. *Ecology* 62: 901–906.

Felger R. S., Lowe C. H. (1976) The island and coastal vegetation and flora of the northern part of the Gulf of California, Mexico. *Natural History Museum of Los Angeles County Contributions in Science* 285: 1–59.

Felger R. S., Moser M. B. (1985) *People of the Desert and Sea: Ethnobotany of the Seri Indians.* University of Arizona Press, Tucson.

Felger R. S., Moser M. B., Moser E. W. (1980) Seagrasses in Seri indian culture. In: Phillips R. C., McRoy C. P. (eds) *Handbook of Seagrass Biology, An Ecosystem Perspective,* pp 260–276. Garland STPM Press, New York.

Felger R. S., Warren P. L., Anderson S. A., Nabhan G. P. (1992) Vascular plants of a desert oasis: flora and ethnobotany of Quitobaquito, Organ Pipe Cactus National Monument, Arizona. *Proceedings of the San Diego Society of Natural History* 8: 1–39.

Franco A. C., Nobel P. S. (1989) Effects of nurse plants on the microhabitat and growth of cacti. *Journal of Ecology* 77: 870–886.

García E. (1973) *Modificaciones al sistema de clasificación climática de Köppen.* Universidad Nacional Autónoma de México, Instituto de Geografía, México DF.

García-Moya E., McKell C. M. (1970) Contribution of shrubs to the nitrogen economy of a desert wash plant community. *Ecology* 51: 81–88.

Gentry H. S. (1942) *Rio Mayo Plants: A Study of the Flora and Vegetation of the Valley of the Rio Mayo in Sonora.* Carnegie Institution of Washington Publication no. 527, Washington, D.C.

Gentry H. S. (1949) Land plants collected by the Velero III, Allan Hancock Pacific expeditions 1937–1941. Allan Hancock Pacific Expeditions 13: 7–245. University of Southern California Press, Los Angeles.

Gibson A. C., Nobel P. S. (1986) *The Cactus Primer.* Harvard University Press, Cambridge.

Glenn E. P., Felger R. S., Búrquez A., Turner D. S. (1992) Ciénega de Santa Clara: endangered wetland in the Colorado River Delta, Sonora, Mexico. *Natural Resources Journal* 32: 817–824.

Grubb P. J. (1977) The maintenance of species-richness in plant communities: the importance of the regeneration niche. *Biological Reviews* 52: 107–145.

Halvorson W. L., Patten D. T. (1975) Productivity and flowering of winter ephemerals in relation to Sonoran Desert shrubs. *American Midland Naturalist* 93: 311–319.

Hastings J. R., Humphrey R. R. (1969) *Climatological Data and Statistics for Sonora and Northern Sinaloa.* Technical Reports on the Meteorology and Climatology of Arid Regions no. 19. University of Arizona Institute of Atmospheric Physics, Tucson.

Hastings J. R., Turner R. M. (1965) *The Changing Mile.* University of Arizona Press, Tucson.

INEGI (1988) *Atlas Nacional del Medio Físico*. Instituto Nacional de Geografía Estadística e Informática, Aguascalientes.

Inouye R. S. (1991) Population biology of desert annual plants. In: Polis G. A. (ed) *The Ecology of Desert Communities*, pp 27–54. University of Arizona Press, Tucson.

Krebs, C. J. (1985) *Ecology: The Experimental Analysis of Distribution and Abundance* (3rd ed.). Harper & Row, New York.

Krizman R. D. (1972) Environment and season in a tropical deciduous forest in northwestern Mexico. Ph.D. dissertation, University of Arizona, Tucson.

Leopold A. (1950) Vegetation zones of Mexico. *Ecology* 31: 507–518.

Lieth H., Whittaker R. H. (eds) (1975) *Primary Productivity of the Biosphere*. Springer Verlag, New York.

Ludwig J. A. (1986) Primary production variability in desert ecosystems. In: Whitford W. (ed) *Pattern and Process in Desert Ecosystems*, pp 5–17. University of Albuquerque, Albuquerque, N.Mex.

Ludwig J. A. (1987) Primary productivity in arid lands: myths and realities. *Journal of Arid Environments* 13: 1–7.

Lumholtz K. (1912) *New Trails in Mexico*. Scribner, New York. (Reprinted by Rio Grande Press, Glorieta, N. Mex. 1971, and University of Arizona Press, Tucson 1990).

MacMahon J. A. (1979) North American deserts: their floral and faunal components. In: Goodall D. W., Perry R. A. (eds) *Arid-land Ecosystems: Structure, Functioning and Management*, pp 21–82. Cambridge University Press, Cambridge.

Martínez-Carretero E. M., Dalmasso A. D. (1992) Litter yield in shrubs of *Larrea* in the Andean piedmont of Mendoza, Argentina. *Vegetatio* 101: 21–35.

Martínez-Yrízar A., Búrquez A., Maass M. (in press) Structure and functioning of tropical deciduous forests in northwestern Mexico. In: Robichaux R. H. (ed) *Tropical Deciduous Forest of the Alamos, Sonora Region*. University of Arizona Press, Tucson.

Martínez-Yrízar A., Núñez S., Búrquez A. (1993) Producción de hojarasca en tres comunidades del Desierto Sonorense. In: *Resumenes del XII Congreso Mexicano de Botánica*. Sociedad Botánica de México, Mérida.

May R. (1973) Resource reconnaissance of the Gran Desierto. Master's thesis. University of Arizona Press, Tucson.

McAuliffe J. R. (1984) Prey refugia and the distribution of two Sonoran Desert cacti. *Oecologia* 65: 82–85.

McAuliffe J. R. (1995) Landscape evolution, soil formation and Arizona's desert grasslands. In: McClaran M. P., Van Devender T. R. (eds) *The Desert Grassland*. University of Arizona Press, Tucson.

Moran R. (1983) Vascular plants of the Gulf Islands. In: Case T. J., Cody M. L.

(eds) *Island Biogeography in the Sea of Cortez,* pp 348–381. University of California Press, Berkeley.

Mosiño P. A., García E. (1981) *Cantidad de lluvia mas frecuente (moda) en la República Mexicana.* Secretaría de Programación y Presupuesto, México DF.

Nabhan G. P., Rea A. M., Reichhardt K. L., Mellink E., Hutchinson C. F. (1982) Papago influences on habitat and biotic diversity: Quitovac oasis ethnoecology. *Journal of Ethnobiology* 2: 124–143.

Nabhan G. P., Suzán H. (1994) Boundary effects on endangered cacti and their nurse plants in and near a Sonoran Desert biosphere reserve. In: Nabhan G. P., Carr J. L. (eds) *Ironwood: An Ecological and Cultural Keystone of the Sonoran Desert,* pp 55–67. Conservation Biology Occasional Papers no. 1, Conservation International, Washington, D.C.

Norris J. N. (1978) Seri Indian seaweed knowledge: taxonomy, ecology and uses. In: *American Philosophical Yearbook 1977.* American Philosophical Society, Philadelphia.

Noy-Meir I. (1985) Desert ecosystem structure and function. In: Evenary M., Noy-Meir I., Goodall D. W. (eds) *Hot Deserts and Shrublands,* pp 93–104. Elsevier, New York.

Patten D. T. (1978) Productivity and production efficiency of an upper Sonoran Desert ephemeral community. *American Journal of Botany* 65: 891–895.

Paulsen H. A. (1953) A comparison of surface soil properties under mesquite and perennial grasses. *Ecology* 34: 727–732.

Petit-Maire N., Casta L. (1977) Un paléolac du nord-ouest Mexicain: La Playa San Bartolo. *Bulletin AFEQ* 50 (Suppl): 303–322.

Polis G. A. (1991) Desert communities: an overview of patterns and processes. In: Polis G. A. (ed) *The Ecology of Desert Communities,* pp 1–26. University of Arizona Press, Tucson.

Raisz E. (1964) *Landforms of Mexico.* 2nd ed. Geography Branch of the Office of Naval Research, Cambridge, Mass.

Rundel P. W., Franklin T. (1991) Vines in arid and semi-arid ecosystems. In: Putz F. E., Mooney H. A. (eds) *The Biology of Vines,* pp 337–356. Cambridge University Press, Cambridge.

Rundel P. W., Nilsen E. T., Sharifi M. R., Virginia R. A., Jarrell W. M., Kohl D. H., Shearer G. B. (1982) Seasonal dynamics of nitrogen cycling for a *Prosopis* woodland in the Sonoran Desert. *Plant and Soil* 67: 343–353.

Rzedowski J. (1978) *La Vegetación de México.* Limusa-Wiley, México DF.

Schmidt R. H. (1989) The arid zones of Mexico: climatic extremes and conceptualization of the Sonoran Desert. *Journal of Arid Environments* 16: 241–256.

Shreve F. (1914) The role of winter temperatures in determining the distribution of plants. *American Journal of Botany* 1: 194–202.

Shreve F. (1951) *Vegetation of the Sonoran Desert.* Carnegie Institution of Washington Publication no. 591. Washington, D.C.

Smith C. M. (1994) Seasonality and synchrony: a site comparison of reproductive phenology in three neotropical figs. Master's thesis. University of Arizona, Tucson.

Steenbergh W. F., Lowe C. H. (1977) *Ecology of the Saguaro: II. Reproduction, Germination, Establishment, Growth, and Survival of the Young Plant.* National Park Service Scientific Monograph Series no. 8. Washington, D.C.

Szarek S. T. (1979) Primary production in four North American deserts: indices of efficiency. *Journal of Arid Environments* 2: 187–209.

Tewksbury J. J., Petrovich C. A. (1994) The influence of ironwood as a habitat modifier species: a case study on the Sonoran Desert coast of the Sea of Cortez. In: Nabhan G. P., Carr J. L. (eds) *Ironwood: An Ecological and Cultural Keystone of the Sonoran Desert,* pp 29–54. Conservation Biology Occasional Papers no. 1, Conservation International, Washington, D.C.

Thames J. L. (1973) Tucson basin validation site report. Research Memorandum 73-3, United States/International Biological Program, Desert Biome, Tucson, Ariz.

Turner R. M. (1990) Long-term vegetation change at a fully protected Sonoran Desert site. *Ecology* 71: 464–477.

Turner R. M., Alcorn S. M., Olin G., Booth J. A. (1966) The influence of shade, soil, and water on saguaro seedling establishment. *Botanical Gazette* 127: 95–102.

Turner R. M., Bowers J. E., Burgess T. L. (1995) *Sonoran Desert Plants. An Ecological Atlas.* University of Arizona Press, Tucson.

Turner R. M., Brown D. E. (1982) Sonoran desertscrub. *Desert Plants* 4: 181–221.

Valiente-Banuet A., Ezcurra E. (1991) Shade as a cause of the association between the cactus *Neobuxbaumia tetetzo* and the nurse plant *Mimosa luisiana* in the Tehuacán Valley, México. *Journal of Ecology* 79: 961–971.

Van Devender T. R. (1990) Late Quaternary vegetation and climate of the Sonoran Desert, United States and Mexico. In: Betancourt J. L., Van Devender T. R., Martin P. S. (eds) *Packrat Middens. The Last 40,000 Years of Biotic Change,* pp 134–165. University of Arizona Press, Tucson.

Webb W. L., Lauenroth W. K., Szarek S. R., Kinerson R. S. (1983) Primary production and abiotic controls in forests, grasslands, and desert ecosystems in the United States. *Ecology* 64: 134–151.

West N. E. (1979) Formation, distribution and function of plant litter in desert ecosystems. In: Perry I. A., Goodall D. W. (eds) *Arid Land Ecosystems: Their Structure, Functioning and Management.* 1: 647–659. Cambridge University Press, Cambridge.

White S. S. (1948) The vegetation and flora of the region of the Río de Bavispe in northeastern Sonora, México. *Lloydia* 11: 220–302.

Whittaker R. H. (1975) *Communities and Ecosystems.* 2nd ed. Macmillan, New York.

Whittaker R. H., Niering W. A. (1975) Vegetation of the Santa Catalina Mountains, Arizona. V. Biomass, production, and diversity along the elevation gradient. *Ecology* 56: 771–790.

Willmott A. P., Búrquez A. (1996) The pollination of *Merremia palmeri* (Convolvulaceae): can hawkmoths be trusted? *American Journal of Botany* 83: 1050–1056.

Wiseman F. M. (1980) The edge of the tropics: the transition from tropical to subtropical ecosystems in Sonora, Mexico. *Geoscience and Man* 21: 141–156.

Yetman D., Búrquez A. (1994) Buffelgrass—Sonoran Desert nightmare. *Arizona Riparian Council Newsletter* 3: 1, 8–10.

Yetman D., Búrquez A. (1995) A tale of two species: speculation on the introduction of *Pachycereus pringlei* in the Sierra Libre, Sonora, Mexico by *Homo sapiens. Desert Plants* 12: 23–32.

3 The Sonoran Desert

Landscape Complexity and Ecological Diversity

Joseph R. McAuliffe

The composition of vegetation in the Sonoran Desert varies greatly from place to place. On a large geographic scale, considerable climatic differences across the region greatly affect plant density and species composition. For example, extremely arid regions of the Lower Colorado River Valley near Yuma, Arizona, annually receive an average of 50 mm or less of precipitation. In contrast, 300 km east of Yuma, semiarid parts of the Sonoran Desert near Tucson, Arizona, receive five to six times that amount. Consequently, desertscrub vegetation cover near Yuma is sparser and contains fewer species than the relatively lush "arboreal desert" in the vicinity of Tucson. Other pronounced vegetation differences among widely separated regions in the Sonoran Desert are in large part historic, involving the evolution and varying dispersal of different species. For example, partial geographic isolation of the Baja California peninsula from the mainland portions of the Sonoran Desert region in Sonora and Arizona contributes to the peninsula's distinct vegetation.

Considerable local variation in vegetation composition is layered on top of these large-scale biogeographic patterns. Over very small distances, often as little as a few meters to tens of meters, species compositions of plant communities may vary substantially in the Sonoran Desert. About a half century ago, Forrest Shreve emphasized that this local compositional variability is largely due to variation in soil characteristics:

> the character of the soil is of primary importance in determining the makeup and distribution of communities. The physical texture of the soil, its depth,

and the nature of its surface are equally important. . . . The profound influence of soil upon desert vegetation is to be attributed to its strong control of the amount, availability, and continuity of water supply. This fundamental requisite of plants is the most effective single factor in the differentiation of desert communities. (Shreve 1951)

Although Shreve may have generally recognized the importance of soils, his studies provided relatively few details of particular plant-soil relationships responsible for the "differentiation" of communities. By 1930, Shreve envisioned a research program at the Desert Laboratory that included investigations of soil moisture and soil-vegetation relationships in the Sonoran Desert. His plan, though, never fully materialized because the economic depression of the 1930s prevented the additions of a soil scientist and plant physiologist to the laboratory research staff (Bowers 1988; McLaughlin and Bowers, chapter 1, this volume). Shreve nevertheless did his best with limited assistance in the 1930s to investigate some aspects of soils and soil-plant relationships, but the few papers he published on this topic (Shreve 1934; Shreve and Mallery 1933; Shreve and Turnage 1936) probably only scratched the surface of the kinds of research problems he recognized.

Since Shreve's death in 1950, other ecologists have also attempted to investigate and explain the factors responsible for local variability in vegetation composition in the Sonoran Desert. However, like Shreve, ecologists have typically had little training and expertise in soil science. This lack has significantly impeded the development of a coherent picture of soil-vegetation relationships in the Sonoran and other warm deserts of North America.

Yang and Lowe (1956) concluded that the vegetation of the bajada (a commonly used term for the gently sloping piedmonts of alluvial fan deposits that flank mountains in the deserts of the southwestern United States) on the northwest side of the Tucson Mountains consisted of different communities determined by soil conditions, echoing the general statements of Shreve (1951). They identified two different communities, creosote bush–bursage and paloverde-saguaro, and related distributions of these to the texture and moisture-retention characteristics of soils. They concluded that the creosote bush–bursage communities of lower elevations on the bajada were associated with fine-textured soils that have higher moisture-retention characteristics, and that the paloverde-saguaro communities were associated with coarser soils of the upper bajada having lower moisture-retention capacity. These conclusions, though, were contrary to earlier results of Yang (1950) that showed that sites where creosote bush (*Larrea tridentata*) was dominant had coarser textures and lower moisture-holding capacities than did adjacent areas where creosote bush was absent. Despite these conflicting interpretations, Yang and Lowe (1956) and Yang (1957) pointed out that a wide area of the bajada actually consisted of a broad transition or ecotone where species compositions varied between these two ideal community types or "edaphic climax communities."

The recognition by Yang and Lowe (1956) of extensive compositional gradation of plant communities reflected the overturn of widely held Clementsian concepts of plant succession and climax plant communities (Clements 1916, 1936) that had recently begun in North American plant ecology (Whittaker 1951). The replacement paradigm for vegetation studies in the 1950s through the present has largely been a view of continuous change in plant community composition due to individualistic responses of species as originally elaborated by Gleason (1926). By the early 1970s, the representation of plant distributions and abundances as responses to ecological gradients became a dominant theme in the interpretation of soil-vegetation relationships in the Sonoran Desert.

Some of the research in the Sonoran Desert during the 1970s associated with the U.S./International Biological Program (US/IBP) Desert Biome project (Barbour and Diaz 1973; Solbrig et al. 1977; Phillips and MacMahon 1978) contributed to a general view of Sonoran Desert bajadas as ideal examples of spatial gradients in soil texture and associated community responses. Like the earlier work of Yang and Lowe (1956), these papers represented the soils of desert bajadas as coarse-textured in the upper bajada and gradually becoming finer in texture with increasing distance from the mountain front. However, the actual patterns of soils and vegetation in bajadas of the Sonoran Desert often show far greater complexity than can be pictured by this simple model.

For example, Solbrig et al. (1977) stated that in the Silver Bell bajada site of the US/IBP, located about 40 km west of Tucson (fig. 3.1), "soil texture becomes coarser with increased elevation." Yet a plot of their published, tabular data on soils for this site (fig. 3.2) shows that the actual pattern of soil texture along the bajada "gradient" is at best extremely irregular. Yang (1957) also reported similar spatial irregularities in soil texture along an

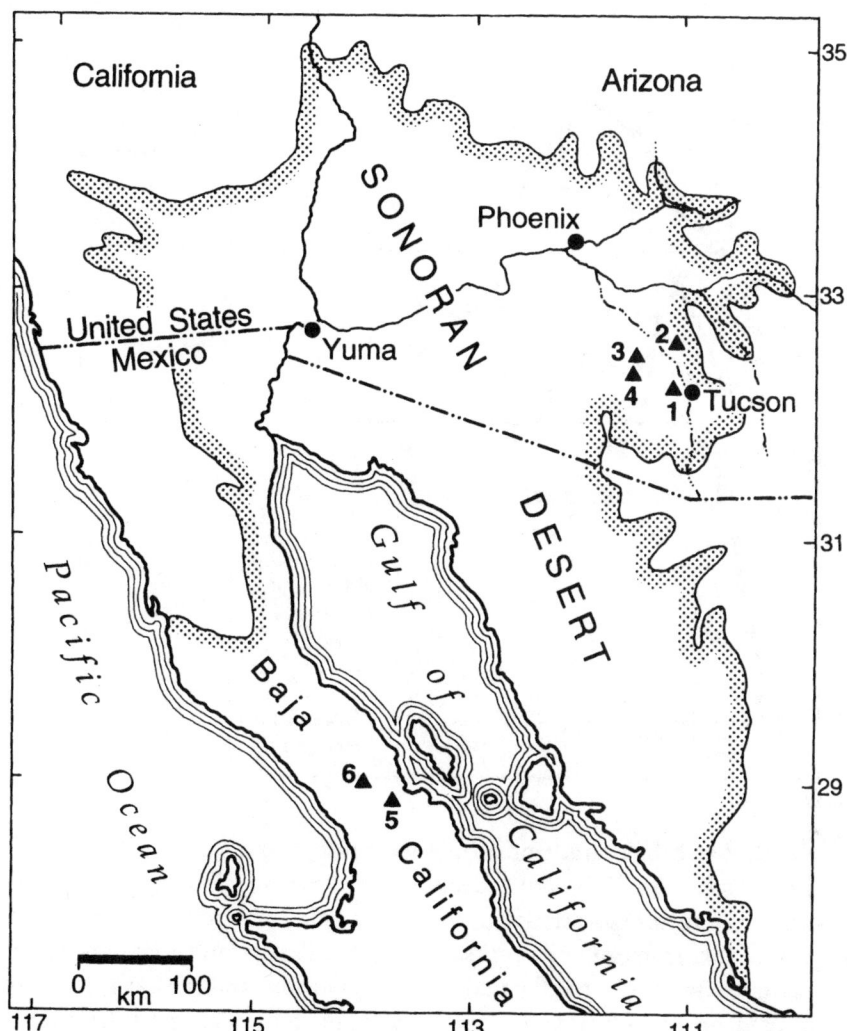

Figure 3.1 Map of the Sonoran Desert, excluding southern Baja California. Numbered triangles indicate study areas discussed in the text: (1) Tucson Mountains; (2) Tortolita Mountains; (3) Silver Bell Mountains; (4) Waterman Mountains; (5) Valle Montevideo; (6) Punta Prieta.

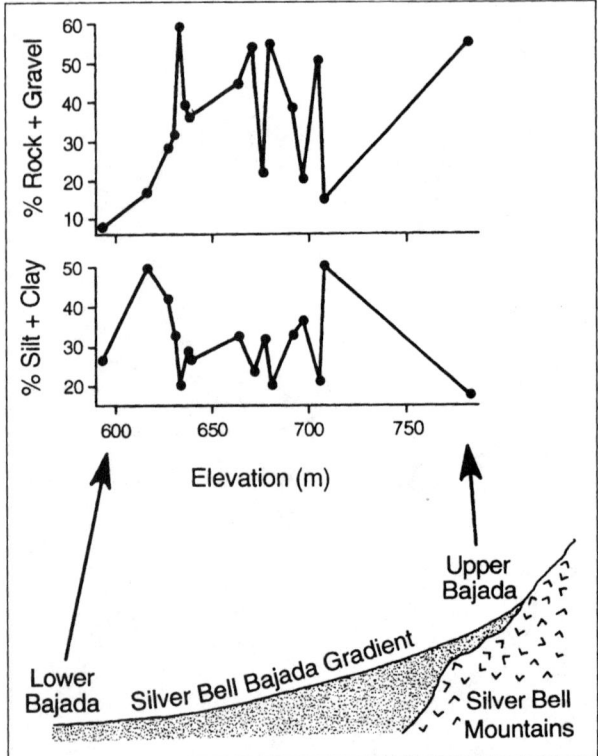

Figure 3.2 Soil attributes along the elevational gradient of the Silver Bell bajada gradient reported in Solbrig et al. (1977).

elevational bajada gradient on the west side of the Tucson Mountains, indicating a complexity not accounted for by the model of regular textural change along the elevational gradient. The gradient model also poorly described highly irregular patterns of plant distribution and abundance. For example, in the Silver Bell bajada IBP site, Barbour and Diaz (1973) commented on considerable variability of vegetation over short distances, where sixfold differences in local abundances of creosote bush made it difficult to find compositionally similar, replicate sampling plots within a limited area.

Ecologists' misunderstandings about soils and the nature of the physical landscape of the Sonoran Desert have been the greatest obstacle to making substantial research progress in deciphering landscape-vegetation relationships in the Sonoran Desert and other warm deserts of North America.

Knowledge about the physical nature of desert landscapes gained through the research efforts of geomorphologists and soil scientists is essential for advancing our understanding of ecological patterns and processes.

The Relevance of Geomorphology Research to Desert Plant Ecology

Knowledge about the geological evolution of alluvial piedmonts, or bajadas, of the Sonoran Desert is essential to ecologists for two reasons. First, episodes of alluvial deposition and erosion that change these landscapes can directly impact some plant populations, especially long-lived plants. Second, characteristics of soils change markedly over time, and knowledge of how and when various parts of the landscape were formed is a prerequisite for understanding and predicting the spatial distribution of different kinds of soils. Learning how to interpret the geological landscape and reconstruct histories of landscape evolution provides considerable new insights into many ecological phenomena. Much of the geological history of alluvial piedmonts in deserts can be reconstructed from readily observed surface features (Peterson 1981; Christenson and Purcell 1985).

For example, contrasting features of the alluvial piedmont flanking the northeast side of the Tucson Mountains provide evidence of how different parts of the landscape have been formed. A 3-km-long transect across the piedmont, paralleling the mountain front, from Wildhorse Wash southward, cuts through two areas that possess very different surface features (fig. 3.3A). Within 1.5 km to the south of Wildhorse Wash, the relatively planar, little dissected landscape contrasts strongly with the deeply dissected terrain of ridges and deep, V-shaped ravines further south (fig. 3.3A). Both parts of the landscape are composed of alluvial fan deposits, but the markedly contrasting surfaces indicate different histories for the two areas.

The deeply dissected terrain to the south is a highly eroded remnant of an ancient alluvial fan that may have been deposited on the order of two million years ago at the time of the Pliocene-Pleistocene transition (Katzer and Schuster 1984). The fan's original surface can be envisioned as a slightly convex plane positioned somewhat above the adjacent, subparallel ridges (fig. 3.3A). At any distance from the mountain front, the ridgeline remnants of this ancient fan deposit are usually higher in elevation than

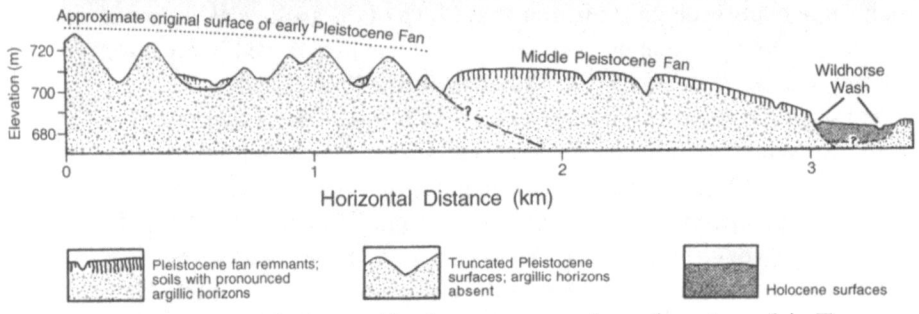

Figure 3.3 Aerial views and landscape cross-sections of portions of the Tucson Mountains piedmont (A, this page) and the Tortolita Mountains piedmont (B, next page). Arrows on aerial photos indicate position of cross-sections. Cross-sections are based on ground survey and elevational data obtained from 1:24000 topographic maps. Cross-section (A) progresses from south to north (left to right). Photographs by Peter Kresan (Tucson Mountains) and Kyle House (Tortolita Mountains).

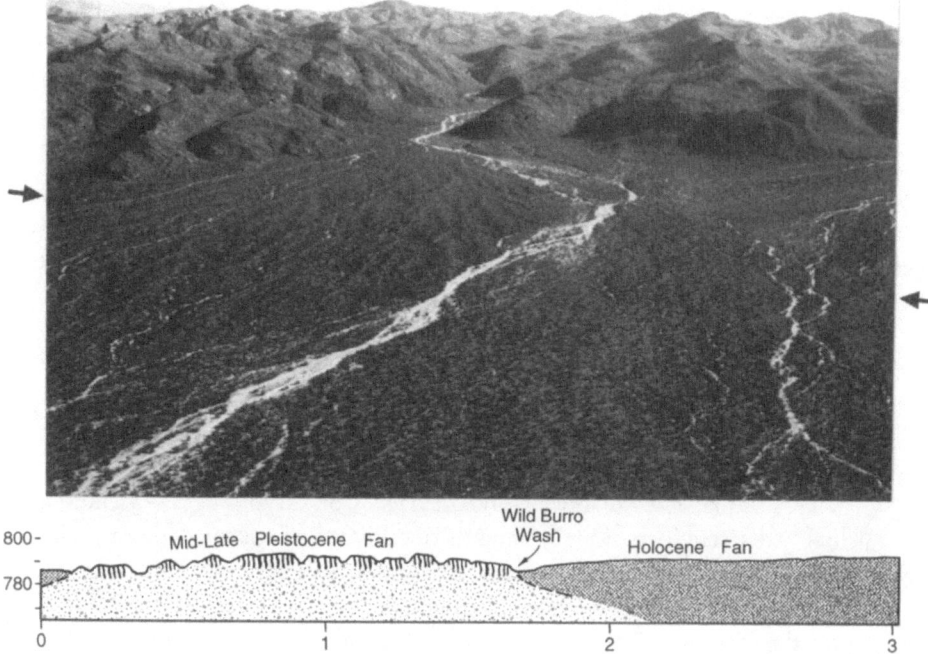

adjacent, younger fan deposits to the immediate north, apparently because the ancient fan was graded to a more elevated base level of the Santa Cruz Valley during the early Pleistocene (Katzer and Schuster 1984).

At a later time, probably during the middle Pleistocene, at least several hundred thousand years ago, another distinct alluvial fan was deposited to the immediate north of the early Pleistocene fan (fig. 3.3A). Active aggradation of the middle Pleistocene alluvial fan ended when the course of Wildhorse Wash switched to a more northerly course and incised into the northern side of the fan, where it remains to this day (Katzer and Schuster 1984). The middle Pleistocene fan retains the slightly convex, little broken surface configuration of the original fan. More recent fan deposits of late Pleistocene age (estimated by Katzer and Schuster [1984] to be 25 000 to 75 000 years old) and Holocene deposits (< 11 000 years old) are inset within the topographic confines of the more elevated, older surfaces to the north near the present course of Wildhorse Wash.

This sequence of distinct alluvial surfaces of varying ages—including highly dissected, erosional remnants of earliest Pleistocene deposits, relatively undissected surfaces of middle- and late-Pleistocene fans, and terraces of Holocene age—occurs throughout the rest of the piedmont flanking the Tucson Mountains, including the alluvial fan deposits to the immediate west of Tumamoc Hill, the site of the Desert Laboratory (McAuliffe 1994).

The morphology and spatial extent of different-aged alluvial fan surfaces varies from piedmont to piedmont within the desert regions of the Basin and Range province (Peterson 1981). Several factors contribute to this variability, including whether or not basin floors have become substantially lowered by erosion or tectonic action (base level changes) and lithology. Relatively rapid (in geological terms) drops in base level due to either marked incision of the basin floor by an external drainage or lowering of the basin floor by tectonic movement typically contributes to deeply incised piedmonts with pronounced, stair-stepped series of alluvial fan remnants of various ages (fig. 3.4A). In such cases, the most recent alluvial deposits (i.e., those of Holocene age, < 11 000 years old) may typically be restricted to narrow terraces inset within the topographic confines of older, more elevated and eroding surfaces. In the Sonoran Desert region of southern Arizona, the piedmont flanking the eastern slopes of the Tucson Mountains, the Upper Gila River Valley near Safford, the lower San Pedro Valley in the vicinity of San Manuel, and the Upper Verde River Valley in

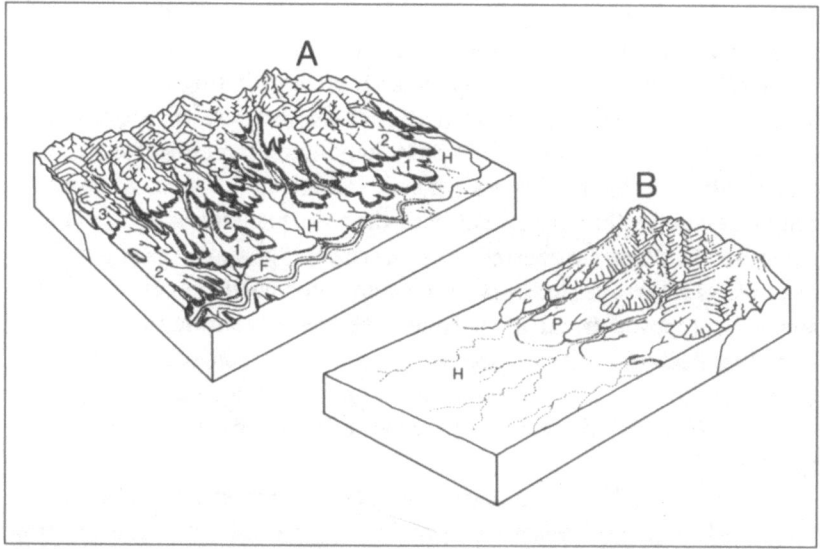

Figure 3.4 Block diagrams of contrasting alluvial piedmonts. (A) Basin exhibiting substantial base-level lowering during the Quaternary and containing stair-stepped sequence of alluvial surfaces with progressively younger surfaces inset within the topographic confines of older, remnant surfaces. Surfaces 1 to 3 represent Pleistocene surfaces (3 = oldest, 1 = youngest), H is a Holocene inset alluvial fan, and F represents floodplain deposits along the axial stream. (B) Alluvial piedmont lacking deep dissection and consisting of relatively small areas of a Pleistocene fan remnant (P) adjacent to the mountain front and more extensive Holocene fan deposits (H) in distal locations. Diagrams adapted from Peterson (1981).

the vicinity of Camp Verde and Cottonwood are good examples of piedmonts containing pronounced, stair-stepped series of elevated, remnant surfaces that are graded to the previously more elevated levels of ancient basin floors.

The types of rocks exposed to weathering and erosion in drainage basins of mountains also greatly influence the morphologies of alluvial piedmonts that flank the mountain slopes. The volume of sediment delivered by fluvial systems from the drainage basins to piedmont surfaces is greatly affected by the weatherability of different kinds of rocks (Bull 1991). Drainage basins with lithologies that weather rapidly to fine-grained particles (e.g., coarse-grained granitic rocks) typically yield large quantities of fine alluvium. Transport of abundant, fine materials derived from rapidly weathered rocks requires less energetic flows than does the transport of coarser, bouldery material derived from more weathering-resistant rocks (e.g., extrusive igneous rocks such as andesite and basalt). Given similar behavior of the basin floor, piedmonts that flank mountains composed of rapidly weatherable rock types typically possess larger areas of more recently deposited (Holocene) surfaces than do piedmonts of mountains composed of more weathering-resistant rocks (fig. 3.4B; Bull 1991). For example, the Tortolita Mountains, located immediately northwest of Tucson, are composed primarily of rapidly weathered granitic rocks, and the Tortolita piedmont is more extensively covered by relatively recent Holocene deposits (fig. 3.3B) than are piedmonts of other nearby mountain ranges composed of more weathering-resistant rhyolite and andesite (McAuliffe 1994).

In summary, alluvial piedmonts flanking the mountain ranges of the Sonoran Desert (and other desert regions of the Basin and Range Province) are mosaics of alluvial deposits of differing ages. Some parts of alluvial piedmonts are geologically young, having been deposited within the last few hundreds or thousands of years. Other surfaces are much older, having been deposited in the Pleistocene, many tens of thousands to more than a million years ago. Depending on their ages, surfaces of these deposits have also been variously modified by erosion.

As originally pointed out, the landscape dynamics of desert piedmonts influence the distributions and abundances of plants in two fundamental ways. First, relatively recent geological changes (either alluvial aggradation or erosion) can directly influence long-lived plants. Second, soil formation is in part dependent on the passage of time. The various ages of alluvial surfaces yield a complex mosaic of different soils. In the next section, I discuss some direct impacts of the most recent geological changes within piedmonts on plant populations and composition of plant communities. This is followed by sections on soil formation, soil moisture, and various soil-plant relationships in some different parts of the Sonoran Desert.

Direct Impacts of Landscape Dynamics on Plant Populations and Community Structure

Although physical changes of the piedmont's surface produced by alluvial aggradation are usually not evident within the time scale of most ecological investigations, they are extremely relevant to the study of long-lived desert plants. For example, individual creosote bush plants exhibit clone-like growth through basal stem splitting and outward propagation in the shape of a circular or elliptical ring (fig. 3.5A, B). Radiocarbon dating of old creosote bush wood in the interior of such rings at a site in the Mojave Desert indicates the clones increase in basal diameter at an average rate of slightly less than 1 mm per year (Vasek 1980). Creosote bush clones with diameters of several meters may have corresponding ages of several thousand years. For plants so long-lived, the sizes of plants and corresponding age structure of populations can directly reflect geological changes in the landscape that have occurred during the Holocene.

In the Silver Bell piedmont west of Tucson, Arizona, and in Valle Montevideo, central Baja California, density and average basal diameter (and probable age) of ringlike clones of creosote bush increase as a function of the ages of Holocene surfaces (McAuliffe 1991, 1994; fig. 3.6). Relative ages of these surfaces were determined by their topographic relationships and additionally, the degree of soil profile development (e.g., structural development of the B horizon and calcium carbonate accumulation). These time-dependent changes in desert soils are discussed later in more detail. Whereas late Holocene surfaces possess relatively low densities of small creosote bush plants, middle Holocene surfaces estimated to be at least a few thousand years old at both sites possess creosote clones up to several meters in basal diameter (McAuliffe 1991, 1994). Vasek (1983) originally pointed out that large clones of creosote bush in the Mojave Desert are common "only on stable surfaces of long duration."

The population and demographic responses of creosote bush to geologically dynamic landscapes ultimately contribute to considerable spatial variation in vegetation composition. In areas of geologically recent alluvial aggradation, creosote bush may be rare or even absent because inadequate time has passed for colonization by this shrub. Short-lived, rapidly dispersing species often predominate in such areas. For example, on the extensive Holocene-aged fan deposits associated with Wild Burro Wash on the southern side of the Tortolita Mountains near Tucson, densities of creosote bush and burrobush (*Hymenoclea salsola*) are strongly negatively correlated (fig. 3.7). Distributions and abundances of the two shrub species differ according to the spatial distribution of geologically recent landscape changes.

The area of the piedmont that was flooded in a record deluge emanating

Figure 3.5 (A, top) Creosote bush clone with basal diameter of approximately 2 m. White stick is one meter in length. Silver Bell piedmont near intersection of Avra Valley Road and Pump Station Road. (B, bottom) Large creosote bush surrounded by 16 people in Organ Pipe Cactus National Monument, Arizona. Basal diameter of this clone is 7.9 m × 5.1 m. April 1992.

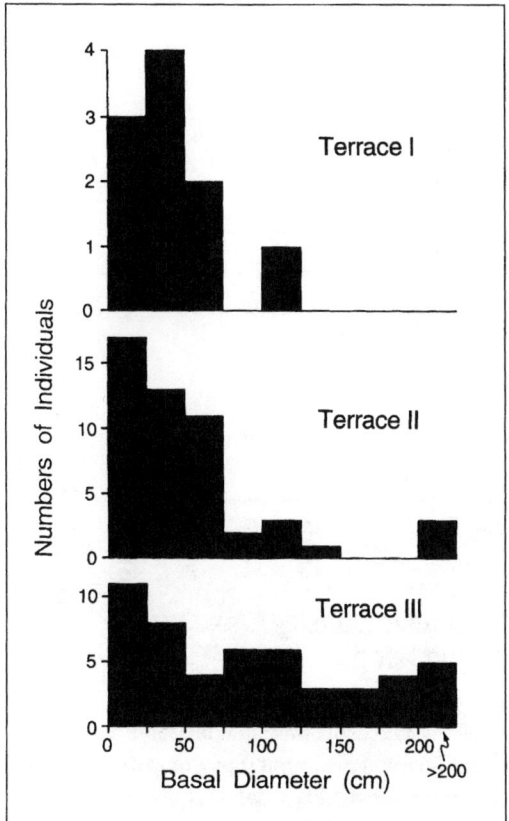

Figure 3.6 Maximum basal diameters of individual creosote bush on three Holocene terraces in Valle Montevideo, Baja California Norte. The scale for numbers of individuals differs for Terrace I because of the extreme rarity of creosote bush on this terrace.

from Wild Burro Canyon in 1988 (Pearthree 1991) closely matched the area in which burrobush is uniformly dominant and creosote bush relatively uncommon or absent (fig. 3.8). Creosote bush was generally absent from this zone even before the 1988 flood (personal observations of site from 1984–1986), indicating that similar episodic flooding and associated aggradation of alluvium over an extended period of recent geological history has largely precluded colonization of these areas by creosote bush. A Hohokam archaeological site containing a pithouse of the Tanque Verde cultural period (1200–1350 A.D.) was excavated near the edge of this broad zone where creosote bush is rare but burrobush is abundant. The pithouse was buried beneath 1.2 m of alluvium (Katzer and Schuster 1984; see fig. 3.8 for location of site) and demonstrates the extent of

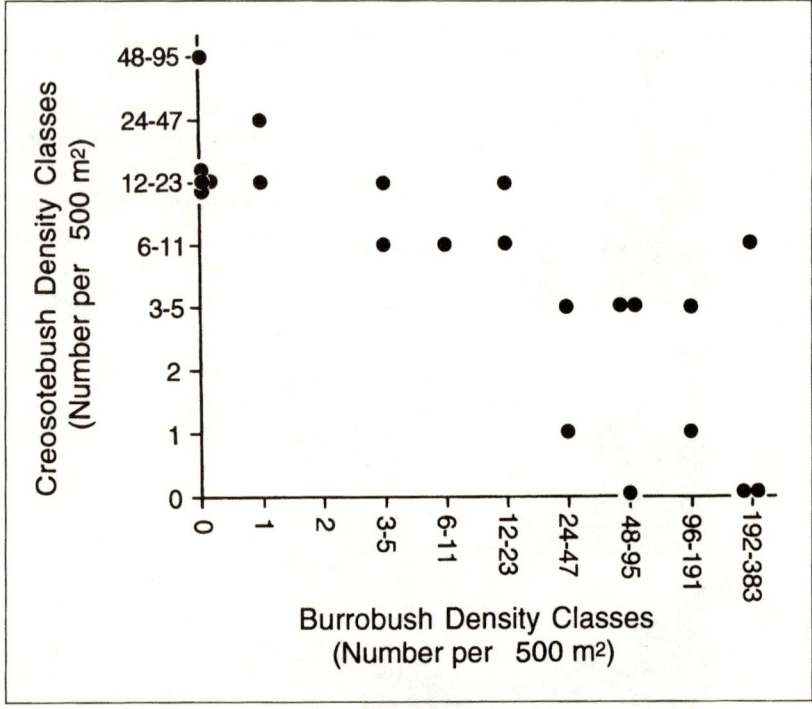

Figure 3.7 Negative correlation between creosote bush and burrobush in 0.1 ha plots located in the lower part of the Tortolita piedmont (Spearman rank correlation = −.82, $p < .01$). Density classes represent a logarithmic scale with geometric midpoints of 1, 2, 4, 8, 16, etc.

alluvial deposition that has occurred within the area over the last six to seven centuries.

Similar responses of creosote bush to geologically recent landscape changes have been identified in other areas. Webb et al. (1988) also documented the absence or rarity of creosote bush and commonness of burrobush in areas subject to recent alluvial aggradation in the Death Valley area of the Mojave Desert. In Valle Montevideo in the Central Desert of Baja California, creosote bush is nearly absent in areas of relatively recent alluvial aggradation. Such areas are instead dominated by short-lived shrub species such as golden-eye (*Viguiera* sp.) and a second species of burrobush (*Hymenoclea monogyra*). On surfaces deposited during the

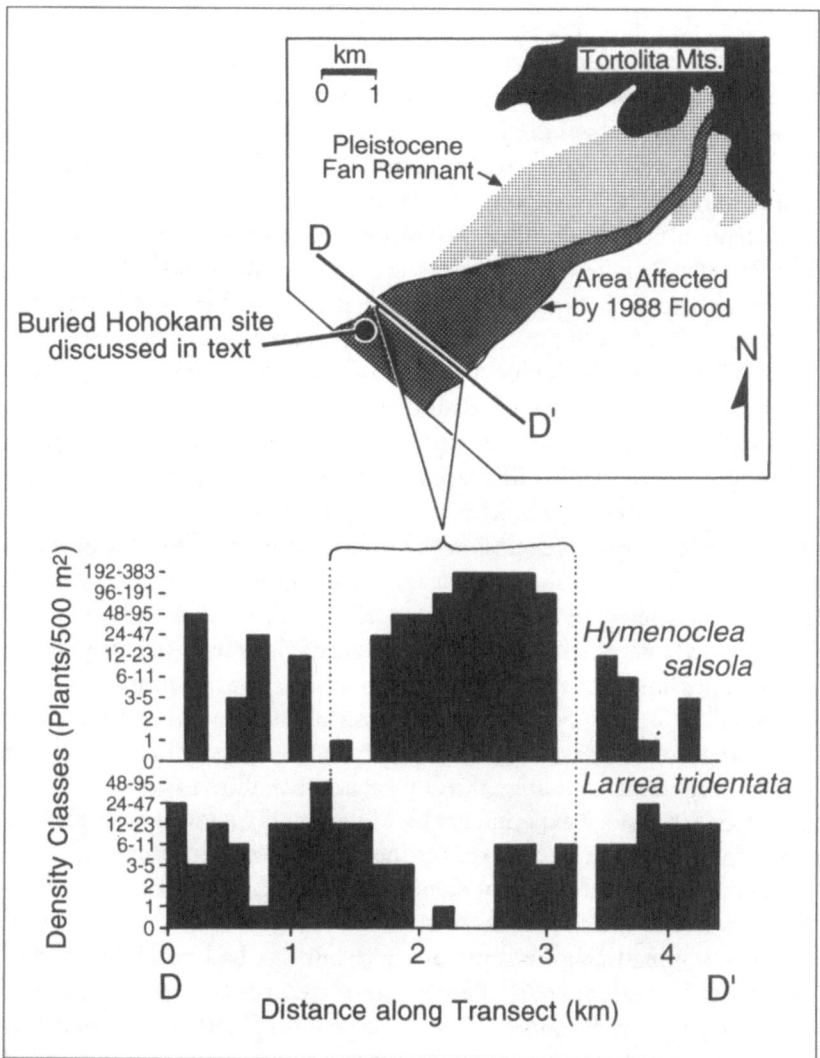

Figure 3.8 Association of high densities of burrobush (*Hymenoclea salsola*) and low densities of creosote bush (*Larrea tridentata*) in the area affected by latest Holocene alluvial aggradation along Wild Burro Wash, Tortolita Mountains piedmont. This area generally corresponds with the area affected by the 1988 flood. The Hohokam archaeological site located in this zone dated at approximately 650 to 800 years old was buried beneath 1.2 m of alluvium.

Holocene that have not experienced aggradation for probably more than a thousand years (estimated independently by means of soil profile development), creosote bush predominates (McAuliffe 1991).

Some of the changes in plant community composition associated with the increases over time in creosote bush densities and sizes of creosote bush clones may be due to competitive interactions. For example, on stable, late Holocene surfaces of the Silver Bell piedmont, creosote bush and triangle-leaf bursage (*Ambrosia deltoidea*) are equally abundant but on mid-Holocene deposits, triangle-leaf bursage is rare, and large, old creosote bush clones account for nearly all perennial plant cover. This reduction in bursage on somewhat older surfaces may be due to strong competition from creosote bush. As individual creosote bush slowly increase in size over long periods of time, they may increasingly monopolize space by inhibiting and eventually eliminating many perennial species (McAuliffe 1991, 1994). Vasek (1979/1980) originally suggested this long-term, slow scenario of competitive exclusion on the basis of an observed lack of establishment of white bursage (*Ambrosia dumosa*) in sites dominated by creosote bush, whereas white bursage rapidly established in adjacent, disturbed sites from which creosote bush was artificially removed. In addition to these possible interspecific competitive interactions, the rarity or complete lack of small, younger creosote bush plants in areas dominated by large creosote bush clones suggests that strong intraspecific competitive interactions prevent additional recruitment (McAuliffe 1994).

Recent laboratory experiments by Mahall and Callaway (1991) demonstrated a possible mechanism for these competitive interactions. Their experiments showed that root elongation of white bursage plants ceased in the presence of active roots of creosote bush, but that this interaction was not symmetrical: the roots of white bursage had no effect on root elongation of creosote bush. Elongation of creosote bush root tips ceases when they approach roots of other creosote bush plants. In environments where creosote bush clones slowly increase in size over long periods of time and their extensive root systems increasingly exploit the soil environment, these root interactions may inhibit establishment and survival of additional creosote bush plants and other species. However, since the changes in demographic and population responses of creosote bush and associated changes in community structure take so long to develop, they become apparent only in light of knowledge about the geological history and dynamics of landscapes over the last several thousand years.

In addition to creosote bush, populations of other long-lived desert plants can also be directly affected by the geological dynamics of Sonoran Desert landscapes. For example, the structure of populations of boojum trees (*Fouquieria columnaris*) and cardón cacti (*Pachycereus pringlei*) differs markedly among three juxtaposed alluvial terraces of Holocene age in Valle Montevideo, Baja California. Maximum heights of individuals of these two species change systematically as a function of increasing ages of the alluvial surfaces (fig. 3.9). On the basis of soil developmental characteristics, the oldest of the three surfaces (Terrace III) is probably within the range of 1000 to 4000 years old (McAuliffe 1991). Terrace III is also the same one supporting large creosote bush clones as previously discussed (fig. 3.6). In contrast, the two younger alluvial surfaces (Terraces I and II) lack any evidence of soil profile development, indicating a substantially lesser age, perhaps of only several centuries. Only on the oldest terrace do boojum trees and cardón cacti reach maximum heights (and ages probably exceeding 500 years; Humphrey 1991; R. M. Turner, personal communication).

In contrast to the great maximum sizes and corresponding ages of boojum trees and cardón cacti on Terrace III, populations of both species on the two younger terraces consist of relatively high densities of young plants. Tall, old plants are altogether lacking (fig. 3.9). Smaller sizes of plants on Terraces I and II cannot be attributed to diminished water availability and slower growth rates since the permeable, sandy soils of both terraces are next to a wash; Terrace III is located further from this source of water. Instead, the complete lack of large, old plants on the two younger terraces indicates that these surfaces have existed for a length of time that is considerably less than the maximum longevities of either boojum trees or cardón cacti (McAuliffe 1991).

These various examples of changes in population and plant community structure over centuries to several millennia are analogous to successional changes that occur in more humid regions (McAuliffe 1994). In arid regions, these changes take a long time to develop because of the extremely infrequent establishment and slow growth of long-lived plants. Ecologists have debated whether or not succession occurs in desert plant communities ever since Forrest Shreve's original disagreements with the writings of Frederick Clements (Shreve 1925, 1951; Shreve and Hinckley 1937; Muller 1940; Wells 1961; Vasek 1979/80, 1983; Zedler 1981; Webb et al. 1987, 1988; McAuliffe 1988; Bowers 1988). Understanding the

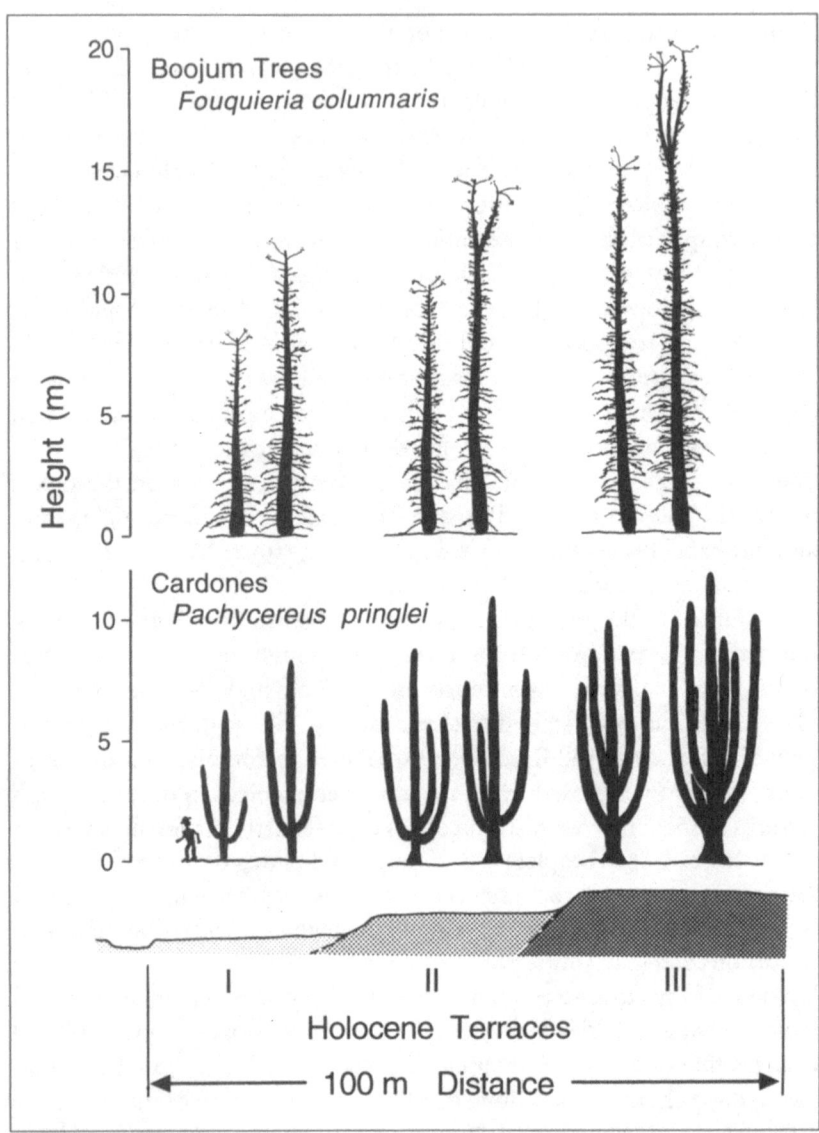

Figure 3.9 Variation in maximum heights attained by boojum trees and cardones on three Holocene alluvial terraces in Valle Montevideo. Each pair of figures represents the height range of the five tallest plants found on each terrace within a 10 ha study area (see McAuliffe 1991).

successional dynamics of many desert plant communities requires an expansion of thinking to time scales far more familiar to geologists (Webb et al. 1988; McAuliffe 1991, 1994).

Spatial Variability of Soils and Soil Moisture

In the previous section, I showed examples of how geological changes in alluvial piedmonts directly affect some plant populations and communities. However, not all of the heterogeneity in vegetation composition within Sonoran Desert piedmonts can be explained by the direct impacts of landscape dynamics. For example, on the east side of the Tucson Mountains (fig. 3.3A), mid-late Pleistocene alluvial fans have provided stable surfaces for more than the duration of the entire Holocene, yet creosote bush is typically very rare or absent on these surfaces. In this case the absence of creosote bush is attributed to soil conditions, not the lack of long-term surface stability (McAuliffe 1994).

Soil Development

Plant ecologists have generally expressed a poor understanding of desert soils, referring to them as "immature" (Solbrig et al. 1977) or lacking significant pedogenic development (Walter 1973). Research by soil scientists and geomorphologists over the last 40 years in the warm deserts of North America provides a far more detailed picture of the characteristics, development, and distribution of soils. Major progress in soils-geomorphic research in the southwestern United States largely began in 1957 with the Desert Soil–Geomorphology Project of the U.S. Soil Conservation Service (commonly known as the "Desert Project"). That project included 16 years of interdisciplinary studies of landform-soil relationships in the northern Chihuahuan Desert near Las Cruces, New Mexico (see Gile 1979; Gile et al. 1981; and references therein). Since those seminal investigations, further soils-geomorphic research has extended and refined the understanding of desert soils and soil-forming processes (Peterson 1981; Weide 1985; McFadden 1988; McFadden et al. 1986, 1989; Wells et al. 1987; McDonald 1994; Ruzicka 1994; and many others). This extensive body of knowledge provides an indispensable foundation for the advancement of ecological research on soil-plant relationships in North American deserts.

Like soils in more humid regions, soils of arid to semiarid regions vary as a function of many different factors, including type of parent material, topographic features, climate regimes, time over which soil development has occurred, and the effects of organisms (Gile 1975a, 1975b). All these factors vary considerably across the geologically diverse and dynamic landscapes of piedmonts in the Basin and Range province of the American southwest, producing mosaics of diverse soil conditions (Peterson 1981).

Soils on alluvial deposits of desert piedmonts vary substantially as a function of deposit age (Gile 1975a). The rocky to gravelly surface horizons of many desert soils often deceptively mask deeper, very strongly developed soil horizons that have formed in older alluvial deposits. Plant ecologists have sometimes relied exclusively on shallow, surface soil samples (< 10 cm depth) to characterize desert soils. Consequently, they have missed the strikingly different features that frequently exist in slightly deeper soil horizons.

Two prominent soil horizons typically form in alluvial fan deposits of arid and semiarid regions. These are an argillic horizon (a zone of clay accumulation) and a calcic horizon (a zone of calcium carbonate accumulation). The accumulation of both materials is time dependent. Depending on the lithology of the alluvium, clay minerals are partly derived from weathering of the coarse parent materials. In arid environments, wind-transported dust and materials deposited by precipitation are additional, major sources of clay minerals and calcium carbonate (Gile et al. 1981; Machette 1985; McFadden and Tinsley 1985; McFadden et al. 1986, 1987; McFadden 1988; Marion 1989).

Soil development over time in the more mesic parts of the Sonoran Desert (specifically in the Arizona Upland subdivision [Shreve 1951; Turner and Brown 1982] where average annual precipitation generally exceeds 200 mm) is similar to that described for areas in southern New Mexico studied in The Desert Project. Holocene fan deposits of noncalcareous alluvium typically exhibit very weak soil development with little or no accumulation of clay and calcium carbonate in distinct horizons (fig. 3.10A). Soils found on surfaces of late Holocene age (less than a few thousand years old) differ only slightly from fresh alluvium. Middle to early Holocene surfaces exhibit somewhat greater accumulations of clay. The accumulation of clays leads to cohesion of particles and structural development of the B horizon (called a structural B or Bw horizon). In such soils, small amounts of accumulated calcium carbonate are leached to somewhat greater depths and appear as thin white coatings on surfaces

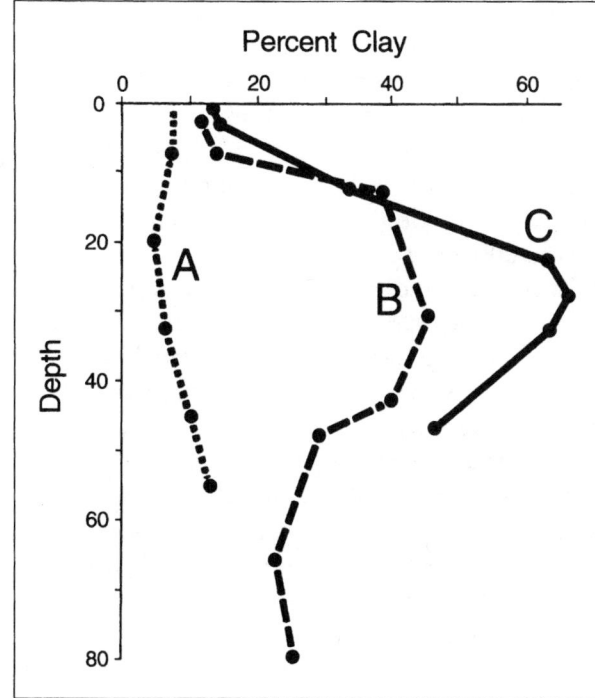

Figure 3.10 Clay accumulation as a function of soil age in alluvial fan deposits of coarse gravelly to rocky alluvium on east-facing piedmont of the Tucson Mountains. (A) Late- to mid-Holocene surface; (B) Middle Pleistocene surface; (C) Early-middle Pleistocene surface. Data from soil profiles "A," "C," and "D" of McAuliffe (1994, appendix C).

of pebbles and gravel particles and also as thin white filaments precipitated in areas of water extraction by fine roots (Gile et al. 1966, 1981; Machette 1985).

In contrast, soils that have been developing for tens to many hundreds of thousands of years on stable Pleistocene fan remnants exhibit very strong profile development with well-developed argillic and calcic horizons. Clay contents of argillic horizons increase as a function of age (fig. 3.10B, C; Gile and Grossman 1968; Gile et al. 1981). Oxidation and accumulation of iron-bearing minerals typically reddens the argillic horizon. Clay particles move downward into the soil in colloidal aqueous suspension, but calcium carbonate moves in solution. Consequently, in semiarid climates, calcium carbonate is transported deeper and accumulates in a distinct calcic horizon below the zone of maximum clay accumulation. Shallower distributions of calcium carbonate in soils of extremely arid environments are discussed later. On some of the older Pleistocene surfaces, the accumulation of calcium carbonate eventually may form a very strongly cemented calcic horizon (commonly known as "caliche").

Knowledge of the history of landform evolution is essential for understanding the complex mosaic of different soils that exists in virtually all alluvial piedmonts within the Basin and Range province (Gile et al. 1981; Peterson 1981). Plant ecologists have just begun to appreciate how these soil mosaics originate and how profoundly they influence plant distributions (McAuliffe 1991, 1994; Parker 1995).

The typical spatial distribution of different-aged surfaces across alluvial piedmonts in the Sonoran Desert's Arizona Upland often yields distributions of soil characteristics opposite to those represented in the bajada gradient model (see fig. 3.2; Solbrig et al. 1977). In most piedmonts, older (Pleistocene) fan remnants are typically located nearer the mountain fronts, whereas younger (Holocene) deposits become increasingly prevalent towards the basin floors (see fig. 3.4; McAuliffe 1994). The extremely high clay content and low gravel and rock content of some soils in the upper part of the Silver Bell piedmont (fig. 3.2) are due to the occurrence of well-developed soils with strong argillic horizons on Pleistocene fan remnants near the mountain front. Holocene fan deposits in the lowermost parts of piedmonts that have undergone considerably less pedogenic alteration have textures that are relatively uniform with depth and typically include loamy sand, sandy loam, and loam textural classes. However, these textures are considerably coarser than are the sandy clay loam, sandy clay, and clay textures of strongly developed argillic horizons in soils of many Pleistocene surfaces. Even though soils on older surfaces in upper parts of piedmonts may contain a large fraction of coarse gravel and cobbles, the argillic horizons of these soils typically have far higher clay contents than do any soil horizons found in Holocene alluvial deposits in lower parts of the piedmont.

In addition to differences in soil development among alluvial surfaces of varying ages, the differential erosion of some parts of the piedmont and the truncation or removal of well-developed soil horizons also contribute to the spatial variability of soils (Gile 1975b). Highly dissected surfaces of great antiquity (e.g., the ridge-and-ravine topography of the earliest Pleistocene fan deposits in the Tucson Mountains piedmont: fig. 3.3A), typically have soils from which the argillic horizons have been completely removed and remnants of calcic horizons are exposed at or near the surface. The relatively great flux of materials in hillslope settings (erosion of upslope areas and accumulation at the bases of slopes) interrupts pedogenesis and inhibits formation of strongly developed horizons. On these surfaces, soils are often highly variable and may be either deep or shallow depending on the degree of erosion and recent accumulation, and the degree to which the calcic horizons have been exposed, weathered, and eroded (Gile et al. 1981; Peterson 1981; McAuliffe 1994).

Soil Characteristics and Soil Moisture

As the quote from Shreve (1951) pointed out at the beginning of this chapter, a more fundamental understanding of plant-soil relationships requires knowledge of how soil characteristics affect the timing, quantity, and depth distribution of soil moisture. Texture has long been recognized as an important factor affecting infiltration and moisture availability in desert soils (Walter 1973). Variation in texture with depth due to soil horizonation leads to additional complexity in the spatial and temporal distributions of soil moisture (McAuliffe 1994, 1995).

Coarse-textured soils lacking argillic horizons are highly permeable, and water can potentially infiltrate and be stored at considerable depth. Greater depth of storage also insulates soil water from rapid evaporative losses to the atmosphere; consequently, relatively deep soil moisture supplies exhibit the least seasonal fluctuation (Noy-Meir 1973; Cable 1975; Monson and Smith 1982; Schlesinger et al. 1987). In contrast, even if a soil containing a strongly developed argillic horizon has a coarse-textured surface horizon that permits rapid initial infiltration, the high moisture-holding capacity of the underlying argillic horizon impedes deeper infiltration. As a consequence, depth of wetting in soils with well-developed argillic horizons is typically considerably less than in coarse textured soils lacking horizonation. Although upper parts of soil profiles with strong argillic horizons may be extremely moist for brief periods, these soils probably exhibit far greater seasonal fluctuations in the moisture availability than do coarser-textured soils of Holocene surfaces lacking strong horizonation (McAuliffe 1994, 1995; McDonald 1994; fig. 3.11A, B). Additionally, because of the infrequency and unpredictability of relatively large rainfall events in desert regions (see Turner 1963; Monson and Smith 1982 for typical annual patterns of individual precipitation events at two Sonoran Desert sites), water may seldom infiltrate beyond the uppermost

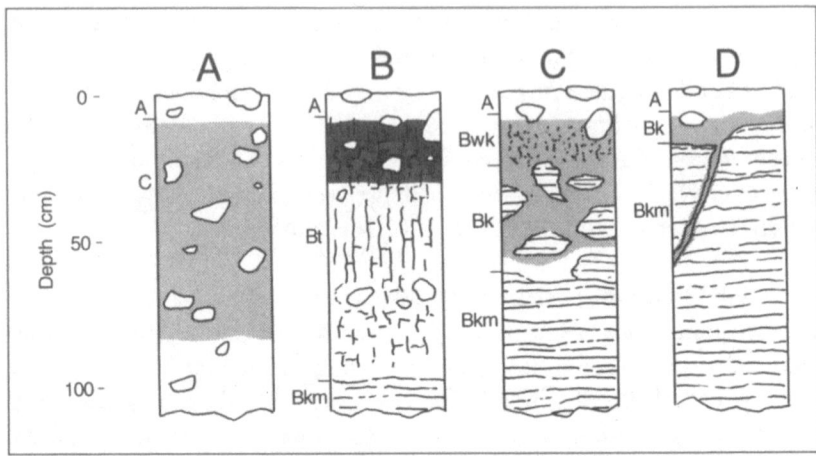

Figure 3.11 Vertical distributions of soil moisture following a moderate rain after the onset of a drying cycle. Stippling represents soil moisture; darker stippling indicates a higher percentage of water at field capacity. Names of soil horizons are on left side of each profile diagram. (A) Holocene surface of gravelly to sandy parent materials, lacking significant soil horizon development ("A" horizon overlying parent material "C"). (B) Pleistocene surface with thick, strongly developed argillic ("Bt") horizons. The "Bkm" horizon is a strongly cemented calcic horizon. (C) Pleistocene-aged surface from which argillic horizons have erosionally truncated and highly weathered calcic horizons serve as parent material. "Bwk" is a structural horizon with abundant carbonate. "Bk" is a calcic horizon containing highly weathered fragments of the former highly cemented calcic horizon. "Bkm" is the underlying highly cemented calcic horizon. (D) Erosionally truncated Pleistocene surface with a very shallow, highly cemented, massive calcic horizon ("Bkm") in which water may be stored in limited numbers of deep cracks.

parts of argillic horizons. Coarse-textured soils lacking strong horizonation would be moistened to greater depths by more frequently occurring, smaller precipitation events.

Some other common soil conditions may permit very little storage of soil moisture under any conditions. Erosionally truncated soils of some older Pleistocene surfaces contain shallow, strongly cemented calcic horizons very near the surface. In these situations, very little fine soil volume may be available for rooting or moisture storage. In these settings, the most persistent, yet limited moisture supply may be that retained in relatively deep cracks and fissures within the otherwise impermeable calcic horizon (fig. 3.11D).

Common Soil-Plant Relationships on Alluvial Piedmonts in the Northeastern Sonoran Desert

The variable quantity, vertical distribution, and seasonality of available soil moisture in different kinds of soils exert a strong control over the predominance of certain plants in the Sonoran Desert. The many different life-forms of desert plants (for example, water-storing succulents, drought-deciduous shrubs and trees, and drought-tolerant, evergreen shrubs) represent various morphological expressions of different modes of water uptake and use. In alluvial piedmonts in the Arizona Upland, where average annual precipitation exceeds 200 mm, there are many striking examples of the relationships between soil characteristics, soil water regimes, and the predominance of different plant life-forms.

Soils with strongly developed argillic horizons on Pleistocene fan remnants are usually dominated by species capable of extracting relatively shallow soil moisture when it is briefly available and then surviving lengthy periods when it is not. These surfaces typically possess a distinct suite of species dominated by triangle-leaf bursage — a small, drought-deciduous shrub — and an abundance of stem succulents, especially staghorn or buckhorn cholla (*Opuntia versicolor*, *O. acanthocarpa*) and occasionally prickly pears (*O. phaeacantha*). The foothill paloverde (*Cercidium microphyllum*), a small, drought-deciduous tree, is typically moderately abundant in these settings with highly seasonal soil moisture (fig. 3.12B; fig 3.13:4; McAuliffe 1994).

Water stored within the argillic horizons of these soils during winter and spring rainy seasons is rapidly exploited by relatively shallow-rooted bursage. With the depletion of these shallow moisture reserves at the

Figure 3.12 Typical vegetation present on different geomorphic surfaces on the east-facing piedmont of the Tucson Mountains. (A, top) Mid-late Holocene surface dominated by creosote bush. (B, center) Mid-late Pleistocene surface with well-developed argillic horizon dominated by triangle-leaf bursage and various *Opuntia* species. Creosote bush is typically absent or extremely rare on these surfaces. (C, bottom) Erosional sideslope of a deeply incised, early Pleistocene surface.

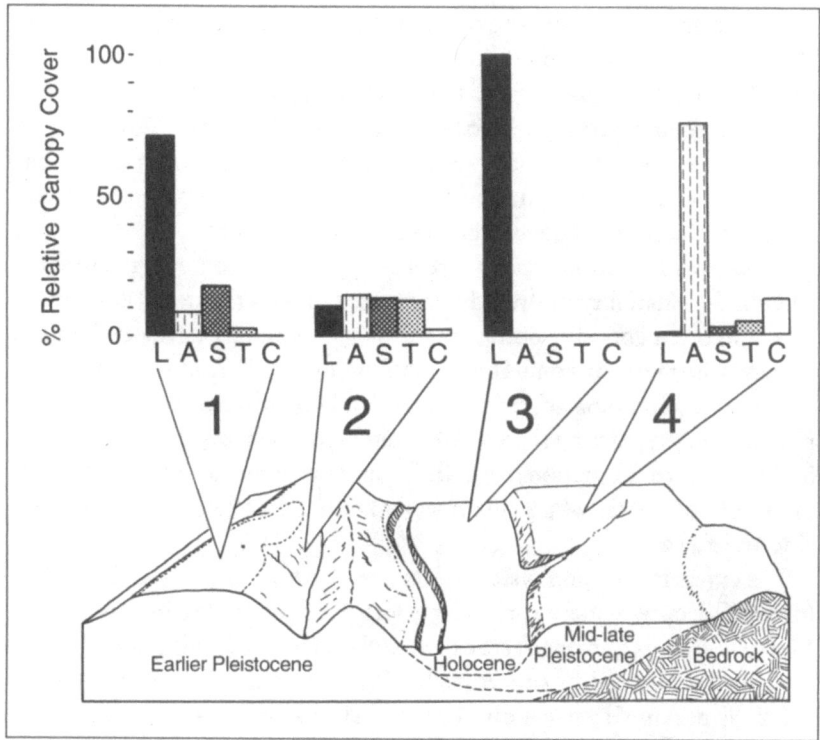

Figure 3.13 Relative percentage canopy cover of some different plant life-forms in three different geomorphic settings in the Tucson Mountains piedmont. "L" = creosote bush (*Larrea tridentata*), a drought-enduring, perennial evergreen; "A" = triangle-leaf bursage (*Ambrosia deltoidea*), a drought-deciduous shrub; "S" = other shrub species; "T" = trees, primarily foothill paloverde (*Cercidium microphyllum*); "C" = cacti, principally staghorn cholla (*Opuntia versicolor*) and prickly pear (*O. phaeacantha*). Site 1 is the relatively level summit of an early Pleistocene surface from which argillic horizons have been truncated but soil is relatively deep due to partial dissolution and weathering of calcic horizon (e.g., fig. 3.11C). Site 2 consists of erosional sideslopes of the same early Pleistocene surface. Site 3 represents mid-late Holocene surfaces with deep, coarse-textured soils. Site 4 represents Pleistocene surfaces with well-developed argillic horizons. Data compiled from 36 sites presented in McAuliffe (1994).

commencement of the two- to three-month-long presummer drought, this shrub loses its leaves and becomes dormant (Halvorson and Patten 1974; Szarek and Woodhouse 1977). The cacti, on the other hand, exhibit considerably greater activity in terms of water uptake from soils during the hot summer rainy period. Storage of water in succulent tissues rather than dormancy enables continuation of photosynthetic activity past the time when soil water no longer can be extracted by the roots.

In contrast, creosote bush, a relatively deep-rooted evergreen shrub lacking the capability of drought dormancy or water storage, is less common and often entirely absent from soils with strongly developed argillic horizons in most Arizona Upland communities. This may be due to a complex combination of factors including lack of water storage at sufficient depth, the extreme seasonal fluctuation of soil moisture, and possible inhibition of root function if argillic horizons become saturated, especially in "El Niño" years, when winter precipitation greatly exceeds the norm (McAuliffe 1994).

The extensive rooting system of creosote bush, consisting of both shallow and deeper, widely spreading roots (Cannon 1911; McAuliffe and McDonald 1995), is one factor that explains its predominance on Holocene surfaces of sandy and gravelly alluvium, where lack of strong soil horizons permits deeper infiltration and storage of water (fig. 3.12A; fig 3.13:3). In soils where restrictive horizons (for example, cemented calcic or strongly developed argillic horizons) inhibit relatively deep and extensive rooting, the size of creosote bush canopies is markedly diminished (Shreve and Mallery 1933; Cunningham and Burk 1973; McAuliffe and McDonald 1995).

Erosional sideslopes of some of the oldest alluvial landforms, such as those of the ridge-and-ravine topography of earliest Pleistocene surfaces of the Tucson Mountains piedmont (fig. 3.3A), contain extremely heterogeneous soil conditions. Strongly developed argillic horizons are usually absent, and soils may either allow or impede deep infiltration and storage of water depending on the presence or absence of shallow, cemented calcic horizons (fig. 3.11C, D). The greater geomorphic instability of these areas due to erosional losses or accumulation of gravity-moved materials (colluvium) during relatively recent geological history can greatly influence populations of long-lived plants like creosote bush, as discussed previously. As a consequence of these varied physical conditions, erosional hillslopes of highly dissected, early Pleistocene fan deposits typically contain some of the most diverse assemblages of desert plants (McAuliffe 1994). Within such hillslope environments in the Tucson area, creosote bush typically occurs in moderate abundance along with triangle-leaf bursage, white ratany (*Krameria grayi*), and foothill paloverde (fig. 3.12C; fig. 3.13:2). In addition, many other species of woody and suffrutescent perennials are found in these settings, including ocotillo (*Fouquieria splendens*), whitethorn acacia (*Acacia constricta*), sangrengado (*Jatropha cardiophylla*), Mexican crucillo (*Condalia warnockii*), desert zinnia (*Zinnia acerosa*), tiquilia (*Tiquilia canescens*), wolfberry (*Lycium* spp.), and paperflower (*Psilostrophe cooperi*).

Although these hillslope environments typically contain a diverse mix of species (fig. 3.13:2), summits of broad ridges directly above slopes are more stable parts of the landscape and are often dominated by large creosote bush clones in areas where argillic horizons have been truncated and highly cemented calcic horizons are not near the surface (fig. 3.13:1). On these summit areas, the predominance of long-lived creosote bush clones may represent the same kind of competitive dominance associated with long-term landscape stability that occurs on some of the older, stable Holocene fan surfaces as discussed earlier (McAuliffe 1994).

The mosaic distribution of different-aged alluvial landforms and their associated soils often produces relatively abrupt discontinuities in vegetation composition (fig. 3.13). Along elevational gradients within Sonoran Desert piedmonts, the variation in vegetation composition at any given elevation can be nearly as great as compositional variation from lower to upper piedmont locations. For example, in the Silver Bell piedmont, much more of the variance in the relative canopy cover of creosote bush among sites is explained by soil differences among various geomorphic surfaces than by position along the elevational gradient. On mid-Holocene surfaces throughout nearly the entire elevational range of the piedmont (600 to 700 m), creosote bush maintains about 80% relative canopy cover (fig. 3.14A). Large, old clonal creosote bush rings are typical in these sites and few other species are present. On more recently deposited, late Holocene surfaces, regardless of position along the gradient, creosote bush is less prevalent (approximately 50% relative canopy cover), large creosote clones are absent, and greater numbers of other species are present (fig. 3.14B). Pleistocene surfaces with strong argillic horizons have the lowest coverage of creosote bush along the entire elevational gradient (fig. 3.14D). Erosional hillslopes of early Pleistocene alluvial surfaces have

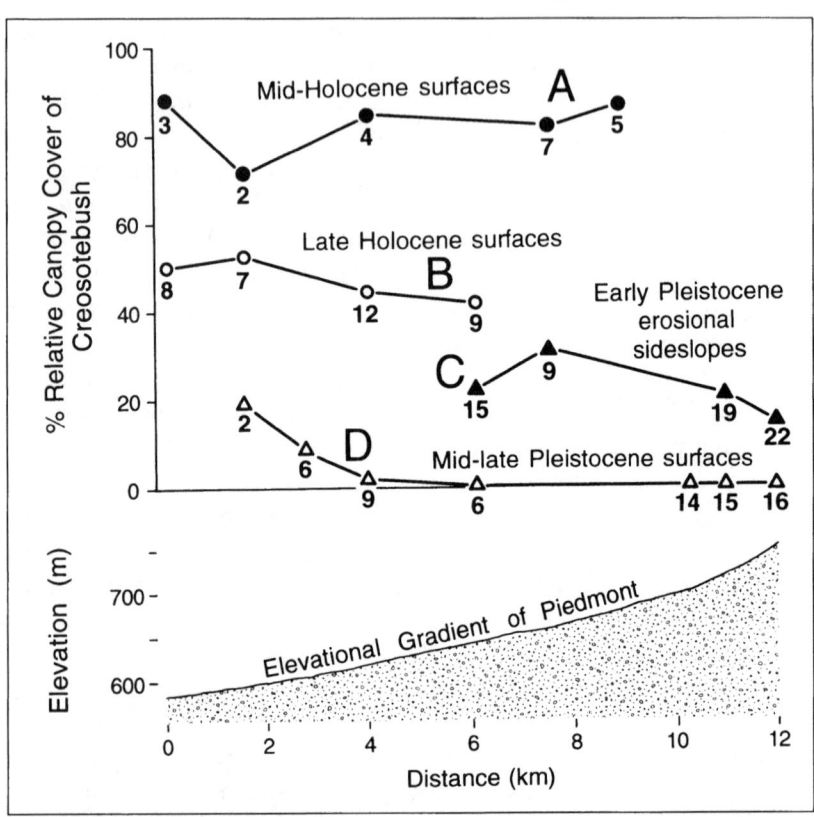

Figure 3.14 Contrasting dominance of creosote bush on four different geomorphic surfaces along the elevational gradient of the Silver Bell Mountains piedmont. (A) mid-Holocene surfaces; (B) late Holocene surfaces; (C) erosional sideslopes of deeply dissected early Pleistocene surfaces; (D) mid-late Pleistocene surfaces with strongly developed argillic horizons. Boldface numbers beneath individual data points indicate the total number of perennial plant species (including creosote bush) recorded within a 500 m^2 sampling area at each site (five 100 m^2 plots).

somewhat higher abundances of creosote bush (15% to 20% relative canopy cover; fig. 3.14C) and, at any given elevation, support greater numbers of species than do any of the other geomorphic surfaces (McAuliffe 1994).

Many details of soil moisture regimes and plant physiological responses responsible for soil-vegetation relationships are yet to be resolved. Nevertheless, a detailed understanding of desert soils and their spatial variability provides an important context for the study of ecophysiological mechanisms responsible for plant performance and distributions.

Mountainside Terrains of the Arizona Upland

The rocky slopes of mountains in the Sonoran Desert typically have extremely complex edaphic settings and associated variation in vegetation composition. This complexity is due to the multitude of combinations of varying slope inclinations, exposures, lithologies, and drainage basin morphologies. These topographically and geologically complex settings provide some of the most difficult challenges for accurately describing and interpreting substrate-vegetation relationships.

One example from the slopes of the Waterman Mountains west of Tucson, Arizona, demonstrates similarities between some of the soil-vegetation relationships in rocky uplands and alluvial piedmonts. On southern aspects in the Waterman Mountains below Waterman Peak, areas dominated by high densities of saguaros (*Carnegiea gigantea*) adjoin areas lacking them (fig. 3.15). Slope inclinations and aspect exposures of the two areas are identical. There are other marked differences between the two areas in vegetation composition. In the area dominated by saguaros, triangle-leaf bursage contributes approximately half of all perennial plant cover and creosote bush is virtually absent. In contrast, bursage is rare ($< 5\%$ relative cover) in the area in which saguaros are rare, while creosote bush is common in this area. The distinct plant assemblages are associated with different underlying lithologies and soils. Quartzite underlies the area with abundant saguaros and triangle-leaf bursage whereas limestone is the substrate harboring creosote bush.

While such stark vegetational differences between calcareous and noncalcareous substrates are well known in the Sonoran Desert region (Whittaker and Niering 1968; Wentworth 1981), the mechanisms underlying these differences have not been satisfactorily explained. At the Waterman site, the quartzite area is mantled by soils with moderate to strong argillic horizons. In contrast, argillic horizons are completely absent from the

Figure 3.15 Hillslopes of limestone (foreground) and quartzite (right background) parent materials located approximately 0.2 km southeast of Waterman Peak. Slope inclines on both sites are 20° and slope aspects are southeast. Elevation is 1040 m. Photograph taken 29 January 1993.

limestone area, since abundant carbonates impede development of clay-rich horizons in arid and semiarid environments (Gile et al. 1981). The manner in which the presence or absence of argillic horizons affects the temporal and depth distributions of soil moisture as previously discussed is probably one of the most important factors controlling vegetation composition in these two areas.

The predominance of different plant life-forms in the two areas is analogous to the relative dominance of different plant life-forms on fan deposits as a function of the presence or absence of argillic horizons. In addition to saguaros, hedgehog cacti (*Echinocereus* spp.), pincushion cacti (*Mammillaria grahamii*), and teddy-bear cholla (*Opuntia bigelovii*) are also abundant in the quartzite area. These cacti are considerably rarer in the limestone area. The predominance of triangle-leaf bursage and the abundance of cacti are indications of highly seasonally variable soil water conditions associated with argillic horizons. The surfaces of the soils in the quartzite area are relatively permeable sandy loams that facilitate infiltration, but underlying sandy clay loam- and clay-textured argillic horizons prevent deeper infiltration and water storage. In contrast, the presence of creosote bush in the limestone area can be attributed to the lack of argillic horizons and the associated extremely different spatial and temporal patterns of water availability (little storage in surface horizons, but less seasonally variable storage in deeper fissures and bedrock joints).

Vegetation Responses to Contrasting Soils in the Vizcaíno Region of Baja California

Throughout widely separated parts of the Sonoran Desert, the flora may change substantially, yet similar relationships between the predominance of various plant life-forms and soil conditions are evident. For example, at a central Baja California site near Punta Prieta (fig. 3.1), in landscapes characteristically occupied by giant boojum trees and cardón cacti, the vegetation composition and relative dominance of various perennial plant life-forms on older (Pleistocene-aged) deposits of fine-grained granitic alluvium differ sharply from the vegetation of adjacent Holocene deposits of the same granitic alluvium. Shallow-rooted century plants (*Agave shawii*) and small, drought-deciduous shrubs including *Ambrosia chenopodifolia*, desert buckwheat (*Eriogonum fasciculatum*), and golden-eye are the dominant occupants of soils in which strong argillic horizons have formed on the Pleistocene surfaces (fig. 3.16A). The predominance of shallow-rooted leaf succulents and small, drought-deciduous shrubs on these soils is analogous to the dominance of cacti and triangle-leaf bursage on Pleistocene surfaces with similar soils in the Arizona Upland near Tucson.

Creosote bush and *Yucca valida,* together with several other shrubs, are far more prevalent on the coarser soils of Holocene alluvial deposits that lack argillic horizons (fig. 3.16B). The more favorable conditions for deep-rooted creosote bush provided on the Holocene surface are indicated by sizes of plants. Creosote bush individuals on Holocene surfaces are considerably taller than are the few scattered individuals present on Pleistocene surfaces. In addition, creosote bush on Holocene surfaces often appear as clonal growth rings exceeding a meter in basal diameter, whereas such basal diameters are never achieved on soils with strong argillic horizons, despite the greater age of the Pleistocene surfaces. Instead, the few creosote bush on Pleistocene surfaces often possess numerous stubs of old, dead basal branches in varying states of decay, which indicate repeated,

Figure 3.16 Vegetation variation near Punta Prieta, Baja California Norte corresponding to presence or absence of argillic horizons. (A, top) Pleistocene-aged surface of fine gravelly granitic alluvium with strongly developed, reddened argillic horizon. Dominant species include century plants and drought-deciduous shrubs. (B, bottom) Holocene alluvial surface lacking argillic horizons where creosote bush and *Yucca valida* are dominant species.

past episodes of stem mortality, evidently caused by more extreme seasonal and interannual fluctuation of soil moisture (personal observations, March 1993).

Soil-Vegetation Relationships in Extremely Arid Regions: The Lower Colorado Subdivision

Materials that contribute to formation of soil horizons, including clay minerals, calcium carbonate, and soluble salts, are all moved downward in the soil by water. With diminished precipitation, these materials accumulate at shallower depths (McFadden and Tinsley 1985; McFadden 1988; Bull 1991). Across a precipitation gradient from semiarid to extremely arid parts of the Sonoran Desert in southern Arizona, soils found on mid-late Pleistocene fan remnants of noncalcareous parent materials exhibit progressively thinner and weaker argillic (Bt) horizons and shallower depth of calcium carbonate accumulation (Bk and Btk horizons) as a function of diminished precipitation (fig. 3.17). Substantial changes in vegetation patterns are associated with these geographic differences in soil characteristics. For example, in areas exceeding approximately 200 mm average annual precipitation, creosote bush is either uncommon or absent from soils with strongly developed argillic horizons on Pleistocene fan surfaces. However, in more arid areas that annually receive less than 200 mm average annual precipitation, creosote bush becomes a more ubiquitous component of vegetation across the entire landscape. Shreve (1951) also pointed out that while soils exerted a strong control on vegetation distributions in the more moist parts of the Sonoran Desert, soil-vegetation relationships "are more weakly manifested" in the driest parts of the region. The more weakly developed argillic horizons in the more arid Lower Colorado subdivision apparently do not inhibit creosote establishment and growth in the same pronounced way as do the exceptionally well-developed argillic horizons of soils in the less arid areas near Tucson (see McAuliffe 1994).

In addition to differences in argillic horizon development, Pleistocene fan remnants in extremely arid parts of the Sonoran Desert often possess distinctive desert pavements devoid of any vegetation (fig. 3.18A). In southern Arizona, desert pavements are generally limited to areas receiving less than 200 mm average annual precipitation and become increasingly pronounced in the driest parts of the state. Desert pavement typically consists of a single layer of tightly packed pebbles and small stones (fig.

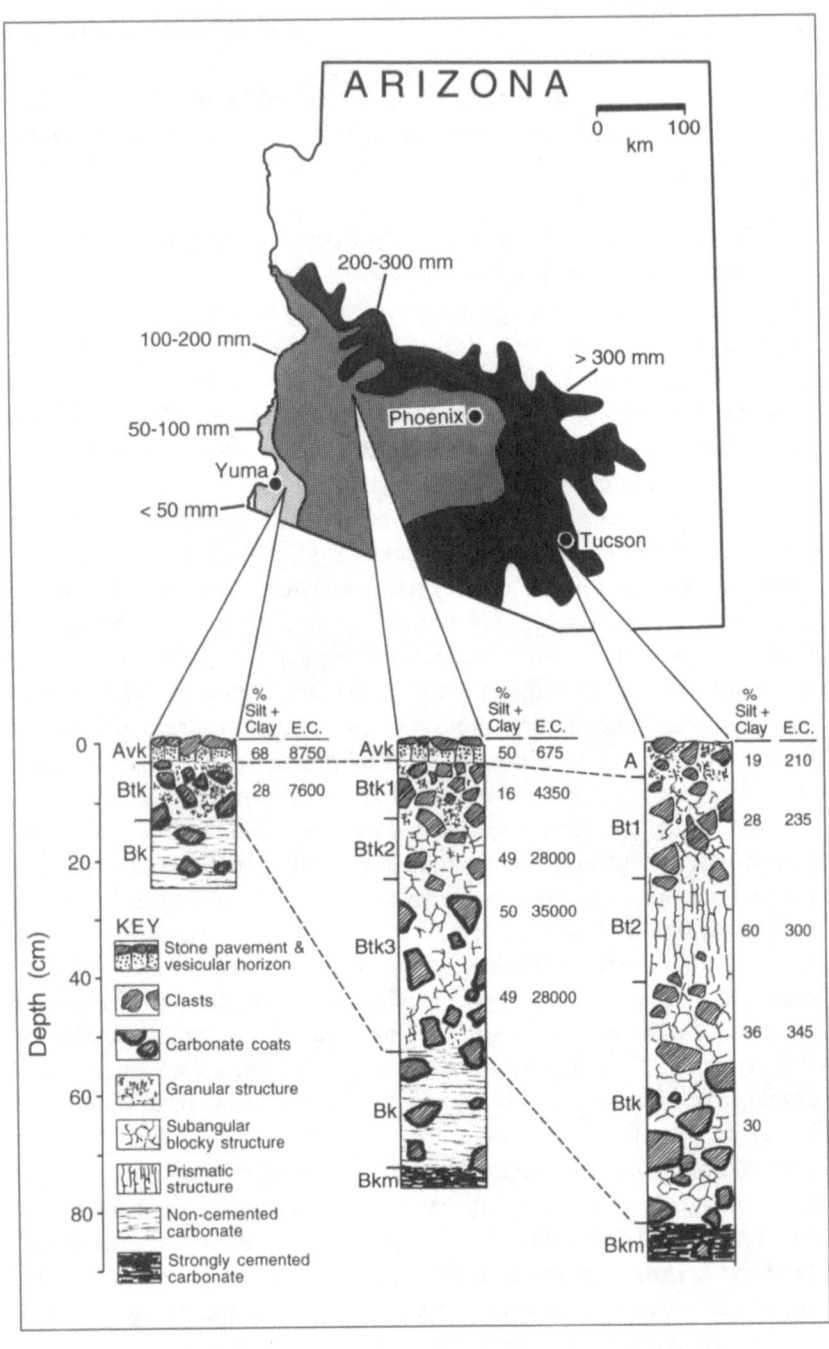

Figure 3.17 *(opposite page)* Variation in soil development as a function of climate on mid-late Pleistocene fan remnants with coarse gravelly to stony parent materials. Data on map of Arizona are average annual precipitation. The shaded areas in the southwest part of the state indicate precipitation zones within the limits of the Sonoran Desert. Precipitation isohyets are redrawn from Turner and Brown (1982). On the soil profile diagrams below the map, horizon names are listed on the left side of each profile. Note the substantial thinning of argillic ("Bt" and "Btk") horizons in extremely arid regions near Yuma versus semiarid regions near Tucson. Dashed lines between soil profile diagrams indicate the change in thickness of argillic horizons. Shallower depths of leaching in the more arid regions leads to calcium carbonate accumulation in all horizons (indicated by "k" suffix on horizon names). Soils in more arid regions possess vesicular surface ("Av") horizons; these horizons are absent in semiarid areas. Data to right side of each profile diagram include percentage silt + clay determined by the hydrometer method and electrical conductivity (E.C.) in micromhos/cm, measured from saturated soil paste extracts.

3.17; fig. 3.18). The exposed, upper surfaces of these pebbles are usually coated with dark, shiny desert varnish; thicker coatings of varnish develop with increasing age of the pavement (see Drake et al. 1993 for a recent discussion of desert varnish composition and formation).

A special soil horizon, the vesicular A (Av) horizon is found directly beneath the pebble pavement surface (fig. 3.17; fig. 3.18B). The Av horizon is extremely fine-grained, composed primarily of silt- and clay-sized particles (McFadden 1988; McFadden et al. 1987). Large vesicles or pores throughout this horizon are attributed to the expansion of soil gas as soil temperatures rapidly increase following wetting by summer storms (Evenari et al. 1974). On rocky or coarse, gravelly parent materials such as those of alluvial fan deposits, the silt and clays comprising the bulk of the Av horizon result from the deposition and entrapment of eolian materials (dust) on the stony surface and downward movement and incorporation of these eolian materials into the Av horizon (McFadden et al. 1987; McFadden 1988). As a consequence of the incorporation and accumulation of fine-grained eolian materials in the Av horizon beneath a thin surface layer of stones, this stone layer is separated from deeper stony parent materials and is maintained on the surface, forming the pavement (McFadden et al. 1987). In the more moist parts of the Sonoran Desert (> 200 mm average annual precipitation), the considerably greater amount of perturbation of

Figure 3.18 (A, top) Area of desert pavement located approximately 160 km west of Phoenix in the zone of 100 to 200 mm average annual precipitation. (B, bottom) Soil profile on Pleistocene-aged surface with strongly developed pavement on the east-facing piedmont of the Gila Mountains in an area receiving between 50 and 100 mm average annual precipitation (soil profile diagrammed on left side of fig. 3.17).

the soil surface through plant establishment and animal burrowing is one factor that may inhibit development of pavements and Av horizons.

Pavements and Av horizons have considerable impacts on soil water movement, the development of other soil features, and ultimately plant responses (McDonald 1994; McDonald et al. 1995, 1996; McAuliffe and McDonald 1995). Tightly packed pebble pavements inhibit infiltration of precipitation and promote runoff. The accumulation of silt and clay within the Av horizon further reduces permeability and infiltration of water, in turn leading to diminished depths of leaching and accumulation of materials such as calcium carbonate and soluble salts at shallow depths (McFadden 1988). In Yuma County, Arizona, where average annual precipitation is approximately 100 mm, Musick (1975) reported electrical conductivity and exchangeable sodium in the surface layers of pavement soils on the order of 30 times that of nonpavement soils. The considerably higher salt contents of soils associated with desert pavements found on older, Pleistocene fan deposits (see electrical conductivity data, fig. 3.17), may be a principal mechanism that prevents many plants from occupying these surfaces. On Pleistocene-aged alluvial fan surfaces with strongly developed desert pavements, vegetation is usually present only in runnels, which represent shallow erosional dissection and destruction of the pavement. Runoff from barren pavement areas to these shallow fluves leads to the amelioration of soil conditions, especially in the reduction of high salinities (Musick 1975).

The harsh edaphic conditions associated with extremely well developed pavements require long periods of geological time to form and are absent or poorly developed on most Holocene surfaces. Additionally, in parts of the landscape where erosional processes continually remove surface materials (e.g., ridge and ravine topography of deeply dissected early Pleistocene fan surfaces), pavements and associated saline soil conditions are eliminated. Consequently, piedmonts in the extremely arid parts of the Sonoran Desert are composed of various surfaces in which perennial plants exhibit markedly different patterns of spatial dispersion and growth responses. Vegetation on Pleistocene surfaces with strongly developed pavements is typically highly "contracted" (Walter 1973) to the narrow areas of shallow runnels that receive runoff and in which harsh soil conditions are ameliorated. In contrast, perennial plants typically exhibit much more diffuse patterns of distribution over erosional hillslopes and Holocene surfaces (fig. 3.19). Whereas creosote bush plants on Holocene alluvial

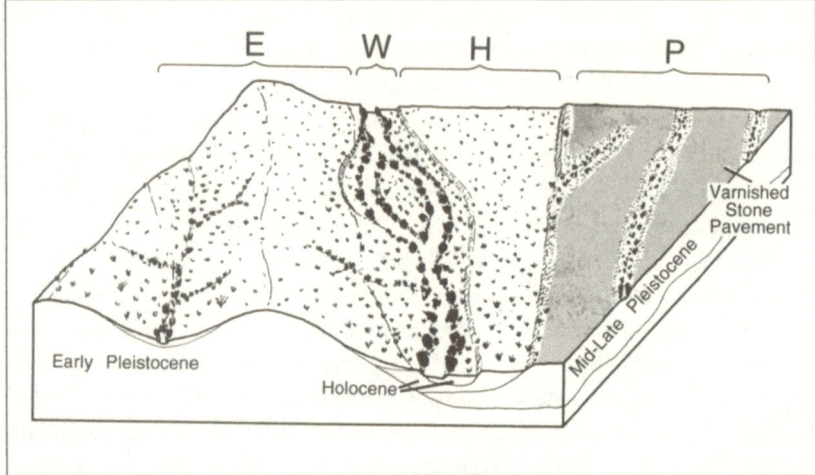

Figure 3.19 Substantial variation in dispersion patterns of perennial plants in more arid regions of the Sonoran Desert due to substantial soil differences among various geomorphic surfaces. Vegetation exhibits a highly "contracted" distribution on mid-late Pleistocene surfaces (P) with strongly developed pavements. Early Pleistocene surfaces (E) from which pavements and argillic horizons have been erosionally truncated and Holocene surfaces (H) lacking substantial pavement development or accumulation of salts in the horizon exhibit much more diffuse distributions of plants. Areas of concentration of runoff along washes (W) contain the densest vegetation.

surfaces typically reach large sizes (McAuliffe and McDonald 1995) and may be regularly dispersed, plants tend to be much smaller and tend towards random to highly aggregated dispersion patterns on erosionally dissected fan deposits and areas of desert pavements. Past comparative studies of size distributions and spatial dispersion of desert plants along precipitation gradients in the Sonoran Desert (e.g., Barbour 1969; Barbour et al. 1977) have overlooked the considerable heterogeneity of individual plant performance and community patterns produced by greatly contrasting soil conditions within a limited area.

Conclusions

Over the last half century, ecologists have repeatedly attempted to apply dominant paradigms and generalized theoretical views of community structure to Sonoran Desert environments without benefit of a solid understanding of the particulars of the history and formation of the physical landscape and soils. The consequence, in some cases, has been a series of rather poor fits between ecological models and ecological reality.

The research focus among many community ecologists during the last 30 years has typically led to focused studies of population dynamics and various ecological interactions, including competition, predation, herbivory, etc. However, before the consequences of these dynamic interactions can be well understood in Sonoran Desert plant communities, the heterogeneous landscape mosaic must first be accurately described and understood, because many biological processes strongly depend on the nature and behavior of the physical setting. For example, the demographic structure of creosote bush populations and potential impacts of this species on community structure make sense only in the context of the geological histories of various parts of the landscape.

The purpose of this chapter is to show the vital importance of the earth sciences to ecological studies of arid environments. Recent advances in geomorphology and soil science prove crucial to interpreting many ecological patterns and processes in the Sonoran Desert. Ecological syntheses that incorporate knowledge from the earth sciences lead to new explanatory hypotheses, productive lines of investigation, and improved models regarding processes that structure desert plant communities.

Acknowledgments

My work on soil-vegetation relationships in the Sonoran Desert began during a year's stay at the Desert Laboratory in 1986–1987. I especially thank Paul Martin for having made that brief stay possible and for encouraging me in the earliest phase of this work. Without that stay, I may never have taken my first steps in forging these links between desert plant ecology and the earth sciences. In retrospect, there could not have been a more historically fitting place for me to embark on these studies than the Desert Laboratory where Forrest Shreve first envisioned a broad research program on soil-plant relationships of the Sonoran Desert. Collaborative work with Leslie D. McFadden and Eric V. McDonald has greatly increased my understanding of soils and geomorphology in the American Southwest and work with Tony L. Burgess has sharpened my thinking about ecological responses of plants. The Desert Botanical Garden in Phoenix, Arizona, has generously supported much of my research since 1990. Comments and suggestions of P. Comas, P. Martin, C. McAuliffe, L. McFadden, and R. Robichaux greatly improved the original manuscript.

Literature Cited

Barbour M. G. (1969) Age and space distribution of the desert shrub *Larrea divaricata*. *Ecology* 50: 679–685.

Barbour M. G., Cunningham G., Oechel W. C., Bamberg S. A. (1977) Growth and development, form and function. In: Mabry T. J., Hunziker J. H., DiFeo D. R. (eds) *Creosotebush, Biology and Chemistry of Larrea in New World Deserts*, pp 48–89. Dowden, Hutchinson, and Ross, Stroudsburg, Pa.

Barbour M. G., Diaz D. V. (1973) *Larrea* plant communities on bajada and moisture gradients in the United States and Argentina. *Vegetatio* 28: 335–357.

Bowers J. E. (1988) *A Sense of Place: The Life and Work of Forrest Shreve*. University of Arizona Press, Tucson.

Bull W. B. (1991) *Geomorphic Responses to Climatic Change*. Oxford University Press, New York.

Cable D. R. (1975) Soil water changes in creosotebush and bursage during a dry period in southern Arizona. *Journal of the Arizona Academy of Science* 12: 15–20.

Cannon W. A. (1911) *The Root Habits of Desert Plants*. Carnegie Institution of Washington Publication no. 131. Washington, D.C.

Christenson G. E., Purcell C. (1985) Correlation and age of Quaternary alluvial-fan sequences, Basin and Range province, southwestern United States. In: Weide D. L. (ed) *Soils and Quaternary Geology of the Southwestern United States*, pp 115–122. Geological Society of America Special Paper 203. Boulder, Colo.

Clements F. E. (1916) *Plant Succession*. Carnegie Institution of Washington Publication no. 242. Washington, D.C.

Clements F. E. (1936) Nature and structure of the climax. *Journal of Ecology* 24: 252–284.

Cunningham G. L., Burk J. H. (1973) The effect of carbonate deposition layers ("caliche") on the water status of *Larrea divaricata*. *American Midland Naturalist* 90: 474–480.

Drake N. A., Heydeman M. T., White K. H. (1993) Distribution and formation of rock varnish in southern Tunisia. *Earth Surface Processes and Landforms* 18: 31–41.

Evenari J., Yaalon D. H., Gutterman Y. (1974) Note on soils with vesicular structures in deserts. *Zeitschrift fur Geomorphologie* 18: 162–172.

Gile L. H. (1975a) Causes of soil boundaries in an arid region; I. age and parent materials. *Soil Science Society of America Proceedings* 39: 316–323.

Gile L. H. (1975b) Causes of soil boundaries in an arid region; II. dissection, moisture, and faunal activity. *Soil Science Society of America Proceedings* 39: 324–330.

Gile L. H. (1979) *The Desert Project Soil Monograph*. U.S. Department of Agriculture, Soil Conservation Service, U.S. Government Printing Office, Washington, D.C.

Gile L. H., Grossman R. B. (1968) Morphology of the argillic horizon in desert soils of southern New Mexico. *Soil Science* 106: 6–15.

Gile L. H., Hawley J. W., Grossman R. B. (1981) *Soils and Geomorphology in the Basin and Range Area of Southern New Mexico—Guidebook to the Desert Project*. Memoir 39, New Mexico Bureau of Mines and Mineral Resources, Socorro.

Gile L. H., Peterson F. F., Grossman, R. B. (1966) Morphological and genetic sequences of carbonate accumulation in desert soils. *Soil Science* 101: 347–360.

Gleason H. A. (1926) The individualistic concept of the plant association. *Bulletin of the Torrey Botanical Garden* 53: 7–26.

Halvorson W. L., Patten D. T. (1974) Seasonal water potential changes in Sonoran Desert shrubs in relation to topography. *Ecology* 55: 173–177.

Humphrey R. R. (1991) Montevideo Valley and its tallest recorded cirio. *Cactus and Succulent Journal (U.S.)* 63: 239–240.

Katzer K. L., Schuster J. H. (1984) The Quaternary geology of the Northern Tucson Basin, Arizona, and its archaeological implications. Master's thesis, University of Arizona, Tucson.

Machette M. N. (1985) Calcic soils of the southwestern United States. In: Weide D. L. (ed) *Soils and Quaternary Geology of the Southwestern United States*, pp 1–21. Geological Society of America Special Paper 203. Boulder, Colo.

Mahall B. E., Callaway R. M. (1991) Root communication among desert shrubs. *Proceedings of the National Academy of Sciences* (U.S.A.) 88: 874–876.

Marion G. M. (1989) Correlation between long-term pedogenic $CaCO_3$ formation rate and modern precipitation in deserts of the American Southwest. *Quaternary Research* 32: 291–295.

McAuliffe J. R. (1988) Markovian dynamics of simple and complex desert plant communities. *American Naturalist* 131: 459–490.

McAuliffe J. R. (1991) Demographic shifts and plant succession along a late Holocene soil chronosequence in the Sonoran Desert of Baja California. *Journal of Arid Environments* 20: 165–178.

McAuliffe J. R. (1994) Landscape evolution, soil formation, and ecological patterns and processes in Sonoran Desert Bajadas. *Ecological Monographs* 64: 111–148.

McAuliffe J. R. (1995) Landscape evolution, soil formation, and Arizona's desert grasslands. In: McClaran M. P., VanDevender, T. R. (eds) *The Desert Grassland*, pp. 100–129. University of Arizona Press, Tucson.

McAuliffe J. R., McDonald E. V. (1995) A piedmont landscape in the eastern Mojave Desert: examples of linkages between biotic and physical components. In: Reynolds R. E., Reynolds J. (eds) *Ancient Surfaces of the East Mojave Desert*, pp 53–63. San Bernardino County Museum Association Quarterly 42(3).

McDonald E. V. (1994) The relative influences of climatic change, desert dust, and lithologic control on soil-geomorphic processes and hydrology of calcic soils forms on Quaternary alluvial-fan deposits in the Mojave Desert, California. Ph.D. dissertation, University of New Mexico, Albuquerque.

McDonald E. V., McFadden L. D., Wells S. G. (1995) The relative influences of climate change, desert dust, and lithologic control on soil-geomorphic processes on alluvial fans, Mojave Desert, California: Summary of results. In: Reynolds R. E., Reynolds J. (eds) *Ancient Surfaces of the East Mojave Desert*, pp 35–42. San Bernardino County Museum Association Quarterly 42(3).

McDonald E. V., Pierson F. B., Flerchinger G. N., McFadden L. D. (1996) Application of a process-based soil-water balance model to evaluate the influence of Late Quaternary climate change on soil-water movement. *Geoderma* 74: 167-192.

McFadden L. D. (1988) Climatic influences on rates and processes of soil development in Quaternary deposits of southern California. In: Sigleo W. R., Reinhardt T. (eds) *Paleosols and Weathering Through Geologic Time*, pp 153–177. Geological Society of America Special Paper 216. Boulder, Colo.

McFadden L. D., Ritter J. B., Wells S. G (1989) Use of multiparameter relative-age methods for age estimation and correlation of alluvial fan surfaces on a desert piedmont, eastern Mojave Desert, California. *Quaternary Research* 32: 276–290.

McFadden L. D., Tinsley J. C. (1985) Rate and depth of pedogenic-carbonate accumulation in soils: formation and testing of a compartment model. In: Weide D. L. (ed) *Soils and Quaternary Geology of the Southwestern United States*, pp 23–41. Geological Society of America Special Paper 203. Boulder, Colo.

McFadden L. D., Wells S. G., Dohrenwend J. C. (1986) Influences of Quaternary climatic changes on processes of soil development on desert loess deposits of the Cima Volcanic field, California. *Catena* 13: 361–389.

McFadden L. D., Wells S. G., Jercinovich M. J. (1987) Influences of eolian and pedogenic processes on the origin and evolution of desert pavements. *Geology* 15: 504–508.

Monson R. K., Smith S. D. (1982) Seasonal water potential components of Sonoran Desert plants. *Ecology* 63: 113–123.

Muller C. H. (1940) Plant succession in the *Larrea-Flourensia* climax. *Ecology* 21: 206–212.

Musick H. B. (1975) Barrenness of desert pavements in Yuma County, Arizona. *Journal of the Arizona Academy of Science* 10: 24–28.

Noy-Meir I. (1973) Desert ecosystems: Environment and producers. *Annual Review of Ecology and Systematics* 4: 25–51.

Parker K. C. (1995) Effects of complex geomorphic history on soil and vegetation patterns on arid alluvial fans. *Journal of Arid Environments* 30: 19–39.

Pearthree P. A. (1991) Geologic insights into flood hazards in piedmont areas of Arizona. *Arizona Geology* 21: 1–5.

Peterson F. F. (1981) Landforms of the Basin and Range Province defined for soil survey. Nevada Agricultural Experiment Station Technical Bulletin 28, Reno.

Phillips D. L., MacMahon J. A. (1978) Gradient analysis of a Sonoran Desert bajada. *Southwestern Naturalist* 23: 669–680.

Ruzicka C. (1994) Soil geomorphic relations of two alluvial fans, Harquahala Mountains, Sonoran Desert, Arizona. Master's thesis, University of New Mexico, Albuquerque.

Schlesinger W. H., Fonteyn P. J., Marion G. M. (1987) Soil moisture content and plant transpiration in the Chihuahuan Desert of New Mexico. *Journal of Arid Environments* 12: 119–126.

Shreve F. (1925) Ecological aspects of the deserts of California. *Ecology* 6: 93–103.

Shreve F. (1934) Rainfall, runoff and soil moisture under desert conditions. *Annals of the Association of American Geographers* 24: 131–156.

Shreve F. (1951) *Vegetation of the Sonoran Desert*. Carnegie Institution of Washington Publication no. 591. Washington, D.C.

Shreve F., Hinckley A. L. (1937) Thirty years of change in desert vegetation. *Ecology* 18: 463–478.

Shreve F., Mallery T. D. (1933) The relationship of caliche to desert plants. *Soil Science* 35: 99–113.

Shreve F., Turnage M. V. (1936) The establishment of moisture equilibria in soil. *Soil Science* 41: 351–355.

Solbrig O. T., Barbour M. A., Cross J., Goldstein G., Lowe C. H., Morello J., Yang T. W. (1977) The strategies and community patterns of desert plants. In: Orians G. H., Solbrig O. T. (eds) *Convergent Evolution in Warm Deserts,* pp 67–106. US/IBP Synthesis Series no. 3. Dowden, Hutchinson and Ross, Stroudsburg, Pa.

Szarek S. R., Woodhouse R. M. (1977) Ecophysiological studies of Sonoran Desert plants. II. Seasonal photosynthesis patterns and primary production of *Ambrosia deltoidea* and *Olneya tesota. Oecologia* 28: 365–375.

Turner R. M. (1963) Growth in four species of Sonoran Desert trees. *Ecology* 44: 760–765.

Turner R. M., Brown D. E. (1982) Sonoran desertscrub. *Desert Plants* 4: 181-221.

Vasek F. C. (1979/1980) Early successional stages in Mojave desert scrub vegetation. *Israel Journal of Botany* 28: 133–148.

Vasek F. C. (1980) Creosote bush: long-lived clones in the Mojave Desert. *American Journal of Botany* 67: 246–255.

Vasek F. C. (1983) Plant succession in the Mojave Desert. *Crossossoma* 9: 1–23.

Walter H. (1973) *Vegetation of the Earth and Ecological Systems of the Geobiosphere.* Springer-Verlag, New York.

Webb R. H., Steiger J. W., Newman E. B. (1988) *The Response of Vegetation to Disturbance in Death Valley National Monument, California.* U.S. Geological Survey Bulletin 1793. U. S. Government Printing Office, Washington, D.C.

Webb R. H., Steiger J. W., Turner R. M. (1987) Dynamics of Mojave desert shrub assemblages in the Panamint Mountains, California. *Ecology* 68: 478–490.

Weide D. L. (ed) (1985) *Soils and Quaternary Geology of the Southwestern United States.* Geological Society of America Special Paper 203. Boulder, Colo.

Wells P. V. (1961) Succession in desert vegetation on streets of a Nevada ghost town. *Science* 134: 670–671.

Wells S. G., McFadden L. D., Dohrenwend J. C. (1987) Influence of late Quaternary climatic change on geomorphic and pedogenic processes on a desert piedmont, eastern Mojave Desert, California. *Quaternary Research* 27: 130–146.

Wentworth T. R. (1981) Vegetation on limestone and granite in the Mule Mountains, Arizona. *Ecology* 62: 469–482.

Whittaker R. H. (1951) A criticism of the plant association and climatic climax concepts. *Northwest Science* 25: 117–131.

Whittaker R. H., Niering W. A. (1968) Vegetation of the Santa Catalina Mountains, Arizona. IV. Limestone and acid soils. *Journal of Ecology* 56: 523–544.

Yang T. W. (1950) The distribution of *Larrea divaricata* in the Tucson area as determined by certain physical and chemical factors of the habitat. Master's thesis, University of Arizona, Tucson.

Yang T. W. (1957) Vegetational, edaphic, and faunal correlations of the western slope of the Tucson Mountains and the adjoining Avra Valley. Ph.D. dissertation, University of Arizona, Tucson.

Yang T. W., Lowe C. H. (1956) Correlation of major vegetation climaxes with soil characteristics in the Sonoran Desert. *Science* 123: 542.

Zedler P. H. (1981) Vegetation change in chaparral and desert communities in San Diego County, California. In: West D. C., Shugart H. H., Botkin D. B. (eds) *Forest Succession: Concepts and Applications*, pp 406–430. Springer-Verlag, New York.

4 Population Ecology of Sonoran Desert Annual Plants

D. Lawrence Venable and Catherine E. Pake

Roughly 50% of the species in local floras in the Sonoran Desert are annuals, with 60% to 80% of these being winter annuals and the rest being summer or nonseasonal annuals (fig. 4.1; Venable et al. 1993). Deserts are ecosystems with a high level of environmental variation driven by rainfall (hot deserts have the highest coefficient of variation in interannual actual evapotranspiration of any of the earth's biomes; Frank and Inouye 1994). Desert annuals are highly responsive to these environmental fluctuations and play an important role in modulating that variation and passing it on to higher trophic levels (Went 1949; Beatley 1967; Patten 1975, 1978; Gutierrez and Whitford 1987). They are most appreciated in "good wildflower years" when the floral displays of the more showy species color the landscape. This often occurs in association with El Niño weather events that tend to result in higher than average winter precipitation in the southwestern United States. Desert annuals spend most of their lives as seeds and in some years may even be inconspicuous during their normal growing season due to low germination or little growth (e.g., Tevis 1958a). Their rapid dynamics appear to make them more sensitive to invasions and extirpations than perennials. Introduced species constitute a higher fraction of annuals than perennials in local floras (fig. 4.2). They may also be more readily lost, judging from the observation that three-fourths of the 27 possible extirpations in the Tucson Mountain flora (species not collected since 1950) are annuals (mostly winter). This is significantly more than expected given that annuals only make up 45% of the flora ($G = 9.87$, $P < .001$; data from Rondeau 1991). The seeds of

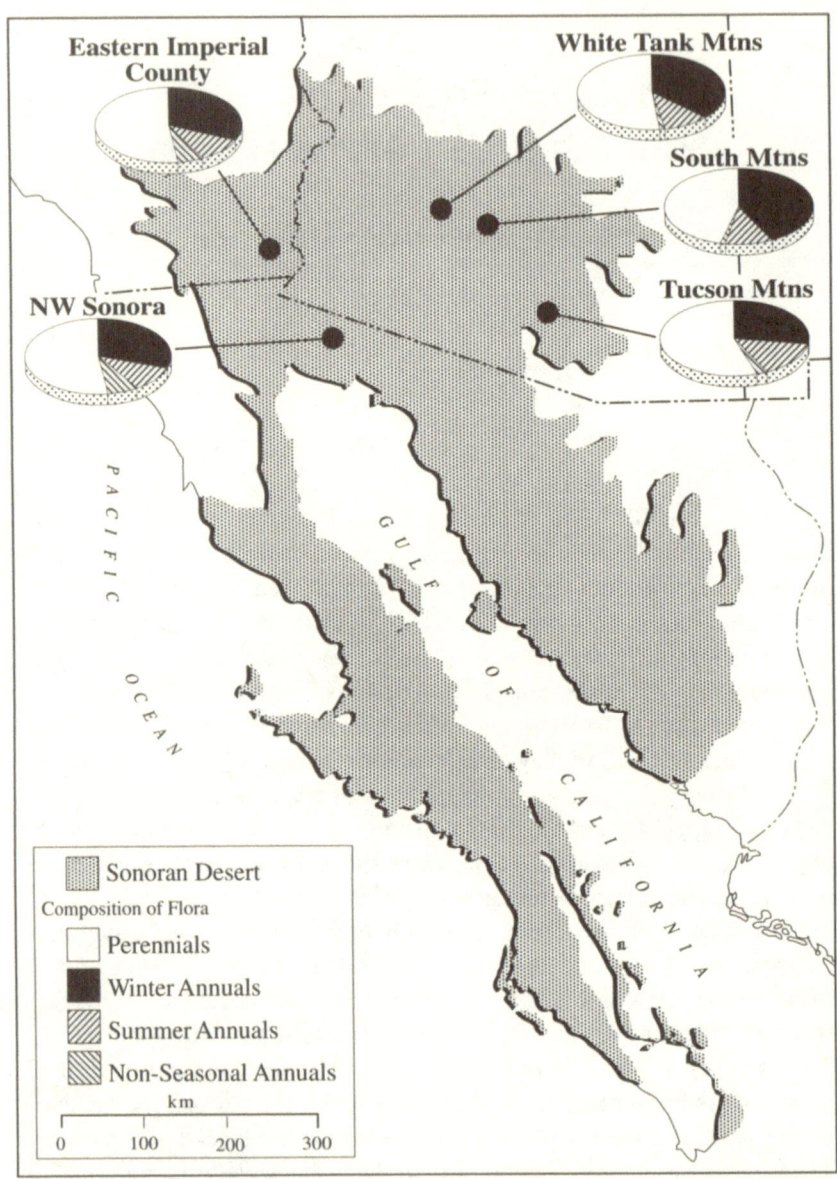

Figure 4.1 Map of the Sonoran Desert showing the proportion of perennials and winter, summer, and nonseasonal annuals in five regional floras. Data taken from Keil (1973), Felger (1980, 1992), McLaughlin et al. (1987), Rondeau (1991), and Daniel and Butterwick (1992).

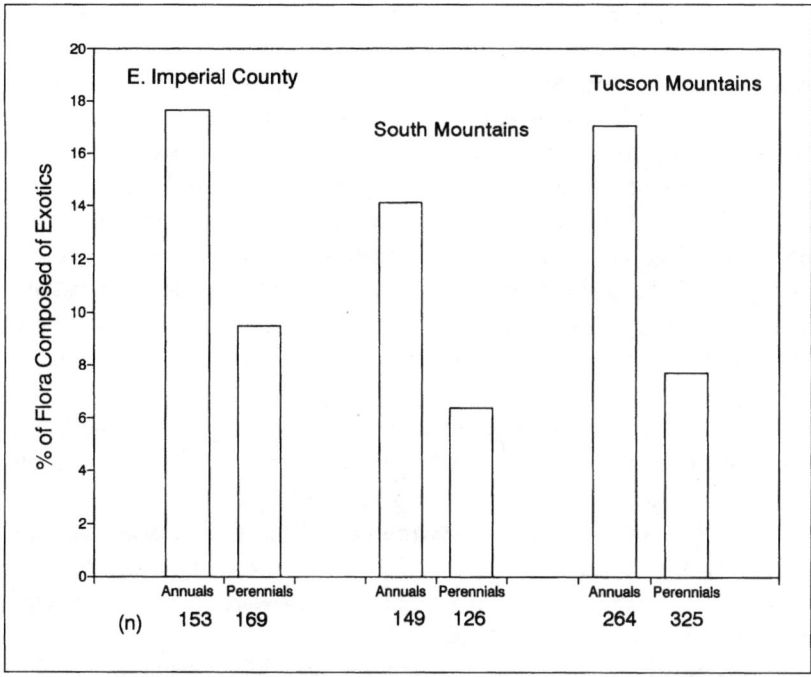

Figure 4.2 Percent of exotic species among the annual and perennial components of three regional floras in the Sonoran Desert. Annuals have a significantly greater proportion of exotic species than perennials in each case (East Imperial County, $G = 4.7$ [$P < .05$]; South Mountains, $G = 19.2$ [$P < .001$]; Tucson Mountains, $G = 12.0$ [$P < .001$]). (n) signifies the number of species.

desert annuals are a primary food source for ants and rodents and their vegetative bodies are important browse for grasshoppers, rabbits and larger grazing animals (Mares and Rosenzweig 1978; Rissing 1986). Annual plants and their fluctuations play a critical role in the population dynamics and species composition of these consumers (Brown et al. 1975, 1979, 1986; Inouye et al. 1980; Davidson et al. 1984, 1985; Samson et al. 1992). Because of the numerical and ecological importance of annual plants in deserts, an understanding of their population dynamics and species interactions is critical to understanding desert ecosystems.

It is widely believed that delayed germination and long-lived seed banks play an important role in the population dynamics of desert annuals. Yet little work has been done to quantify nongerminating seed fractions of desert annuals in natural environments (but see Nelson and Chew 1977; Reichman 1984; Price and Reichman 1987 for documentation of seed-bank densities). Long-lived seeds and species-specific germination requirements have been hypothesized to explain observed fluctuations in population densities, especially the sudden appearance of species following one or more years of absence (Went 1949; Juhren et al. 1956; Tevis 1958a, 1958b; Shreve and Wiggins 1964). More recently, the germination of a few species has been studied in well-controlled tests in growth chambers. These studies have demonstrated that a fraction of viable seeds usually remains dormant, even under apparently ideal germination conditions (Baskin et al. 1993; Philippi 1993a, 1993b; S. Adondakis and Venable, unpublished data).

Desert annuals and their seeds have played an important role in the development of theories about adaptation to variable environments, the population dynamic functions of dispersal and dormancy, and variance-mediated species coexistence. Environmental variability, dispersal, and dormancy are generally recognized to be widespread in plant communities, biologically significant in a variety of contexts, and not very amenable to experimental investigation.

In this paper we review some results on population dynamics of desert annuals from our research conducted at the Desert Laboratory. We have documented the conditions under which desert winter annuals have emerged since 1982. We have also documented the survival and reproduction of emerging seedlings over this period. In this research we have measured population dynamic properties critical to evaluating some theoretical issues for which desert annuals have been used as examples. We have documented germination fractions for different species in several years. Investigations of shifts in interspecific interactions, in concert with information on seed-bank dynamics, have shed light on mechanisms of species coexistence in this community. We have also partitioned the sources of seedlings into local reproduction, seed bank, and dispersal and measured seed dispersal distances.

The Desert Laboratory Site

The study area for the long-term plots and competition studies is a gently sloped alluvial plain at 725 m elevation northwest of Tumamoc Hill. Other results come from work done further east, from the northeastern base of

Tumamoc Hill extending around to the northwest side. The Desert Laboratory property has been ungrazed by livestock since 1907 (Bowers and Turner 1985; Burgess et al. 1991), and our study sites are dominated by *Larrea tridentata* (creosote bush), which may be joined by *Ambrosia deltoidea* (triangle-leaf bursage), *Krameria grayi* (white ratany), *K. parviflora* (range ratany), *Opuntia fulgida* (jumping cholla), *O. phaeacantha* (prickly pear), and *Fouquieria splendens* (ocotillo) (Bowers and Turner 1985).

The Long-Term Plots

In the autumn of 1982, research was initiated on the demography of desert winter annuals along a 250-m transect through the previously mentioned gently sloped creosote flat. Fifteen permanent plots (0.10 m^2 each) were placed in random positions in the open along this transect. Plots have been visited each year following each rainfall event, and censuses have been taken if germination occurred. Generally, plots had censuses taken four to six times a season for emergence, survival, and reproduction. This sampling scheme gave sample sizes of hundreds (occasionally thousands) of individuals for most species in most years. Data were collected by mapping on acetate sheets placed with fixed coordinates over plexiglass mapping tables. Maps were later digitized to record spatial coordinates, germination and death dates, and fecundity of individual plants. Estimates of seed-bank dynamics were obtained from soil cores (28 per year) collected from 1982–1986 and 1989–1996 (180 cores per year).

More than 30 species of winter annuals have been found on these plots with the most common species being *Plantago patagonica* (Indian wheat) and *P. insularis* (Plantaginaceae), *Schismus barbatus* (Poaceae; introduced from the arid Middle East), *Erodium texanum* (native) and *E. cicutarium* (filaree; introduced from the Mediterranean; Geraniaceae), *Evax multicaulis*, *Stylocline micropoides*, *Monoptilon bellioides* (Mohave desert star), *Eriophyllum lanosum* (woolly daisy; Asteraceae), and *Pectocarya recurvata* (Boraginaceae).

We now understand many basic aspects of the population dynamics of these species. Germination typically occurs from October through January (fig. 4.3). Plants grow and reproduce until late March to early May when a combination of high temperatures ($> 35°C$) and drought results in death (fig. 4.3). High temperatures can be associated with either mortality or high rates of plant growth, depending on rainfall. Thus mortality or

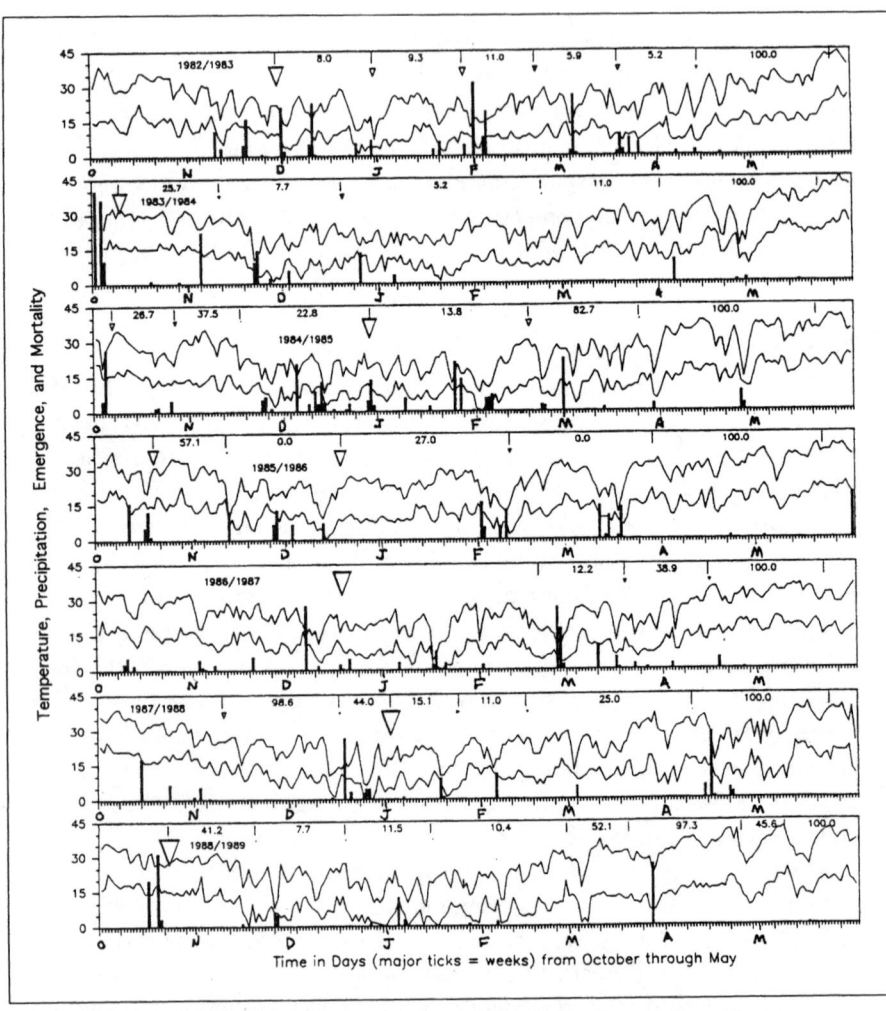

Figure 4.3 Weather, emergence, and mortality of winter annuals on the long-term plots at the Desert Laboratory from 1982–1989. Solid lines indicate daily max/min temperatures (°C). Solid bars indicate daily rainfall (mm). Down-pointing arrows indicate emergence, with the size of the arrow being proportional to the percentage of the total year's emergence that occurred at that census. The numbers across the top of each year's graph indicate the average per capita mortality between the dates indicated by the ticks, expressed on a per month basis, for plants alive at the beginning of each interval.

growth is sometimes high in the autumn (if plants emerge early) or in the spring, but neither is high in the cool winter months (fig. 4.3). These winter annuals do well in terms of growth and reproduction in El Niño years (1982/83, 1986/87, 1990/91, 1991/92, and 1994/95) and worse in dry years. However, exceptions exist and species have individualistic responses to temporal variation (fig. 4.4, Venable et al. 1993). The total number of seedlings of all species emerging has varied by two orders of magnitude over the 15-year period (1982–1997). Populations tend to increase following El Niño events (fig. 4.5), though the details vary from species to species (fig. 4.6). Desert annual populations were high during the early 1980s but crashed in 1989/90, the third year of the late '80s winter drought (1987/88–1989/90). Populations recovered in the period from 1991/92 to 1992/93 after low-density, high-fecundity seasons in 1989/90 and 1990/91 (fig. 4.4; fig 4.5; fig 4.6). They remained high through 1994/95 but dropped again during the mid-1990's drought due to low germination survival and fecundity in 1995/96 and 1996/97.

It is interesting to contemplate how these patterns observed at the Desert Laboratory might vary over broad geographic areas. Personal observations and preliminary results from a lower-elevation, drier study site near Gila Bend, Arizona, plus published accounts from other studies suggest that there are overriding regional patterns, probably produced by regional winter frontal storm systems. Published accounts of year-to-year variation in annual plant densities, biomass, or reproduction for sites in Nevada or southeastern California (Nelson and Chew 1977; Beatley 1969) correspond fairly well to winter precipitation records taken at the Desert Laboratory in Tucson (Venable, personal observation). Yet many details of desert annual population dynamics appear to vary, sometimes dramatically, on a scale of tens to hundreds of miles or along short elevational, slope, or aspect gradients. Frontal systems may drop different amounts of rain at sites only a few miles apart resulting in sites with missing or extra germination cohorts (Venable, personal observation). Also, late cohorts may be missing from some sites where a heavy early rain apparently depleted the seed bank of most nondormant seeds. Plant populations that reliably experience different weather regimes are also likely to have ecotypic differences. Germination and flowering times have been documented to be more flexible in dry Lower Colorado Valley or Mojave Desert sites than at our Desert Laboratory site. For example, rains before September or after January result in only negligible germination at the Desert Laboratory, but

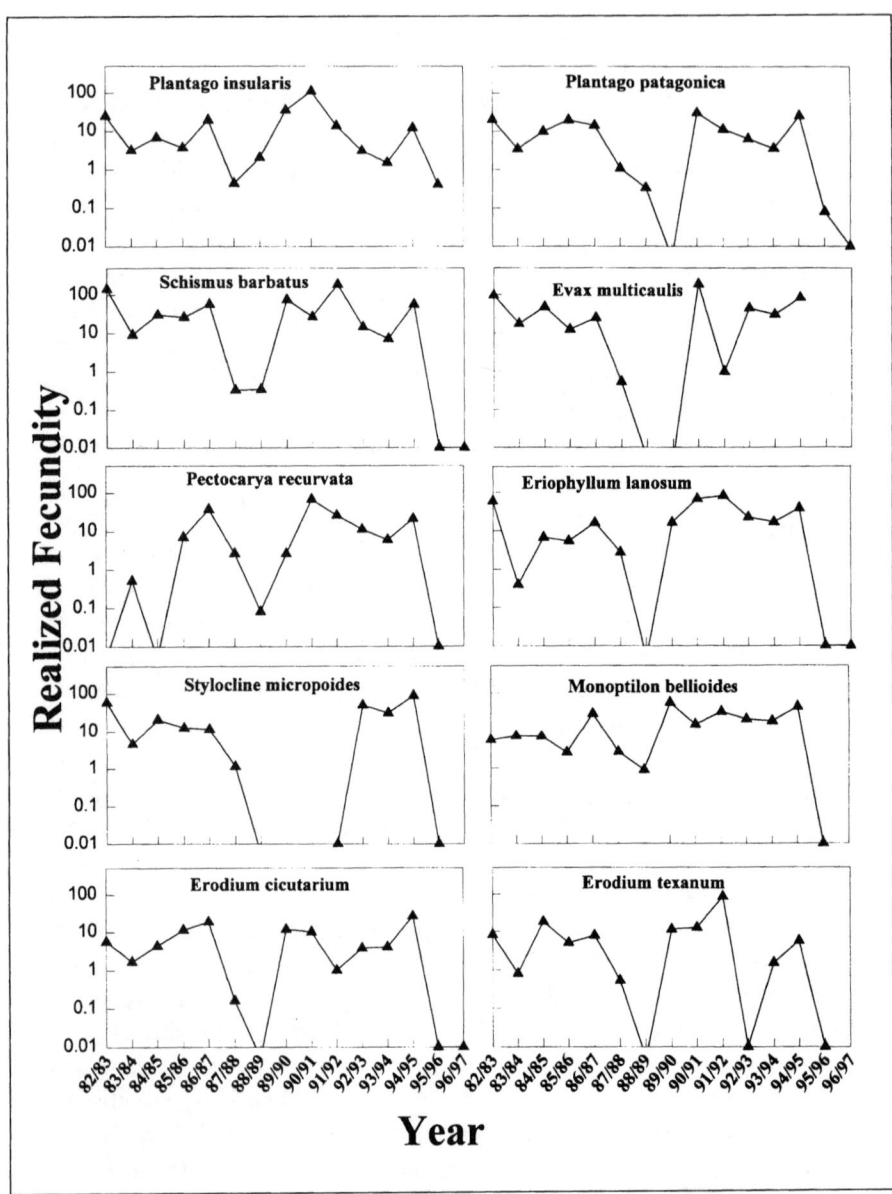

Figure 4.4 Realized fecundity (average survival [from emergence to maturity] × average fecundity [seed per adult]) for 10 species of winter annuals on the long-term plots at the Desert Laboratory for each year from 1982/83–1996/97.

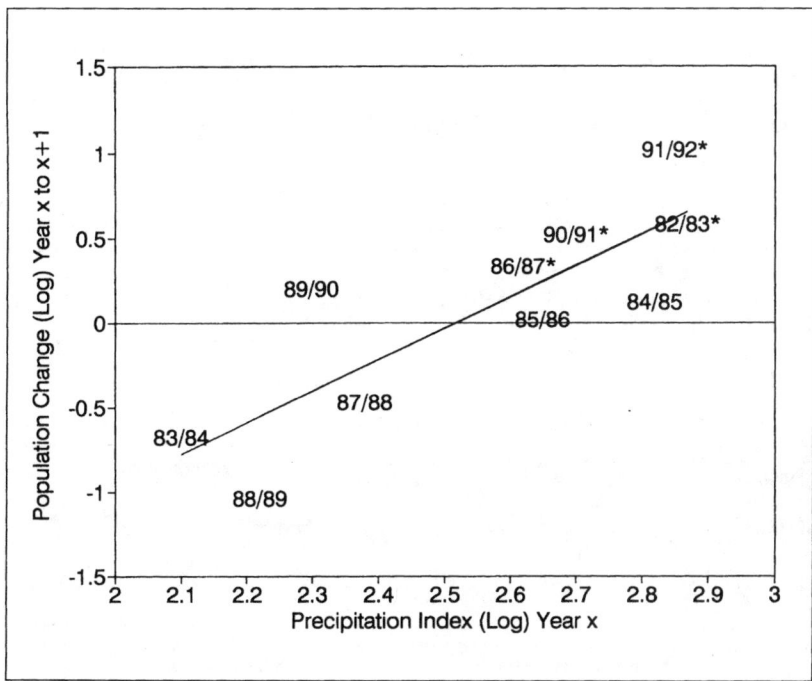

Figure 4.5 The total population of all desert annuals on the long-term plots at the Desert Laboratory tends to increase following years of higher winter precipitation. The precipitation index is total precipitation from November through February of each year. Population change is calculated as $\log(N_{x+1}) - \log(N_x)$, where N_x is the number of seedlings emerging in year x. Asterisks indicate the population increases that followed El Niño events. $R^2 = 0.68$, $P < .0025$.

may produce large cohorts at drier, more variable sites (e.g., Tevis 1958a; Beatley 1967; Venable, unpublished data from Gila Bend).

Germination Fractions, Seed Size, and Environmental Variation

Seed-bank dynamics and germination behavior of desert annuals in natural habitats are relevant to a variety of ecological and evolutionary issues beyond desert ecology per se. Germination behavior of desert annuals has been used as a model system for understanding adaptations to variable

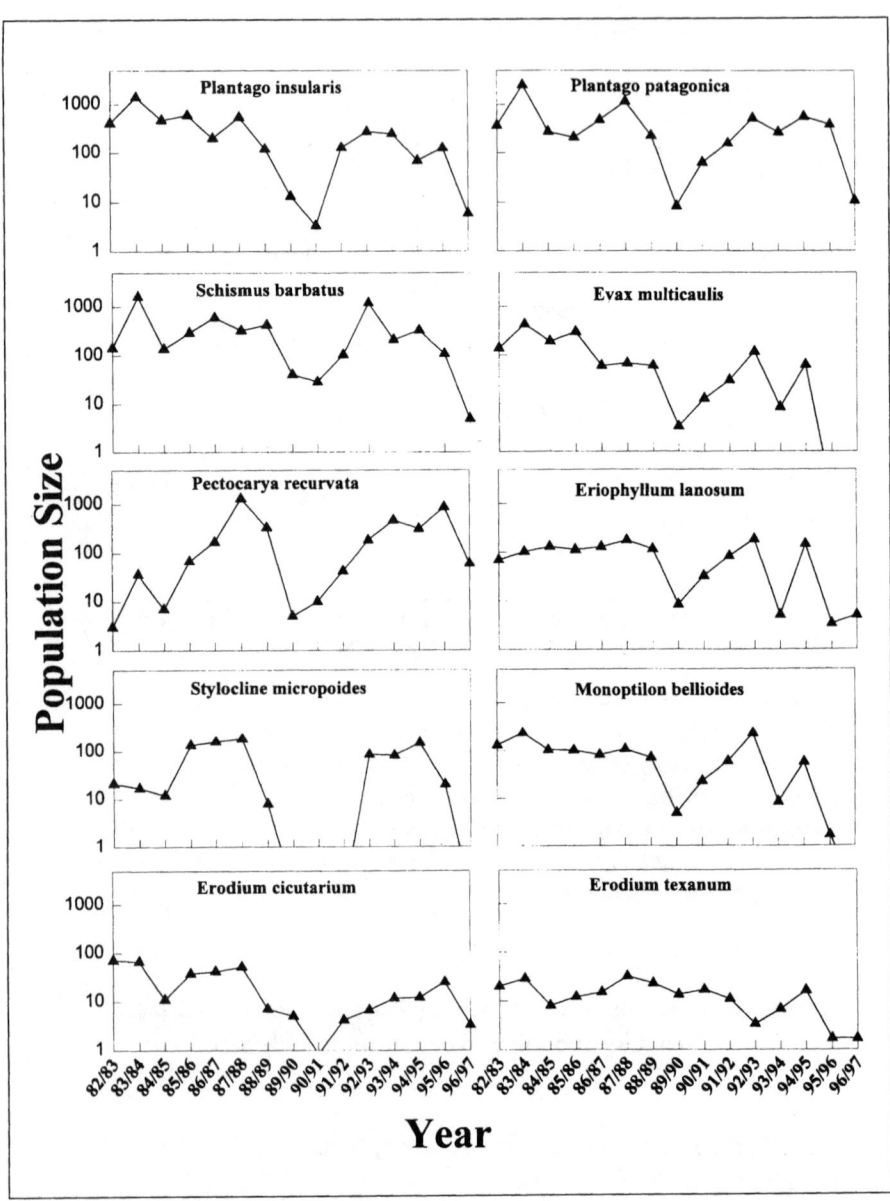

Figure 4.6 Population size (number of seedlings emerged) for each of 10 species of winter annuals on the long-term plots at the Desert Laboratory for each year from 1982/83–1996/97.

environments. Theoretical studies have shown that fractional germination for desert annuals can result in the adaptive reduction of temporal variation in fitness commonly called bet hedging (Cohen 1966; MacArthur 1972; Venable and Lawlor 1980; Ritland 1983; Bulmer 1984; Leon 1985; Philippi and Seger 1989; Venable and Brown 1988; Venable 1989). Delayed germination of a fraction of a plant's progeny buffers it from the consequences of near or complete reproductive failure in unfavorable years, but it may also reduce success in favorable years by sacrificing opportunities to reproduce. While desert annuals have been useful to evolutionary ecologists for developing theories about how adaptation should occur in variable environments, very little empirical data has been collected in a way suitable for testing such ideas.

To that end, we have quantified the fractions of viable seed banks that germinated in each of three years for 17 species at the Desert Laboratory. We used these data to look for the predicted negative correlation of germination fraction with demographic variance (Venable et al. 1993; Pake and Venable 1996). This was done by sifting and counting the viable seeds from soil samples (180 each year, collected with a stratified randomization scheme, each 5.4 cm diameter × 2.5 cm deep; viability determined by visual inspection of embryo/endosperm as in Pake and Venable 1996). These samples were always collected after the winter germination season but before new seeds dispersed in the spring. Densities of emerging seedlings were obtained from 48 nearby quadrants so that germination fractions could be calculated as seedlings/m^2 ÷ (seeds + seedlings)/m^2. We calculated fitness variance as the year-to-year variance in reproductive success of seeds that germinate (survival × mean fecundity of survivors for each of nine species using 10 years of demographic data). We found that the species whose germinating seeds experienced greater variance in per capita reproductive success had lower germination fractions (fig. 4.7). This is the first study to use long-term data on population dynamic variance to test the prediction from bet-hedging theory that selection should lead to lower germination fractions in species with higher variance in success (Cohen 1966). Phillipi (1993b) found a similar but less-direct result. Seeds of *Lepidium lasiocarpum* collected from sites with lower mean rainfall had lower germination fractions in growth chamber experiments.

Other aspects of desert annual life histories may also reduce variance in ways that should increase fitness. Seed size is one such trait. A larger-seeded plant is likely to produce fewer seeds in favorable years than a

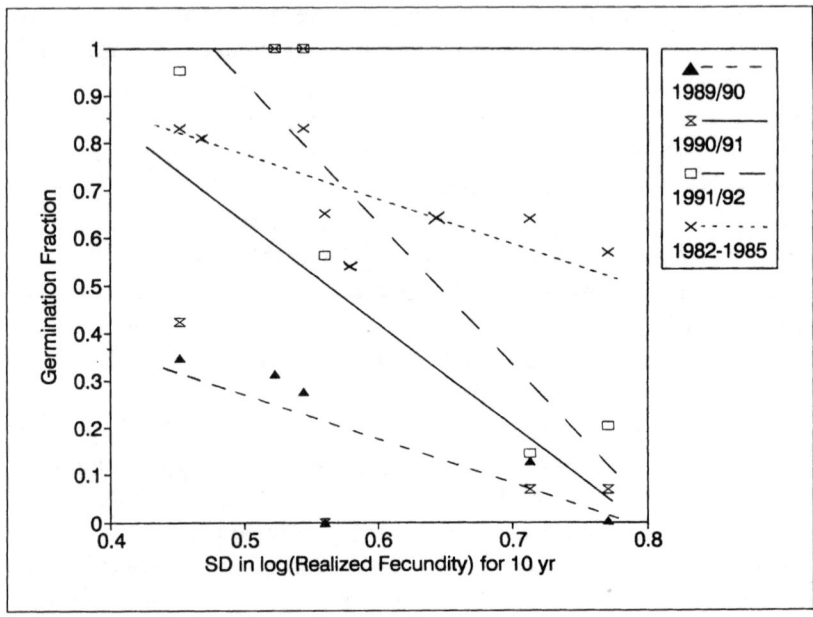

Figure 4.7 Germination fractions of desert winter annuals were lower for species that experienced greater year-to-year variance in reproductive success. Germination fractions are graphed for field data collected at the Desert Laboratory in 1989/90, 1990/91, and 1991/92 and for the average of data collected at the Desert Laboratory from 1982–1985.

small-seeded one (given similar amounts of nutrients and energy available for reproductive allocation). However, large seed size means more maternal provisioning, which may buffer seeds from some negative effects of dry years, as originally posited by Baker (1972; see Venable and Brown 1988 for the theoretical arguments regarding bet hedging). In our system, we found that smaller-seeded species have a higher variance in reproductive success (Spearman $r = -0.7448$, $P < .0001$; Pake and Venable 1996). Thus our results are consistent with the idea that larger seed size buffers plants from variance in arid environments (see also Venable and Brown 1988).

By reducing the variance experienced by a desert annual, larger seed size should also modify the strength of natural selection for delayed germination. Thus theoretical studies have suggested a positive correlation between germination fraction and seed size (Venable and Brown 1988), and this pattern is also highly significant in our data (Pake and Venable 1996). A positive correlation between germination fraction and seed size has also been reported in a comparative study of species of British plants (Rees 1993).

Species Coexistence

Another line of inquiry for which the population dynamics of desert annuals has had an impact on general ecological concepts is the theory of species coexistence mechanisms promoted by environmental variance. Desert annuals are frequently cited as an example of a system in which temporal variation may promote species coexistence (Shmida and Ellner 1984; Ellner 1987; Chesson and Huntley 1988, 1989; Chesson 1994). Again, few empirical results are available to test the ideas. Since the pioneering works of Hutchinson (1961) and Grubb (1977), ecologists have been aware that temporal environmental variation may promote species coexistence of organisms with different "temporal niches." More recently, the importance of resistant life-history stages, such as seed banks, for temporal-variance-mediated species coexistence has been recognized (Chesson and Huntley 1988). Temporal heterogeneity is a factor that might be expected to be very important in promoting coexistence in desert annuals, since deserts are the biome with the greatest coefficient of variation among years in actual evapotranspiration (Frank and Inouye 1994).

One frequently suggested annual-plant scenario with the population dynamic elements necessary for variance-mediated coexistence involves persistent seed banks and species-specific germination responses to environmental variation (Chesson and Huntley 1989). Persistent seed banks are necessary to buffer populations against extinction in unfavorable conditions (i.e., to "store" population inputs from favorable years). Species-specific germination responses provide opportunities for species to bounce back from rarity by occasionally escaping competition with abundant species. Another related scenario (Venable et al. 1993; Pake and Venable 1995) involves germination fractions that are correlated with reproductive success (predictive germination). Predictive germination, with year-to-year variation in survival and fecundity that is not completely correlated among species, increases the coexistence-promoting properties of variable environments. These theoretical scenarios make specific predictions about

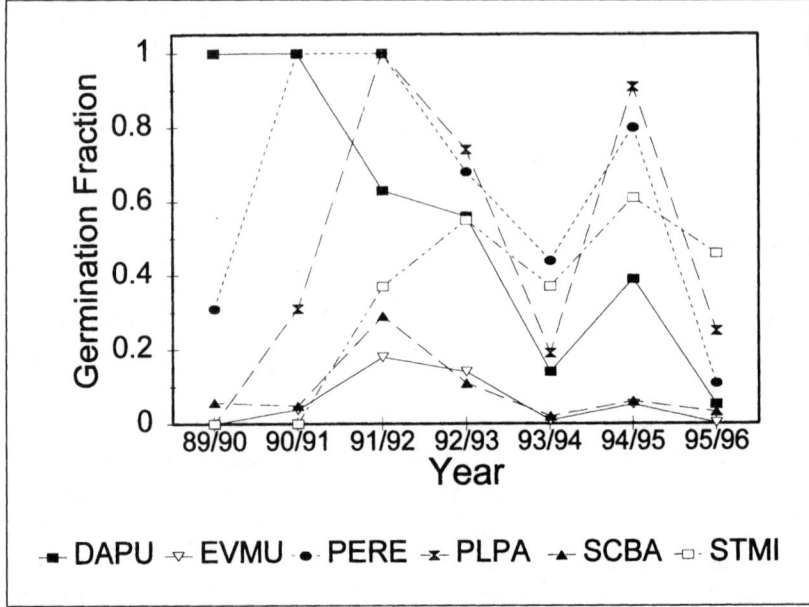

Figure 4.8 Graph of germination fraction of six species of desert winter annuals on permanent plots at the Desert Laboratory in seven years, illustrating species-by-year interaction. DAPU = *Daucus pusillus*; EVMU = *Evax multicaulis*; PERE = *Pectocarya recurvata*; PLPA = *Plantago patagonica*; SCBA = *Schismus barbatus*; STMI = *Stylocline micropoides*.

the dynamics of seeds in the soil and about shifts in the relative abundance and relative performance of species over time. Such predictions can be tested with data from desert annuals (Venable et al. 1993).

Persistent seed banks are produced by most (probably all) members of our desert annual community (Pake and Venable 1996). Statistical analysis of the germination fraction data described above shows that germination fraction varied significantly among years, with individual species responding differently in different years (species × year interaction; fig. 4.8; Pake and Venable 1995). Thus, the basic conditions for variance-mediated coexistence seem to be met in this community. Also, we have shown that germination fractions were higher in years when reproductive success was higher, suggesting that germination fraction is cued to conditions corre-

Table 4.1 Some statistically significant differences ($p < .05$) in species' performances that reveal shifts in hierarchies associated with temporal variation (either natural or simulated year-types).

Plots in different years,
all at low density
    1990/91:   PERE > SCBA, PERE > PLPA
    1991/92:   SCBA > PERE, PLPA > PERE, SCBA > PLPA
Plots in year-types simulated by varying density,
all under dry conditions
    Low density: PLPA > PERE
    Medium density: PERE > PLPA, PERE > SCBA
    High density: SCBA > PERE, SCBA > PLPA

Source: Adapted from Pake and Venable (1995).
Note: PERE = *Pectocarya recurvata*; PLPA = *Plantago patagonica*; SCBA = *Schismus barbatus*.

lated with future prospects for reproduction (Pake and Venable 1996; cf. Cohen 1967 and Rice 1985). As mentioned above, a positive relationship between germination fraction and reproductive success suggests a role for the variation experienced by germinated seedlings in species coexistence.

To test for such shifts in reproductive success of species among years, we conducted a demographic field experiment with three winter annuals: *Plantago patagonica, Pectocarya recurvata,* and *Schismus barbatus* (Pake and Venable 1995). In this experiment we determined the extent to which species performance hierarchies shifted among years, among simulated year-type factors (water addition/removal and density manipulation), and between shrub and open habitats (the major contributor to local spatial variation). Seedlings were mapped in replicated plots in shrub and open habitats, and in different years, densities, and moisture levels, and their per capita survival and reproductive success were determined. The relative performance of these species shifted significantly as we experimentally varied the factors that are important components of the natural year-to-year environmental variation (table 4.1). For example, under low-density, dry conditions, *Plantago* outperformed *Pectocarya;* at medium-density, dry conditions, *Pectocarya* outperformed both *Plantago* and *Schismus;* while at high-density, dry conditions, *Schismus* outperformed both *Pectocarya* and *Plantago.* In 1990/91 at low density, *Pectocarya* outperformed both *Schismus* and *Plantago,* yet in 1991/92 at low density, *Plantago* and *Schismus* outperformed *Pectocarya,* and *Shismus* outperformed *Plantago*. Thus, the relative success of competing species shifts under real and simulated temporal variation. Significant shifts in performance hierarchies were not found between shrub and open habitats, suggesting that this spatial component of environmental patchiness may be less important than the temporal components, at least under the conditions investigated here. These experiments represent an attempt to use population dynamic approaches to investigate plant competition in undisturbed natural habitats. Our results so far indicate that variance-mediated coexistence mechanisms are likely to be important in the species diversity of desert annuals (Venable et al. 1993; Chesson 1994; Pake and Venable 1995, 1996).

Community ecologists are well aware that a great variety of factors may potentially contribute to species coexistence in a given system (Tilman and Pacala 1993). In the research described above, we have attempted to explore the population dynamic components required by one potentially important factor for desert annuals: temporal variance. Spatial heterogeneity is also an important source of desert annual species diversity, with the standard gradients of slope, aspect, soil type, moisture, and nutrients contributing at a variety of scales (Shreve and Wiggins 1964)[5]. The mosaic pattern of open area and perennial cover, as well as microtopographic variation, also contribute at a local scale (Shmida and Whittaker 1981; Samson 1986). Another likely factor contributing to the coexistence and species diversity of desert annuals is the species-specific behavior of grazers and seed predators (Brown et al. 1979; Pacala and Crawley 1992).

Partitioning of Seedling Sources

While understanding the role of seed banks is clearly important to our understanding of desert annual population dynamics, we would also like to know the role of seed dispersal. Seed dispersal is another understudied aspect of plant population dynamics with many important consequences, including persistence and coexistence in variable environments. Since seed dispersal and seed-bank dynamics are both difficult to measure in natural populations, little good population dynamic data exists on their consequences for any plant species. To attack this problem, we have performed a variety of removal experiments in natural populations of desert annuals at the Desert Laboratory. The goal of these experiments was to determine the proportion of the seedlings emerging in each of three years that came from *in situ* reproduction the previous year, delayed germination from prior years, or dispersal.

In the spring of 1991 (and in each subsequent year), we used a herbicide to inhibit reproduction of desert annuals either solely under shrubs, solely in open sites, both, or neither. This was carried out in replicated randomized-block removal experiments with replicated removal treatment areas roughly 10 m in diameter. During the following germination season (autumn 1992), we measured seedling emergence in small plots within the larger removal areas. These plots were either "natural" (uncovered and unbordered) or "seed bank" (small, bordered plots from which plants were removed and which were covered with a fine-mesh organza cloth during the seed dispersal season). Pairs of natural and seed-bank plots were located in shrub and open habitats in each of the larger removal areas representing the factorial combinations of reproduction inhibition. Comparison of seed bank and natural plots in the control areas (with no removal) determines the relative contribution from first-year germination and delayed germination. Comparison of shrub or open removal areas to control areas (with no removal) indicates the role of seed dispersal between the two habitat types. This process was repeated in each of the subsequent two years, but was not quite finished for the second year at the time of this writing.

During the first year in which we measured the results of these manipulations, 1992/93, the Sonoran Desert experienced the third favorable winter of high precipitation following the late 1980s drought (fig. 4.4; fig 4.5; fig. 4.6). During the second year, 1993/94, germination season rainfall was low and early. Some noteworthy shifts occurred in seedling abundances between these years (data taken from unmanipulated plots). Two annual grass species increased considerably (*Schismus barbatus* and *Bromus rubens*) while the forbs either remained constant (*Plantago patagonica, Pectocarya recurvata*) or declined (*Stylocline micropoides* and *Eriastrum diffusum* [miniature wool star]; fig. 4.9). Data from seed-addition plots indicate that the dramatic decline in *Eriastrum* seedling density was due to low germination in the second year (0.5% germination vs. 26% the previous year). This was probably due to the low precipitation during the autumn germination season. Despite these shifts in abundance of species, partially due to differences in germination response, within species the relative densities in shrub and open sites did not shift between years. The

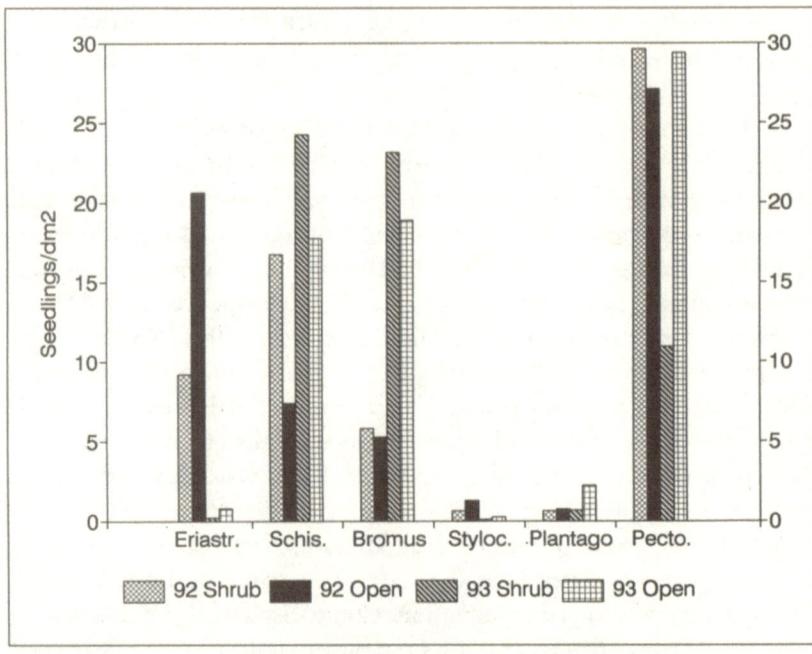

Figure 4.9 Densities of seedlings emerging in shrub and open habitats for two years in the seedling source experiment at the Desert Laboratory. Eriastr. = *Eriastrum diffusum;* Schis. = *Schismus barbatus;* Bromus = *Bromus rubens;* Styloc. = *Stylocline micropoides;* Plantago = *Plantago patagonica;* Pecto. = *Pectocarya recurvata.*

two grasses were consistently more abundant under the shrubs. *Eriastrum, Stylocline,* and *Plantago* were somewhat more abundant in the open in both years (and *Pectocarya* was exceptional in shifting from more abundant under shrubs in the first year to more abundant in the open the second year; fig. 4.9).

Dispersal was usually a less important source of seedlings than delayed germination (as predicted for desert annuals by Ellner and Shmida [1981]; fig. 4.10). Nonetheless, the shrub and open habitats have a source-sink relationship that differs in direction for different species: the habitat in which a species occurs at lower density usually has a greater fraction of its seedlings originating from dispersed seeds (e.g., lower density and more immigration under the shrubs for *Eriastrum* and *Stylocline* and in the

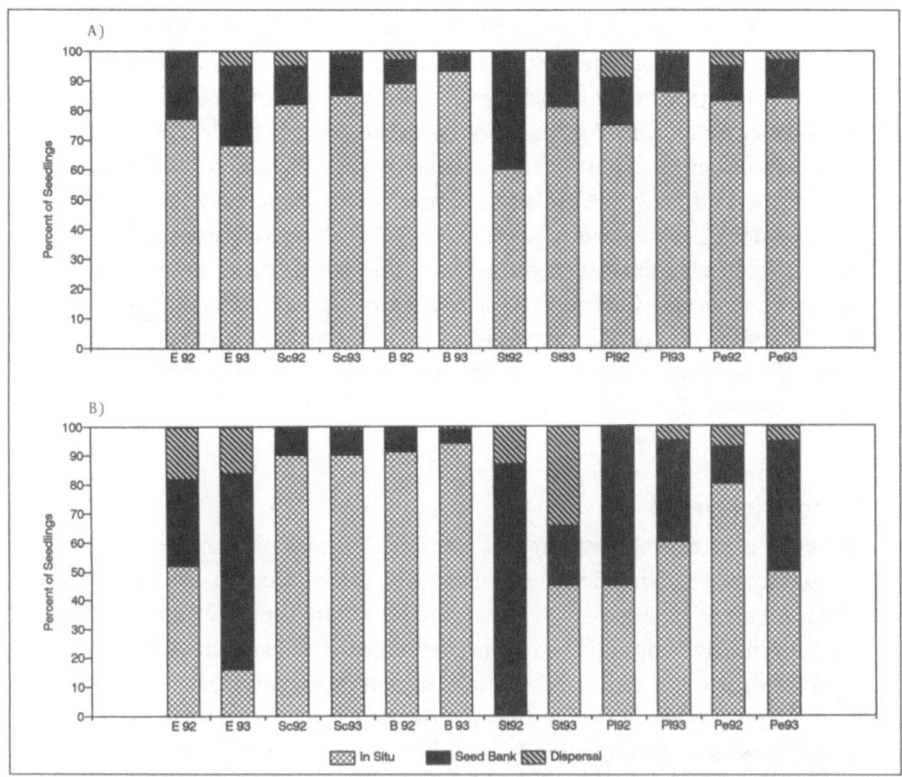

Figure 4.10 Partition of seedling origin into *in situ* reproduction, delayed germination, and immigration in two years in the seedling source experiment at the Desert Laboratory. (A) Open habitats; (B) Under shrubs. E = *Eriastrum diffusum*; Sc = *Schismus barbatus*; B = *Bromus rubens*; St = *Stylocline micropoides*; Pl = *Plantago patagonica*; Pe = *Pectocarya recurvata*.

open for the two grasses [fig. 4.10A vs. 4.10B; cf. Shmida and Ellner 1984; Pulliam 1988]). Compared with the forbs, the two annual grasses had little contribution from delayed germination or dispersal in either year. Also, as compared to seedlings in open sites, a greater fraction of the seedlings under shrubs came from delayed germination (cf. solid black bars, fig. 4.10A vs. 4.10B). These experiments have enabled us to decompose the densities we see in a given year into *in situ* production (in the local shrub or open habitat), delayed germination (seeds more than one year old), and dispersal (immigration from the other habitat type: shrub for the open sites or open for the shrub sites). In this way some interesting, difficult-to-measure attributes of the population dynamics of this system have been revealed.

Seed Dispersal

We have also taken advantage of the large annual plant removal areas created for the previously discussed experimental system to obtain a more direct measure of dispersal distance for desert annuals (Flores and Venable, unpublished data). The boundaries of these removal areas have been marked with spray paint. The density of seedlings coming from the seed bank inside removal areas has been estimated from "seed-bank plots" as described above. Seedling densities within the large removal areas that are greater than the background seed-bank density are due to seed dispersal across the borders. To estimate dispersal distances, we have taken censuses of seedlings in transects that run perpendicular across these borders. A blurred, sigmoidal gradient of seedling densities is found along such transects (e.g., fig. 4.11). Densities along these transects represent dispersal curves for the population of plants outside the removal border. These can be converted to dispersal curves for individual plants using the fact that the population dispersal curve is the integral (or sum) of the dispersal curves of all the plants from the border outward. The dispersal curve in figure 4.11 is for all annual plants together and indicates a rapid decline in seed number with distance. The estimated rate of decline in seed numbers is -4.0 (seeds $\times$ seed^{-1} $\times$ meter^{-1}). This is quite low among published estimates of dispersal rate (the average exponential dispersal rate for herbaceous plants without dispersal mechanisms was -2.75 in the studies reviewed by Willson [1993]).

This predictable and low rate of seed dispersal is not the whole story for

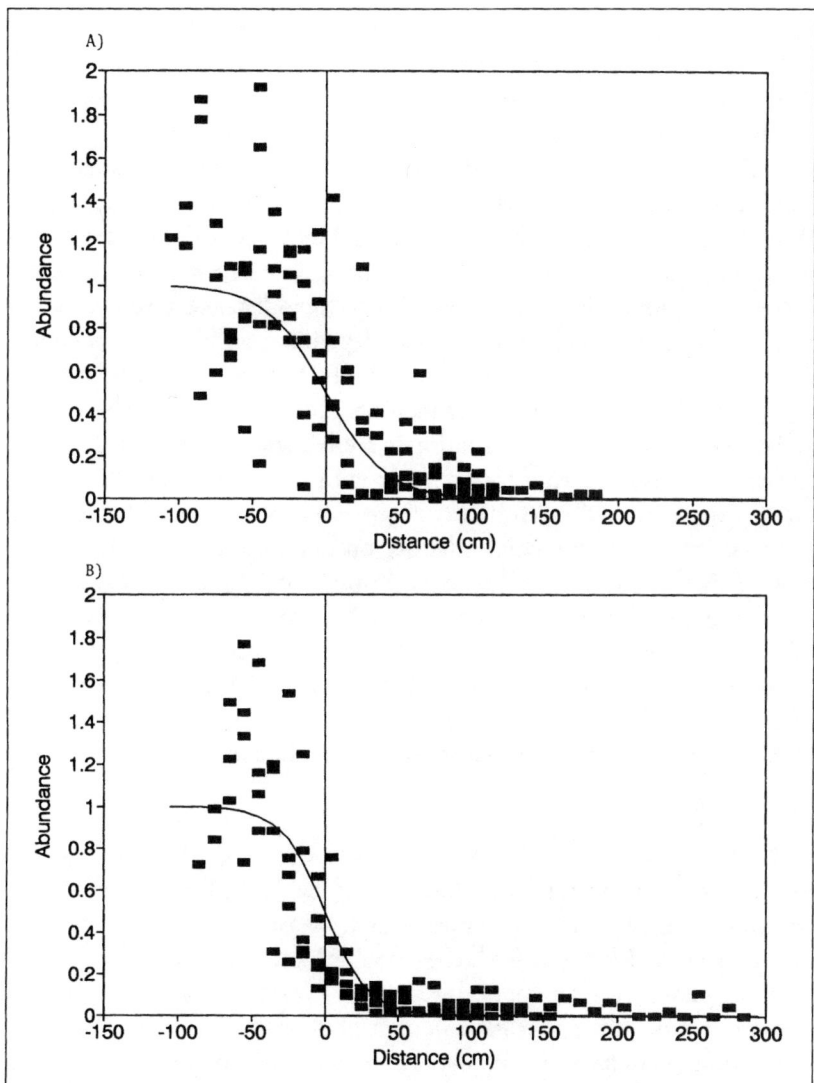

Figure 4.11 Seed dispersal across an annual plant removal boundary. The abundance of emerging seedlings of all species is plotted against distance from the boundary. (A) Dispersal down a gently sloped bajada; (B) Dispersal up a gently sloped bajada.

desert winter annuals. High densities of desert annual seedlings can be found where tiny fragments of plant litter accumulate in clumps or even lines that represent the edges of rainfall sheet flow. It would appear from the distribution of such patches of seeds and seedlings that dispersal in sheet flow can take seeds considerable distances. We find seed-laden litter deposits on the downhill side of 10 m wide total-removal plots. At our Desert Laboratory site, such dramatic seed movement does not seem to occur frequently in most plots. Yet it appears to occur regularly in a few exceptional plots. The importance of such long-distance sheet wash dispersal appears to vary highly in space, depending on local runoff conditions determined by surface microtopography and soil water penetration. Occasional very heavy rains may increase the importance of sheet wash and generate high temporal variation in seed dispersal distances. The conclusions of these experiments, together with those from the experiment that partitioned seedling sources, suggest that seed dispersal is often a minor source of local seedling density on our Desert Laboratory study plots. This is consistent with prior suggestions of limited seed dispersal in desert annuals (Ellner and Shmida 1981). Yet, another point of view in the literature suggests that seed dispersal via sheet wash (or wind) is widespread (Mott and McComb 1974; Reichman 1984). It seems from our studies that both points of view are partially correct, but that there may be considerable spatial and temporal variation in their applicability.

Conclusions

While desert annuals represent around half the flora of the Sonoran Desert, they are easily overlooked during many seasons and some years. Yet they are a critical resource for higher trophic levels and their wide year-to-year fluctuations are an important source of variation for the organisms that depend on them. We have documented 15 years of desert annual population fluctuations in an ongoing study of the ecology and evolution of desert annuals. Besides gaining a better understanding of how desert annuals respond to variation in weather between years, we have documented germination fractions of desert annuals in natural field sites. Delayed germination is greater for those species that experience more year-to-year demographic variation, as predicted by life history theory, and larger-seeded species experience less year-to-year variation in reproductive success. These are the first long-term plant population dynamic data to confirm these theoretical predictions for bet hedging. Desert annuals also appear to have many properties necessary for temporal-variance-mediated species coexistence. These include delayed germination, with germination fractions varying among years in different ways for different species. Also, there is a correlation between a higher germination fraction for a given species and the years in which it has higher reproduction. Competitive hierarchies shift among years, densities, and moisture levels, which should contribute to temporal-variation-based coexistence. Dispersal may be a less important source of seedlings than delayed germination for desert annuals. Both vegetation cover and open habitats appear to provide population sources and sinks for different species that have a habitat preference for one or the other. While seed dispersal of desert annuals appears often to be quite limited, at certain places and times seeds may travel considerable distances in sheet flow during heavy rains.

Desert annuals provide a tractable system for studying many aspects of plant population and community ecology in variable environments. Since they are small, working with many of them is easy. By virtue of being annuals, their vegetative life cycle is short and they have the many convenient properties that attract researchers to short-lived plants. They occur as persistent dominant members of relatively undisturbed natural communities. Thus, their responses to temporal variation can be easily studied in nature. Also, since water availability is an important limiting factor, causes of population fluctuations may be less obscure than for some plant populations and communities. The long-term availability of protected research stations like the Desert Laboratory is critical to our growing understanding of the population and community ecology of desert annuals.

Acknowledgments

This work was supported by NSF grants BSR 91107323, DEB-9419905 and BSR 9520888. Special thanks go to Anthony C. Caprio, Donna Conlan, Tani Hubbard, Michelle Stubbs, and Sandy Adondakis.

Literature Cited

Baker H. G. (1972) Seed mass in relation to environmental conditions in California. *Ecology* 53: 997–1010.

Baskin C. C., Chesson P. L., Baskin J. M. (1993) Annual seed dormancy cycles in two desert winter annuals. *Journal of Ecology* 81: 551–556.

Beatley J. C. (1967) Survival of winter annuals in the Northern Mojave Desert. *Ecology* 48: 745–750.

Beatley J. C. (1969) Biomass of desert winter annual plant populations in southern Nevada. *Oikos* 20: 261–263.

Beatley J. C. (1974) Phenological events and their environmental triggers in Mojave desert ecosystems. *Ecology* 55: 856–863.

Bowers J. E., Turner R. M. (1985) A revised flora of Tumamoc Hill, Tucson, Arizona. *Madroño* 32: 225–252.

Bowers M. A. (1987) Precipitation and the relative abundances of desert winter annuals: a 6 year study in the northern Mojave Desert. *Journal of Arid Environments* 12: 141–149.

Brown J. H., Davidson D. W., Munger T. C., Inouye R. S. (1986) Experimental community ecology: the desert granivore system. In: Diamond J., Case T. J. (eds) *Community Ecology*, pp 41–61. Harper and Row, New York.

Brown J. H., Grover J. J., Davidson D. W., Lieberman G. A. (1975) A preliminary study of seed predation in desert and montane habitats. *Ecology* 56: 987–992.

Brown J. H., Reichman O. J., Davidson D. W. (1979) Granivory in desert ecosystems. *Annual Reviews of Ecology and Systematics* 10: 201–227.

Bulmer M. G. (1984) Delayed germination of seeds: Cohen's model revisited. *Theoretical Population Biology* 26: 367–377.

Burgess T. L., Bowers J. E., Turner R. M. (1991) Exotic plants at the Desert Laboratory, Tucson, Arizona. *Madroño* 38: 96–114.

Chesson P. L. (1994) Multispecies competition in variable environments. *Theoretical Population Biology* 45: 227–276.

Chesson P. L., Huntley N. (1988) Community consequences of life-history traits in a variable environment. *Annales Zoologici Fennici* 25: 5–16.

Chesson P. L., Huntley N. (1989) Short-term instabilities and long-term community dynamics. *Trends in Ecology and Evolution* 4: 293–298.

Cohen D. (1966) Optimizing reproduction in a randomly varying environment. *Journal of Theoretical Biology* 12: 119–129.

Cohen D. (1967) Optimizing reproduction in a randomly varying environment, when a correlation may exist between the conditions at the time a choice has to be made and the subsequent outcome. *Journal of Theoretical Biology* 16: 1–14.

Daniel T. L., Butterwick M. L. (1992) Flora of the South Mountains of south-central Arizona. *Desert Plants* 10: 99–119.

Davidson D. W., Inouye R. S., Brown J. H. (1984) Granivory in a desert ecosystem: experimental evidence for indirect facilitation of ants by rodents. *Ecology* 1965: 1780–1786.

Davidson D. W., Samson D. A., Inouye R. S. (1985) Granivory in the Chihuahuan desert: interactions within and between trophic levels. *Ecology* 66: 486–582.

Ellner S. (1987) Alternate plant life history strategies and coexistence in randomly varying environments. *Vegetatio* 69: 199–208.

Ellner S., Shmida A. (1981) Why are adaptations for long-range seed dispersal rare in desert plants? *Oecologia* 51: 133–144.

Felger R. S. (1980) Vegetation and flora of the Gran Desierto, Sonora, Mexico. *Desert Plants* 2: 87–114.

Felger R. S. (1992) Synopsis of vascular plants of Northwestern Sonora, Mexico. *Ecologica* 2: 11–44.

Frank D., Inouye R. S. (1994) Temporal variation in climate of terrestrial ecosystems: patterns and ecological implications. *Journal of Biogeography* 21: 401–411.

Grime J. P. (1977) Evidence for the existence of three primary strategies in plants and its relevance to ecological and evolutionary theory. *American Naturalist* 111: 1169–1194.

Grubb P. J. (1977) The maintenance of species-richness in plant communities: the importance of the regeneration niche. *Biological Reviews* 52: 107–145.

Gutierrez J. R., Whitford W. G. (1987) Response of Chihuahuan Desert herbaceous annuals to rainfall augmentation. *Journal of Arid Land Environments* 12: 127–139.

Halvorson W. L., Patten D. T. (1975) Productivity and flowering of winter ephemerals in relation to Sonoran Desert shrubs. *American Midland Naturalist* 93: 311–319.

Hutchinson G. E. (1961) The paradox of the plankton. *American Naturalist* 95: 137–145.

Inouye R. S. (1980) Density-dependent germination response by seeds of desert annuals. *Oecologia* 46: 235–238.

Inouye R. S. (1982) Population biology of desert annual plants. Ph.D. dissertation, University of Arizona, Tucson.

Inouye R. S. (1990) Population biology of desert annual plants. In: Polis R. (ed) *The Ecology of Desert Communities*, pp 25–44. University of Arizona Press, Tucson.

Inouye R. S., Byers G. S., Brown J. H. (1980) Effects of predation and competition on survivorship, fecundity, and community structure of desert annuals. *Ecology* 61: 1344–1351.

Juhren M., Went F. W., Phillips E. (1956) Ecology of desert plants. IV. Combined field and laboratory work on germination of desert annuals at Joshua Tree National Monument, California. *Ecology* 37: 318–330.

Keil D. J. (1973) Vegetation and flora of the White Tank Mountains Regional Park, Maricopa County, Arizona. *Journal of the Arizona Academy of Science* 8: 35–48.

Klikoff L. G. (1966) Competitive response to moisture stress of a winter annual of the Sonoran Desert. *American Midland Naturalist* 75: 383–391.

Leon J. A. (1985) Germination strategies. In: Greenwood P. J., Harvey P. H., Slatkin M. (eds) *Evolution: Essays in Honour of John Maynard Smith*, pp 129–143. Cambridge University Press, Cambridge.

MacArthur R. H. (1972) Appendix: Delayed germination of desert annuals. *Geographic Ecology*, pp 165–168. Harper and Row, New York.

Mares M. A., Rosenzweig M. L. (1978) Granivory in North and South American deserts: rodents, birds, and ants. *Ecology* 59: 235–241.

McLaughlin S. P., Bowers J. E., Hall K. P. (1987) Vascular plants of eastern Imperial County, California. *Madroño* 34: 359–378.

Mott J. J., McComb A. J. (1974) Patterns in annual vegetation and soil microrelief in an arid region of Western Australia. *Journal of Ecology* 62: 115–126.

Nelson J. F., Chew R. M. (1977) Factors affecting seed reserves in the soil of a Mojave desert ecosystem, Rock Valley, Nye County, Nevada. *American Midland Naturalist* 97: 300–320.

Pacala S. W., Crawley M. J. (1992) Herbivores and plant diversity. *American Naturalist* 140: 243–260.

Pake C. E., Venable D. L. (1995) Environmental variation and Sonoran Desert annual plants: is coexistence mediated by temporal variability? *Ecology* 76: 371–391.

Pake C. E., Venable D. L. (1996) Seed banks in desert annuals: implications for persistence and coexistence in a variable environment. *Ecology* 77: 1427–1435.

Patten D. T. (1975) Phenology and function of Sonoran Desert annuals in relation to environmental changes. In: USABP *Desert Biome Research Memorandum 75010*, pp 109–116. Utah State University, Logan.

Patten D. T. (1978) Productivity and production efficiency of an upper Sonoran Desert ephemeral community. *American Journal of Botany* 65: 891–895.

Philippi T. (1993a) Bet-hedging germination of desert annuals: beyond the first year. *American Naturalist* 142: 474–487.

Philippi T. (1993b) Bet-hedging germination of desert annuals: variation among populations and maternal effects. *American Naturalist* 142: 488–507.

Philippi T., Seger J. (1989) Hedging one's evolutionary bets, revisited. *Trends in Ecology and Evolution* 4: 41–44.

Price M. V., Reichman O. J. (1987) Spatial and temporal heterogeneity in Sonoran Desert soil seed pools, and implications for heteromyid rodent foraging. *Ecology* 68: 1797–1811.

Pulliam H. R. (1988) Sources, sinks and population regulation. *American Naturalist* 132: 652–661.

Rees M. (1993) Trade-offs among dispersal strategies in British plants. *Nature* 366: 150–152.

Reichman O. J. (1984) Spatial and temporal variation of seed distributions in Sonoran Desert soils. *Journal of Biogeography* 11: 1–11.
Rice K. J. (1985) Responses of *Erodium* to varying microsites: the role of germination cuing. *Ecology* 66: 1651–1657.
Rissing S. W. (1986) Indirect effects of granivory by harvester ants: plant species composition and reproductive increase near ant nests. *Oecologia* 68: 231–234.
Ritland K. (1983) The joint evolution of seed dormancy and flowering time in annual plants living in variable environments. *Theoretical Population Biology* 24: 213–243.
Rondeau R. J. (1991) Flora and vegetation of the Tucson Mountains, Pima County, Arizona. Master's thesis, University of Arizona, Tucson.
Samson D. A. (1986) Community ecology of Mojave Desert winter annuals. Ph.D. dissertation, University of Utah, Salt Lake City.
Samson D. A., Philippi T. E., Davidson D. W. (1992) Granivory and competition as determinants of annual plant diversity in the Chihuahuan Desert. *Oikos* 65: 61–80.
Shmida A., Ellner S. (1984) Coexistence of plant species with similar niches. *Vegetatio* 58: 29–55.
Shmida A., Whittaker R. H. (1981) Pattern and biological microsite effects in two shrub communities, southern California. *Ecology* 62: 234–251.
Shreve F., Wiggins I. L. (1964) *Vegetation and Flora of the Sonoran Desert*. Stanford University Press, Stanford.
Tevis L. Jr (1958a) Germination and growth of ephemerals induced by sprinkling a sandy desert. *Ecology* 39: 681–688.
Tevis L. Jr (1958b) A population of desert ephemerals germinated by less than one inch of rain. *Ecology* 39: 688–695.
Tevis L. Jr (1958c) Interrelation between the harvester ant *Veromessor pergandei* (Mayr) and some desert ephemerals. *Ecology* 39: 695–704.
Tilman D., Pacala S. (1993) The maintenance of species richness in plant communities. In: Ricklefs R. E., Schluter D. (eds) *Species Diversity in Ecological Communities: Historical and Geographical Perspectives*, pp 13–25. University of Chicago Press, Chicago.
Venable D. L. (1989) Modeling the evolutionary ecology of seed banks. In: Leck M. A., Parker V. T., Simpson R. L. (eds) *The Ecology of Soil Seed Banks*, pp 67–87. Academic Press, San Diego.
Venable D. L., Brown J. S. (1988) The selective interactions of dispersal, dormancy, and seed size as adaptations for reducing risk in variable environments. *American Naturalist* 131: 360–384.
Venable D. L., Lawlor L. (1980) Delayed germination and dispersal in desert annuals: escape in space and time. *Oecologia* 46: 272–282.

Venable D. L., Pake C. E., Caprio A. C. (1993) Diversity and coexistence of Sonoran Desert winter annuals. *Plant Species Biology* 8: 207–216.

Went F. W. (1942) The dependence of certain annual plants on shrubs in southern California deserts. *Bulletin of the Torrey Botanical Club* 69: 100–114.

Went F. W. (1948) Ecology of desert plants. I. Observations on germination in the Joshua Tree National Monument, California. *Ecology* 29: 242–253.

Went F. W. (1949) Ecology of desert plants. II. The effect of rain and temperature on germination and growth. *Ecology* 30: 1–13.

Willson M. F. (1993) Dispersal mode, seed shadows, and colonization patterns. In: Fleming T. H., Estrada A. (eds) *Frugivory and Seed Dispersal: Ecological and Evolutionary Aspects,* pp 261–280. Kluwer Academic Publishers, Dordrecht.

5 Form and Function of Cacti

Park S. Nobel and Michael E. Loik

Perhaps no other image of the Sonoran Desert is as enduring as that of the saguaro cactus (*Carnegiea gigantea*). The saguaro and other cacti are uniquely adapted to the characteristic droughts and thermal extremes of the region. Precipitation rarely exceeds 300 mm per year, and a particular location may not receive any rainfall for many months. High temperatures can reach 70°C at the soil surface, and temperatures below 0°C can occur for up to 30 days per year. Various animals can escape the harsh extremes of the physical environment of the Sonoran Desert, but plants are sessile and therefore must tolerate the extended periods of drought and extremes of temperature.

The approximately 140 species of cacti native to the Sonoran Desert are exclusively perennial stem succulents (Shreve and Wiggins 1964). Their primary physiological adaptation to drought is the utilization of Crassulacean acid metabolism (CAM), in which stomatal opening and CO_2 uptake occur mainly at night, when air and tissue temperatures are cooler than during the daytime (Nobel 1988). This results in less water loss and a higher water-use efficiency (mass of CO_2 taken up per mass of water lost) than for C_3 or C_4 plants (Nobel 1991). Morphological adaptations of cacti to drought and temperature extremes include stem succulence and orientation, ribs and tubercles, spines and apical pubescence, shallow roots, and association with "nurse plants." This chapter will review such features related to the form of cacti and how they function to increase survival in the often harsh environment of the Sonoran Desert.

Stem Morphology

The cacti of the Sonoran Desert vary in size and shape (fig. 5.1). They range from the approximately spherical *Mammillaria tetrancistra*, about 12 cm in height (fig. 5.1A), to the tallest of the American cacti, *Carnegiea gigantea* (fig. 5.1B), which can reach heights of 15 m (Benson 1982; Nobel 1988). Many species have stems that are columnar (e.g., *C. gigantea* and *Lophocereus schottii* [senita]), cylindrical (e.g., chollas such as *Opuntia bigelovii* [teddy-bear cholla]; fig. 5.1C), or flattened (e.g., prickly pear cacti such as *O. phaeacantha*; fig. 5.1D). Whereas most species occur as solitary stems, some species, such as those in the genus *Echinocereus*, can contain hundreds of stems that form mounds.

Variations in morphology directly influence how a cactus interacts with its environment. For example, stem size can influence the temperature of cactus tissues. For *Carnegiea gigantea*, maximal temperatures near the top of the stem decrease, and minimum temperatures increase, as stem diameter increases from 10 to 60 cm (Nobel 1978). Therefore, the apex of *C. gigantea* will tend to be warmer in winter and cooler in summer for thicker stems. Because the meristem producing the cell division necessary for stem growth occurs at the apex, moderation of apical temperatures is crucial for the growth and survival of columnar cacti as well as barrel cacti.

Stem height can also influence the ability of an individual to compete for light, usually referred to as the photosynthetic photon flux density (PPFD; wavelengths of 400 to 700 nm that can be absorbed by photosynthetic pigments). For example, stem height of *Stenocereus gummosus* (pitahaya agria) increases from northern to southern Baja California as the height of the neighboring species increases (Nobel 1980b). Such increases in the height of *S. gummosus* are necessary for interception of adequate levels of PPFD in different habitats.

A key to drought survival that is related to overall stem size and shape is the ability to store water. The water storage capacity of the stems of desert cacti can be quantified using the ratio of stem volume to stem area (V/A), representing the volume for water storage divided by the area across which water can be lost by transpiration. The units for V/A are length^3/length^2, or length; in particular, V/A indicates the average depth in the stem for water storage. For a medium-sized barrel cactus, *Ferocactus acanthodes*, V/A is 6.3 cm (Nobel 1977), but V/A is much less, 0.53 cm, for the prickly pear *Opuntia phaeacantha* (fig. 5.1D; Conde 1975). For the pencil cholla, *O. leptocaulis*, V/A is only 0.14 cm, whereas the thicker

Figure 5.1 Representative shapes of cacti from the Sonoran Desert: (A, upper left) *Mammillaria tetrancistra,* known as the fish-hook cactus; (B, upper right) *Carnegiea gigantea,* the saguaro; (C, lower left) *Opuntia bigelovii,* the teddy-bear cholla; and (D, lower right) *Opuntia phaeacantha,* a representative pear cactus.

stems of the tree cholla, *O. imbricata*, have a V/A of 0.40 cm (Conde 1975). A one-week-old, approximately spherical seedling of *F. acanthodes* with a diameter of 2 mm has a volume/area ratio of 0.05 cm, but a year-old seedling can have a V/A of 0.55 cm (Gibson and Nobel 1986). The ability of older seedlings to survive drought is determined largely by increases in their V/A.

The stems of cacti have two major tissue types, as is readily apparent from their coloration when a stem is cut open. The photosynthetic chlorenchyma, which contains the enzymes of the CAM pathway, constitutes an outer green layer that is 2 to 5 mm thick. The inner yellowish-white tissue is the water-storage parenchyma, which ranges from 0.3 cm in thickness for *Opuntia fragilis* (little prickly pear) to about 45 cm for *Ferocactus acanthodes* (Benson 1982; Loik and Nobel 1993). The water-storage parenchyma stores water in large cells that contain few or no chloroplasts. During drought, water moves out of such storage and into the chlorenchyma, where it helps sustain biochemical reactions and is used to support stomatal opening and CO_2 uptake. Water supplied by the water-storage parenchyma of *F. acanthodes* can support stomatal opening for up to 40 days after water uptake from the soil ceases (Nobel 1977). For the sympatric CAM succulent *Agave deserti* (desert agave), which has a V/A of about 1 cm, such stomatal opening occurs for only 8 days after the soil dries. The large V/A associated with thick stems influences the physiology of various cacti, such as supplying stored water that extends their period of net CO_2 uptake by days to weeks.

Stem Orientation

Some species of Sonoran Desert cacti have growth patterns leading to orientations that affect stem temperature. One of the best known examples is the equatorial tilting of the compass barrel, *Ferocactus wislizenii*, native to southern Arizona and to Sonora, Mexico (Shreve and Wiggins 1964; Benson 1982). This orientation can raise the minimum temperature of the apical meristem as well as the temperature of the areoles (lateral buds that can produce spines and flowers) during the spring when flower production and development occur (Nobel 1981; Ehleringer and House 1984). Earlier flowering can increase the reproductive potential of individuals of this species by allowing more time for seed maturation.

The orientation of stems can also influence the PPFD intercepted, which

Table 5.1 Net CO_2 uptake over 24 hours by cladodes of pear cacti at various latitudes in the Sonoran Desert.

Time of Year	Directions Faced by Cladode	Net CO_2 Uptake (% of Maximum)		
		Northern (35° N)	Central (30° N)	Southern (25° N)
Summer solstice	East-west	99	98	97
	North-south	52	56	60
Equinox	East-west	96	95	93
	North-south	57	59	59
Winter solstice	East-west	61	49	39
	North-south	55	54	54

Note: Data are for unshaded vertical cladodes on clear days. See Nobel (1986) for further details.

in turn affects net CO_2 uptake and hence growth (table 5.1). The cladodes (flattened stem segments) of many species of *Opuntia*, known as platyopuntias, are oriented in a way that maximizes PPFD interception during the wet part of the year when most growth takes place. For example, *O. chlorotica* (pancake pear) occurs in the Sonoran Desert of Arizona, where it receives summer and winter rainfall, and in the southern Mojave Desert of California, where it receives primarily winter rainfall. In Arizona, its terminal, vertical, unshaded cladodes tend to face east-west, whereas more cladodes face north-south for plants in the Mojave Desert (Nobel 1981). Facing east-west maximizes summertime light interception in the morning and afternoon, when the sun is lower in the sky. Indeed, facing east-west leads to more PPFD interception at all latitudes and times of the year, except in the winter at latitudes at least 30° from the equator, where facing north-south is advantageous for net CO_2 uptake (table 5.1).

Local topographic features such as slope and aspect, canyon walls, and nearby mountains can also influence the orientation of terminal cladodes of pear cacti. For *Opuntia erinacea* (grizzly bear cactus) growing in Mitchell Caverns State Nature Preserve in eastern California, terminal cladodes tend to face southeast-northwest for plants shaded on the west and northwest sides (Nobel 1982). For *O. chlorotica* in a north-south canyon of the Granite Mountains of California, cladodes tend to face north-south because of shading by the canyon walls. Facing toward the direction of prevailing light leads to more PPFD interception and thus more CO_2 uptake (table 5.1). Such cladodes tend to produce more "daughter" cladodes. Because daughter cladodes tend to face in the same direction as the underlying "mother" cladodes on which they originate, these daughter cladodes also have the same favorable orientation. This interesting interplay between morphology and physiology leads to greater growth for prickly pear cacti than would occur with random orientation of their cladodes.

The pattern of branching for cacti is also influenced by the prevailing direction of the sunlight during the period of stem growth. For the chollas *Opuntia acanthocarpa* (buckhorn cholla) and *O. echinocarpa* (silver cholla), the upper surface of the stem canopy tilts equatorially, which maximizes PPFD interception. The canopy formed by their cylindrical stems tilts an average of 52° toward the light source when illuminated from one side in the laboratory (Geller and Nobel 1987). For three Sonoran Desert species of *Ferocactus*, plants in the field tilt an average of 12° from the vertical, and plants illuminated with a light source at 10° above the horizontal in the laboratory tilt 19° from the vertical (Nobel 1981; Geller and Nobel 1987). For *Carnegiea gigantea* in the Saguaro National Monument near Tucson, Arizona, the majority of branches ("arms") occur on the south-facing half of the stems (Geller and Nobel 1986). Such a branching pattern can increase the amount of surface area for PPFD interception, but at the same time some shading of adjacent stems and the main trunk occurs. Indeed, computer simulations show that nocturnal CO_2 uptake per unit stem area decreases as the number of stems, stem diameter, and self-shading increase (Geller and Nobel 1986).

Ribs and Tubercles

Much of the unique appearance of cacti is due to the presence of ribs or tubercles. Ribs and tubercles contain areoles, the lateral buds from which spines emerge. Tubercles (fig. 5.2A) are leaf bases and appear as projections or raised areas on the surface of the stem. Tubercles can be virtually absent, such as for *Opuntia leptocaulis,* or very prominent, such as for *Mammillaria microcarpa* of southern Arizona. Ribs (fig. 5.2B) are formed from the fusion of tubercles. For cacti from the Sonoran Desert, ribs can number from four or five for some species of *Echinocereus* to over 30 for those in the genus *Ferocactus* (Benson 1982). The number of ribs tends to increase with age as the stems increase in girth (Gibson and Nobel 1986).

Some of the earliest plant research at the Desert Laboratory was con-

Figure 5.2 Surface features of cactus stems, showing (A, left) tubercles of *Opuntia spinosior* (cane cholla; photograph courtesy of Arthur C. Gibson) and (B, right) ribs of *Ferocactus acanthodes*. Some of the apical spines of the barrel cactus have been removed to show the apical pubescence.

ducted by Effie Southworth Spalding and Daniel MacDougal relating ribs to the water relations of cacti, especially *Carnegiea gigantea*. Their pioneering work showed that the ribs of *C. gigantea* become thinner and closer together as water is lost during drought (Spalding 1905; MacDougal and Spalding 1910). Upon rainfall or rewatering, the ribs become shallower, wider, and further apart, similar to the expansion of the pleats of an accordion.

Ribs and tubercles can also increase the surface area of the stem across which CO_2 can be taken up. This manner of increasing surface area allows the plant to maintain a spherical or cylindrical shape with a large water-storage capacity. Shading of one rib or tubercle by another will reduce PPFD interception and therefore reduce CO_2 uptake at the shaded location. Rib and tubercle size and shape also affect air movement and thereby influence stem temperature. Ribs and tubercles thus represent a series of trade-offs among water storage, increased area for CO_2 uptake, reduced PPFD interception through shading, and modification of the thermal environment at the stem surface (Nobel 1988).

The effect of rib size and number can be quantified using the fractional rib depth and the perimeter ratio (Geller and Nobel 1984). Fractional rib depth equals the rib depth (radial distance from a rib crest to the trough between adjacent ribs) divided by the radius of the stem from its center to the crest of the rib. For Sonoran Desert cacti, fractional rib depth can vary from 0.14 for *Carnegiea gigantea* to 0.35 for *Lophocereus schottii* (Geller and Nobel 1984). Perimeter ratio is the total perimeter, from crest to trough to crest for all ribs, divided by the stem diameter. For a cylinder of diameter d, the perimeter is πd, and so the perimeter ratio is π. As rib number increases, the perimeter ratio generally increases.

For many species, fractional rib depth and perimeter ratio vary with direction. For example, the ribs on the south-facing side of *Carnegiea gigantea* and *Ferocactus acanthodes* are about 20% closer together than are those on the north-facing side, resulting in a fractional rib depth that is 24% and 5% greater, respectively, for the south side (Geller and Nobel 1984). Such changes in fractional rib depth and perimeter ratio influence PPFD interception and CO_2 uptake. Computer simulations in which perimeter ratio is varied, along with time of the year and the percentage of surface area shaded, show that CO_2 uptake is maximal for a perimeter ratio of about 6.5 in the summer and near 5 in the winter. A perimeter ratio of 5 is ideal for CO_2 uptake for stems that are shaded 10%, but for stems shaded 30% the ideal perimeter ratio is 4, and for 50% shading the ideal perimeter ratio is essentially equal to π. The actual perimeter ratio for Sonoran Desert cacti is often above 4, which maximizes CO_2 uptake when shading by spines, neighboring plants, and topographical features as well as cloudiness are taken into account (Geller and Nobel 1984).

Spines and Pubescence

In 1911 Forrest Shreve, one of the original researchers at the Desert Laboratory and later its director, noted that the southern extent of freezing temperatures is the most limiting feature in the northern distribution of tropical species, including cacti (Shreve 1911). In the Sonoran Desert, only three columnar cacti (*Carnegiea gigantea*, *Lophocereus schottii*, and *Stenocereus thurberi* [organ pipe cactus]) occur where freezing temperatures are common. A key to the survival of episodic freezes by these three species, as well as by barrel cacti (fig. 5.2B), is the covering of the sensitive apical region of their stems by spines and pubescence.

Spines, which are leaves that have become modified during evolution, can be an antiherbivory adaptation but also provide an essential means of modifying the thermal environment adjacent to the stem (fig. 5.3). Spines can be long, thin, and hairlike, as for *Opuntia erinacea*, or they can be thick, stout, and relatively inflexible, as for most species in the genus *Ferocactus*, such as *F. acanthodes* (fig. 5.2B). Many species of Sonoran Desert cacti have a special class of modified spines called pubescence, a soft, fuzzy covering over the apical meristem (fig. 5.2B). In the genus *Opuntia*, another type of modified spine occurs — glochids. Glochids are tiny spines with barbs that facilitate their attachment to skin and fur. Because they are so small, they are difficult to see and to remove from the skin. Spines can help distribute some cactus species, such as *O. fulgida* (jumping cholla) and *O. bigelovii*. However, spines represent costs in terms of biomass invested and reduced PPFD interception by the stem, with its consequential reduction in CO_2 uptake.

An advantage of spines and pubescence is their ability to insulate the apical region from direct exposure to the cold nighttime sky (Nobel 1991). The thickness of the apical pubescence and the percentage cover by spines vary among and within species, especially over regions that differ in low wintertime temperatures. For example, *Carnegiea gigantea* at 30° N in Sonora, Mexico, has an apical pubescence depth of 10 mm and 9% shading of the apex by spines. Nearby, *Lophocereus schottii* has essentially no apical pubescence but apical spine shading of 20%. *Stenocereus thurberi* also has no apical pubescence but apical spine shading of 60% (Nobel 1980d). To the north, the apical pubescence of all three species is unchanged. However, shading of the apex by spines increases to 41% for *C. gigantea* at 35° N, is unchanged for *L. schottii* at 32° N, and increases to 71% for *S. thurberi* at 32° N (Nobel 1980d). Its lack of apical spines may restrict the distribution of *L. schottii* in the United States, where it is native only to Organ Pipe Cactus National Monument in southernmost Arizona (Benson 1982). On the other hand, the thick apical pubescence of *C. gigantea*, together with moderate shading by spines, substantially raises the minimum temperatures of its apical region (fig. 5.3), allowing it to occur in the northern half of Arizona (Nobel 1980d).

A similar interspecific comparison can be made among barrel cacti of the Sonoran Desert (Nobel 1980a). In particular, *Ferocactus acanthodes* (fig. 5.2B) can have a thick layer of apical pubescence (9 mm), apical spine shading of over 90%, and a distribution that includes the coldest habitats

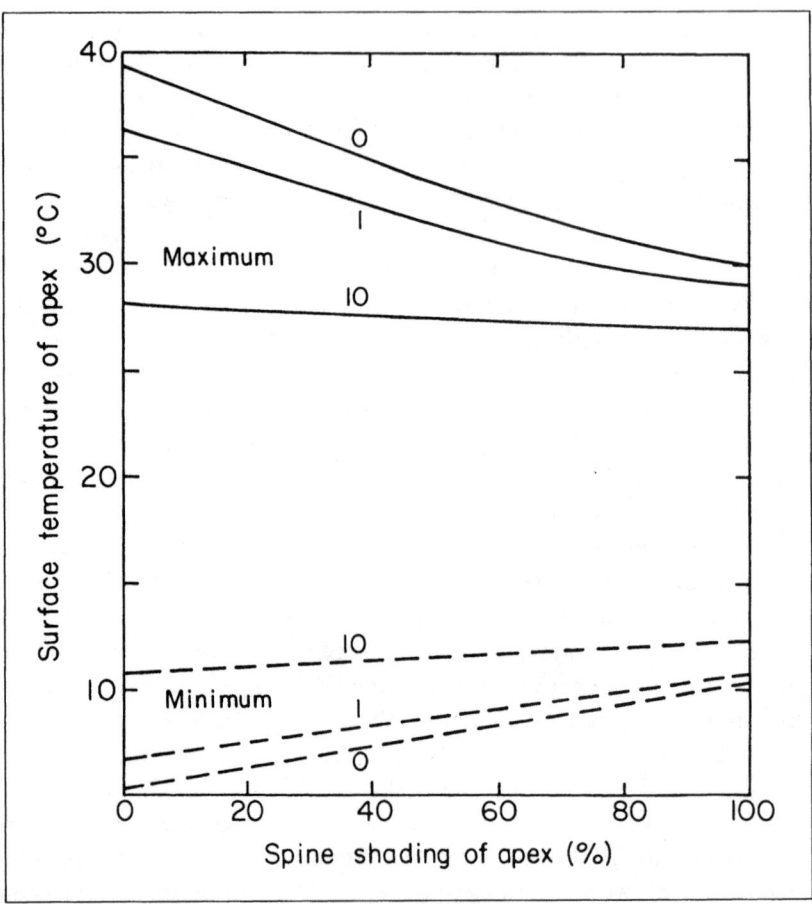

Figure 5.3 Influence of spine shading of the apex and thickness of apical pubescence (indicated in mm next to the curves) on the surface temperature of the stem apex for *Carnegiea gigantea* (saguaro). Used by permission from Nobel 1988. © Cambridge University Press. See Nobel (1980d) for further details.

of the Sonoran Desert. *Ferocactus wislizenii* also has thick apical pubescence (8 mm) and a 50% spine shading of the apex; it can occur as far north as 33° N in central Arizona. *Ferocactus covillei* (emory barrel) has a similar depth of apical pubescence but only up to 16% shading of the apex; its northernmost distribution is limited to southern Arizona (Benson 1982; Nobel 1980a).

Another effect of spines is to reduce the amount of light intercepted by cactus stems. Light absorption varies with time of day, season, location on the stem, stem orientation, neighboring vegetation, topography, and cloud cover. Taking into account stem dimensions and orientation, the average total daily PPFD incident on the spines of *Opuntia bigelovii* (fig. 5.1C) is about 21 mol m^{-2} d^{-1} over the course of a year (Nobel 1983). Nocturnal acid accumulation for this species is 90% of maximal at a total daily PPFD of 23 mol m^{-2} d^{-1} with spines but at only 16 mol m^{-2} d^{-1} when the spines are removed. Therefore, spines cause a considerable reduction in PPFD interception, resulting in lower rates of nocturnal acid accumulation, CO_2 uptake, and thus growth. In fact, spine removal results in a 60% greater increase in stem volume for *O. bigelovii* over 2½ years compared with plants with intact spines (Nobel 1983).

In addition to their effects on abiotic factors, spines also deter many animals from eating cacti. Antelope, bighorn sheep, wild horses, rodents, and the desert tortoise are all consumers of cacti, especially when the spines are absent or are removed, such as by burning (Steenbergh and Lowe 1977; Stelfox and Vriend 1977; Norman and Martin 1986; Nobel 1994). Sonoran Desert bighorn sheep will butt and paw barrel cacti to break the stem open, which they readily eat and vigorously defend as an important source of water, especially during the drier times of the year (Warrick and Krausman 1989). A spineless prickly pear, *Opuntia phaeacantha* var. *laevis,* grows under trees and on rocky ledges where it is inaccessible to most grazing mammals except small rodents (Benson 1982). Because spines reduce PPFD interception and provide thermal protection necessary for survival only at the distribution extremes of certain cacti, the biomass invested in spines, which can be 8% of the dry mass for *Ferocactus acanthodes* and 16% for *O. bigelovii,* must most often be justified as a deterrent to herbivory (Nobel 1983).

Some species of cacti benefit from an association between animals and spines. Spines can help to disseminate some cholla cacti of the Sonoran Desert, such as *Opuntia bigelovii* and *O. fulgida,* which reproduce mainly asexually. The terminal segments of their stems are easily detached when the barbed spines stick to the skin or fur of a passing animal. A detached segment can dislodge from the animal at some new location, where it takes root and develops into a plant (Benson 1982; Holthe and Szarek 1985). Both *Ferocactus acanthodes* and *O. acanthocarpa* possess extrafloral nectaries in the areoles. Nectar is produced in large quantities in the spring, and the ants it attracts may act as a defense against herbivory during the flowering season (Pickett and Clark 1979; Ruffner and Clark 1986).

Roots

Roots of cacti, like those of other plants, anchor the plant in the soil, take up water, and absorb nutrients. Cacti grow new roots when soil water is available following rainfall, yet their roots are able to resist water loss to the soil as drought commences. Roots use respiration to derive energy for growth and for nutrient uptake. Respiration requires oxygen and produces carbon dioxide, both of which must diffuse through the soil. Hence, soil structure and porosity have implications for root metabolism.

In 1911 William Cannon, also working at the Desert Laboratory, reported on the root systems of many Sonoran Desert plants, including cacti. The roots of *Ferocactus wislizenii* radiate away from the plant base to a distance of over 3 m, are thin along most of their length (5 mm at 1 m from the plant base), and average only 9 cm in depth (Cannon 1911). A few years later, MacDougal and Spalding noted that the deepest roots of *F. wislizenii* occur at 20 cm, which is still quite shallow (MacDougal and Spalding 1910; Nobel 1988). Other species of cacti, such as *Opuntia phaeacantha* (fig. 5.1D), have shallow roots that are associated with rocks. On the other hand, *O. versicolor* (staghorn cholla) has a vertical tap root that extends to 30 cm deep and unbranched lateral roots within the top 5 cm of the soil (Cannon 1911). Cactus roots are generally shallower than those of other species, whose roots are rarely found in the same zone as the cacti, suggesting that cacti preempt the wetting front of water as it seeps into the desert soil.

A rainfall of 13 mm causes stems of *Opuntia phaeacantha* to regain turgor, as does watering for a desiccated plant (Cannon 1911). Such immediate regaining of turgor is due to water uptake by existing roots (Nobel and Sanderson 1984). Rewetting also induces the growth of new "rain roots" (Nobel 1988). For *Ferocactus acanthodes* (fig. 5.2B), new roots are visible 8 h after rewetting and are 2 to 6 mm in length after 24 h (Nobel and Sanderson 1984). Rewetting causes water uptake and the swelling of stems of *Carnegiea gigantea* and *O. basilaris* (beavertail cactus) within 24 to 36 h (Szarek et al. 1973; Nobel and Sanderson 1984). Even though stem water content is correlated with the induction of new roots, computer simulations have shown that water uptake during the first few days after drought is due primarily to existing roots. The area of new roots, which is initially zero, rapidly increases but does not lead to major water uptake until about 4 days after rewetting of the soil (Nobel and Sanderson 1984).

The distribution and branching patterns of roots of many Sonoran Desert cacti, such as *Opuntia phaeacantha* and *O. versicolor,* are affected by underground rocks (Cannon 1911). For example, their roots proliferate under relatively shallow rocks but run alongside rocks deeper than 25 cm. Also, soil at a depth of 25 cm can be wet under rocks but dry 1 m laterally away. For *Echinocereus engelmannii* (strawberry hedgehog), the root length per unit soil volume is threefold higher within 1 cm of rocks compared to 5 cm away (Nobel et al. 1992). The length of lateral roots per unit length of main roots is 4.2 for *Ferocactus acanthodes* near rocks but only 0.8 away from rocks. The roots of *O. acanthocarpa* form fine nets adjacent to rocks, which may help it obtain water to support its relatively high transpiration rates (Nobel et al. 1991). Rocks can block the upward movement of water in the soil, and water may condense on their relatively cool lower surfaces. The preferential association of roots of Sonoran Desert cacti with rock surfaces apparently reflects the wetter soil in such regions (Nobel et al. 1992).

The roots of Sonoran Desert cacti usually occur in sandy, porous soils (Nobel 1988). One possible reason for this is that silty or clayey soils retain water longer, facilitating the growth of fungi and bacteria that may act as root pathogens. An alternative hypothesis is that the large air phases typical of porous soils are required for an efficient diffusion of oxygen toward roots and carbon dioxide away from roots. Roots of *Ferocactus acanthodes* can survive several days of soil anoxia (Nobel 1990). However, elevated CO_2 levels decrease the viability of cells in its roots. Thus, restriction of roots of cacti to well-aerated soils may be a consequence of CO_2 toxicity rather than a requirement for a certain level of O_2 (Nobel 1990).

As soil dries during drought, the soil water potential can drop to -10 MPa in 1 month (fig. 5.4), but the tissues of cacti, including their roots, generally remain above -0.6 MPa (Nobel 1988). In such a situation, water would be expected to move toward lower water potentials — from the shoot to the roots to the dry soil. Cacti have evolved a number of

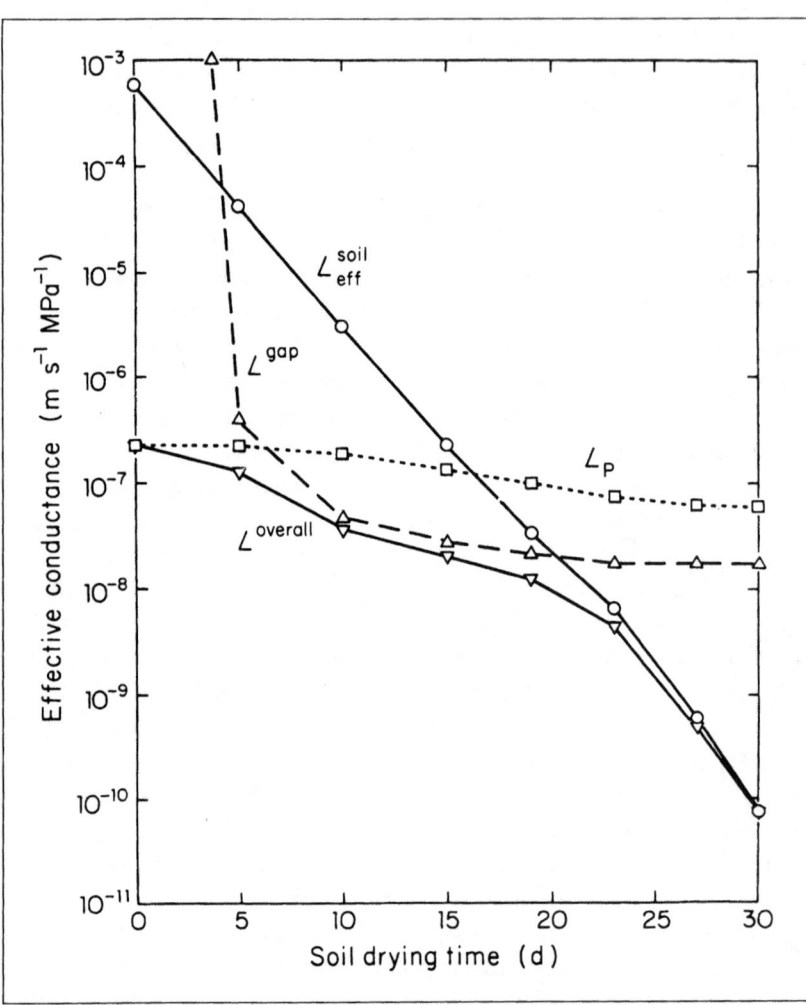

Figure 5.4 Influence of drought duration on the three hydraulic conductances of the root-air gap-soil system. The hydraulic conductance of the root is represented by L_P, that of the air gap by L^{gap}, and that of the soil by L^{soil}_{eff}. The three conductances add as reciprocals to give the overall conductance, $L^{overall}$, for water movement from the soil to the root xylem, or vice versa. At the onset of drought, the bulk soil water potential was -0.01 MPa, decreasing to -1 MPa at 15 days and to -10 MPa at 30 days. See Nobel and Cui (1992), from which the figure is used by permission, for details. © Oxford University Press.

features that reduce the amount of water lost to a drying soil. For example, the roots of some species of *Ferocactus* and *Opuntia* exude a fibrous mucopolysaccharide that binds soil particles to the root hairs and forms a sheath of soil around the root. As the soil dries, the soil sheath reduces water loss (North and Nobel 1992).

The root hydraulic conductance (L_P), which quantifies the ease of water movement into or out of a root, decreases as the soil dries (fig. 5.4). The development of cortical gaps (lacunae), embolism of the root xylem, and suberization accompany the reduction in L_P. The hydraulic conductance of the junctions between main roots and stems as well as between main roots and lateral roots also decreases during drought, further limiting water loss from the roots to a drying soil (Ewers et al. 1992; North et al. 1992). The dehydration of the root cortex leading to formation of cortical lacunae causes the root to shrink radially and hence to pull away from the adjacent cylinder of soil. Such shrinkage produces an air gap between the root and the soil across which water must move as a vapor (Nobel and Cui 1992). Upon rewetting, L_P recovers for young roots (as does the structure of cortical cells), xylem embolism is reversed, but suberization remains (North and Nobel 1992). Rewetting also causes the root to re-swell, so the air gap disappears.

When soil water is plentiful, such as after rainfall, water uptake by roots is favored and is limited primarily by the root L_P (fig. 5.4). As the soil dries, roots shrink and the air gap develops. Young roots of *Ferocactus acanthodes* shrink about 10% as the soil water potential decreases to −1 MPa during 15 days of drought. At this time, the air gap is the most limiting factor in the movement of water from roots to the drying soil (fig. 5.4). As the soil dries further, the low water content of the soil becomes the limiting factor for water movement. Indeed, the decrease in the soil hydraulic conductance over a 30-day period causes the overall conductance from the root, across the air gap, and into the soil to decrease by a factor of 3000 (fig. 5.4). Therefore, water loss from a cactus to a dry soil is mainly determined by the low hydraulic conductance of the soil.

Nurse Plants

Many species of Sonoran Desert cacti germinate under the canopy of an adult plant of another species, often referred to as a "nurse plant." Shreve (1931) noted that *Cercidium microphyllum* (foothill paloverde) reduces air and soil temperatures and is crucial for the germination and the survival of seedlings of *Carnegiea gigantea*. Almost all seedlings of *C. gigantea* in unshaded microsites die, even when supplied with supplemental water (Turner et al. 1966; Despain 1974).

Nurse plants moderate the thermal environment by lowering air temperatures as well as soil temperatures, which in the Sonoran Desert can reach 70°C (Nobel 1988). Furthermore, convective cooling of seedlings by winds is reduced near the ground. Field measurements and computer simulations show that a 70% reduction in shortwave radiation by a nurse plant reduces the maximal plant temperature by 17°C for a 2-cm-tall seedling of *Ferocactus acanthodes* and by 9°C for a 5-cm-tall seedling, which can be crucial for survival (Nobel 1984; Nobel et al. 1986). In addition, the shading and hence cooling of the soil surface reduce water evaporation, which can further benefit the young seedling.

Nurse plants also moderate the thermal environment for seedlings during the winter and can thereby increase the distribution of cacti into colder regions. During episodic freezes, the branches and leaves of nurse plants protect the apical meristem from the cold nighttime sky, analogous to the effects of spines and pubescence on apical stem temperatures. For instance, 1-cm-tall seedlings of *Carnegiea gigantea* had survival rates in the winter of 1967/68 of 58% in exposed locations and 80% under the canopy of *Cercidium microphyllum* (Steenbergh and Lowe 1976). In January 1971, survival was 78% for seedlings protected by screens compared to 35% for unprotected plants. For *C. gigantea* less than 50 cm tall, 14% exhibited freezing damage in exposed locations but only 6% did so under nurse plants (Nobel 1980c). As the plants grow taller (e.g., greater than about 2 m, the average height of the surrounding vegetation), the protection from the nurse plants diminishes and also is generally not needed. Specifically, minimum apical temperatures of *C. gigantea* then increase due to the effects of stem height, stem diameter, shading by spines, and apical pubescence (Nobel 1988).

Nurse plants reduce the PPFD intercepted by the cactus stems, which decreases nocturnal CO_2 uptake and growth of cacti. The two common nurse plant species for *Carnegiea gigantea*, *Cercidium floridum* (blue paloverde) and *C. microphyllum*, when leafless can reduce the PPFD incident on the stems of cacti by 31% near the winter solstice (Lowe and Hinds 1971). For the CAM succulent *Agave deserti* of the Sonoran Desert, shading by its nurse plant, *Pleuraphis rigida* (big galleta; formerly *Hilaria rigida*), reduces incident PPFD by 70% for a 4-cm-tall seedling (Franco and Nobel 1988). The roots of *A. deserti* and *P. rigida* overlap, resulting in competition for soil water. However, the soil nitrogen content under the nurse plants can be 60% higher compared to exposed sites (Franco and Nobel 1988). Even though the PPFD is always lowered and the soil water can be lowered by the association of Sonoran Desert succulents with nurse plants, the reduced CO_2 uptake and reduced growth can be partially offset by the greater nutrient availability under the nurse plant compared with exposed sites.

Small cacti, such as *Mammillaria microcarpa* and *Echinocereus engelmannii*, often germinate in the canopy of *Opuntia fulgida*. *Opuntia fulgida* reproduces by dropping terminal stem segments to the ground, which readily take root. In addition to the moderation of the thermal environment for the seedlings of the two smaller species, the accumulation of the spiny stems of *O. fulgida* on the ground apparently keeps small herbivores from eating the tiny seedlings of *M. microcarpa* and *E. engelmannii* (McAuliffe 1984). Some nurse plants protect cactus seedlings by providing camouflage. Other cactus species may be involved in a cyclical relationship with their nurse plants. Seeds of *O. leptocaulis* are disseminated by birds that use *Larrea tridentata* (creosote bush) as a perch and by rodents that use it as a refuge (Yeaton 1978). Because of competition by the shallow roots of the cactus for water combined with the high water-use efficiencies of cacti, *O. leptocaulis* can overgrow *L. tridentata*, which eventually dies. When *O. leptocaulis* dies, the nutrients of the dead cactus return to the soil and apparently provide a preferential site for the germination of seedlings of *L. tridentata*, and the cycle begins anew (Yeaton 1978).

Conclusions

The forms of cacti have profound influences on their function and their survival in the often harsh environment of the Sonoran Desert. The massiveness of the stems and especially the insulation of the apical meristem by spines and pubescence moderate the extremes of tissue temperature. Nurse plants also moderate the thermal environment at the soil surface, again at a cost for PPFD interception and hence growth. Even though ribs and tubercles can greatly expand the surface area of cacti, PPFD is generally limiting for their net CO_2 uptake. Thus, orientation of platyopuntia cladodes toward the direction of the highest PPFD increases their net CO_2 uptake and hence their growth. The roots of cacti can proliferate near underground rocks, which influence soil water content. During drought, the roots of desert cacti undergo anatomical changes that reduce water loss, although substantial water loss is prevented mainly by a decrease in the hydraulic conductance of the soil. The interplay of many factors of structure and function determines the establishment, survival, growth, and reproduction of the remarkable and highly visible Sonoran Desert cacti.

Acknowledgments
Financial support by the Environmental Sciences Division, Office of Health and Environmental Research, U.S. Department of Energy is gratefully acknowledged.

Literature Cited
Benson L. (1982) *The Cacti of the United States and Canada.* Stanford University Press, Stanford.
Cannon W. A. (1911) *The Root Habits of Desert Plants.* Carnegie Institution of Washington Publication no. 131, Washington, D.C.
Conde L. F. (1975) Anatomical comparisons of five species of *Opuntia* (Cactaceae). *Annals of the Missouri Botanical Garden* 62: 425–473.
Despain D. G. (1974) The survival of saguaro (*Carnegiea gigantea*) seedlings on soils of differing albedo and cover. *Journal of the Arizona Academy of Science* 9: 102–107.
Ehleringer J., House D. (1984) Orientation and slope reference in barrel cactus (*Ferocactus acanthodes*) at its northern distribution limit. *Great Basin Naturalist* 44: 133–139.
Ewers F. W., North G. B., Nobel P. S. (1992) Root-stem junctions of a desert monocotyledon and a dicotyledon: hydraulic consequences under wet conditions and during drought. *New Phytologist* 121: 377–385.
Franco A. C., Nobel P. S. (1988) Interactions between seedlings of *Agave deserti* and the nurse plant *Hilaria rigida*. *Ecology* 69: 1731–1740.
Geller G. N., Nobel P. S. (1984) Cactus ribs: influence on PAR interception and CO_2 uptake. *Photosynthetica* 18: 482–494.
Geller G. N., Nobel P. S. (1986) Branching patterns of columnar cacti: influences on PAR interception and CO_2 uptake. *American Journal of Botany* 73: 1193–1200.
Geller G. N., Nobel P. S. (1987) Comparative cactus architecture and PAR interception. *American Journal of Botany* 74: 998–1005.
Gibson A. C., Nobel P. S. (1986) *The Cactus Primer.* Harvard University Press, Cambridge.

Holthe P. A., Szarek S. R. (1985) Physiological potential for survival of propagules of Crassulacean acid metabolism species. *Plant Physiology* 79: 219–224.

Loik M. E., Nobel P. S. (1993) Freezing tolerance and water relations of *Opuntia fragilis* from Canada and the United States. *Ecology* 74: 1722–1732.

Lowe C. H., Hinds D. S. (1971) Effect of paloverde (*Cercidium*) trees on the radiation flux at ground level in the Sonoran Desert in the winter. *Ecology* 52: 916–922.

MacDougal D. T., Spalding E. S. (1910) *The Water-Balance of Succulent Plants.* Carnegie Institution of Washington Publication no. 141, Washington, D.C.

McAuliffe J. R. (1984) Prey refugia and the distributions of two Sonoran Desert cacti. *Oecologia* 65: 82–85.

Nobel P. S. (1977) Water relations and photosynthesis of a barrel cactus, *Ferocactus acanthodes*, in the Colorado Desert. *Oecologia* 27: 117–133.

Nobel P. S. (1978) Surface temperatures of cacti – influences of environmental and morphological factors. *Ecology* 59: 986–996.

Nobel P. S. (1980a) Influences of minimum stem temperatures on ranges of cacti in southwestern United States and central Chile. *Oecologia* 47: 10–15.

Nobel P. S. (1980b) Interception of photosynthetically active radiation by cacti of different morphology. *Oecologia* 45: 160–166.

Nobel P. S. (1980c) Morphology, nurse plants, and minimum apical temperatures for young *Carnegiea gigantea*. *Botanical Gazette* 141: 188–191.

Nobel P. S. (1980d) Morphology, surface temperatures, and northern limits of columnar cacti in the Sonoran Desert. *Ecology* 61: 1–7.

Nobel P. S. (1981) Influences of photosynthetically active radiation on cladode orientation, stem tilting, and height of cacti. *Ecology* 62: 982–990.

Nobel P. S. (1982) Orientations of terminal cladodes of platyopuntias. *Botanical Gazette* 143: 219–224.

Nobel P. S. (1983) Spine influences on PAR interception, stem temperature, and nocturnal acid accumulation by cacti. *Plant, Cell and Environment* 6: 153–159.

Nobel P. S. (1984) Extreme temperatures and the tolerances for seedlings of desert succulents. *Oecologia* 64: 310–317.

Nobel P. S. (1986) Form and orientation in relation to PAR interception by cacti and agaves. In: Givnish T. J. (ed) *On the Economy of Plant Form and Function*, pp 83–103. Cambridge University Press, New York.

Nobel P. S. (1988) *Environmental Biology of Agaves and Cacti.* Cambridge University Press, New York.

Nobel P. S. (1990) Soil O_2 and CO_2 effects on apparent cell viability for roots of desert succulents. *Journal of Experimental Botany* 41: 1031–1038.

Nobel P. S. (1991) *Physicochemical and Environmental Plant Physiology.* Academic Press, San Diego.

Nobel P. S. (1994) *Remarkable Agaves and Cacti*. Oxford University Press, New York.

Nobel P. S., Cui M. (1992) Hydraulic conductances of the soil, the root-soil air gap, and the root: changes for desert succulents in drying soil. *Journal of Experimental Botany* 43: 319-326.

Nobel P. S., Geller G. N., Kee S. C., Zimmerman A. D. (1986) Temperatures and thermal tolerances for cacti exposed to high temperatures near the soil surface. *Plant, Cell and Environment* 9: 279-287.

Nobel P. S., Loik M. E., Meyer R. W. (1991) Microhabitat and diel tissue acidity changes for two sympatric cactus species differing in growth habit. *Journal of Ecology* 79: 167-182.

Nobel P. S., Miller P. M., Graham E. A. (1992) Influence of rocks on soil temperature, soil water potential, and rooting patterns for desert succulents. *Oecologia* 92: 90-96.

Nobel P. S., Sanderson J. (1984). Rectifier-like activities of roots of two desert succulents. *Journal of Experimental Botany* 35: 727-737.

Norman F., Martin C. E. (1986) Effects of spine removal on *Coryphantha vivipara* in central Kansas. *American Midland Naturalist* 116: 118-124.

North G. B., Ewers F. W., Nobel P. S. (1992) Main root-lateral root junctions of two desert succulents: changes in axial and radial components of hydraulic conductivity during drying. *American Journal of Botany* 79: 1039-1050.

North G. B., Nobel P. S. (1992) Drought-induced changes in hydraulic conductivity and structure in roots of *Ferocactus acanthodes* and *Opuntia ficus-indica*. *New Phytologist* 120: 9-19.

Pickett C. H., Clark W. D. (1979) The function of extrafloral nectaries in *Opuntia acanthocarpa* (Cactaceae). *American Journal of Botany* 66: 618-625.

Ruffner G. A., Clark W. D. (1986) Extrafloral nectar of *Ferocactus acanthodes* (Cactaceae): composition and its importance to ants. *American Journal of Botany* 73: 185-189.

Shreve F. (1911) The influence of low temperatures on the distribution of the giant cactus. *The Plant World* 14: 136-146.

Shreve F. (1931) Physical conditions in sun and shade. *Ecology* 12: 96-104.

Shreve F., Wiggins I. L. (1964) *Vegetation and Flora of the Sonoran Desert*. Vol. II. Stanford University Press, Stanford.

Spalding E. S. (1905) Mechanical adjustment of the sahuaro (*Cereus giganteus*) to varying quantities of stored water. *Bulletin of the Torrey Botanical Club* 32: 57-68.

Steenbergh W. F., Lowe C. H. (1976) Ecology of the saguaro. I. The role of freezing weather in a warm-desert population. In: *Research in the Parks*, pp 49-92. National Park Service Symposium Series no. 1. U.S. Government Printing Office, Washington, D.C.

Steenbergh W. F., Lowe C. H. (1977) *Ecology of the Saguaro. II. Reproduction, Germination, Establishment, Growth, and Survival of the Young Plant.* National Park Service Scientific Monograph Series no. 8. U.S. Government Printing Office, Washington, D.C.

Stelfox J. G., Vriend H. G. (1977) Prairie fires and pronghorn use of cactus. *Canadian Field Naturalist* 91: 282–285.

Szarek S. R., Johnson H. B., Ting I. P. (1973) Drought adaptation in *Opuntia basilaris*. Significance of recycling carbon through Crassulacean acid metabolism. *Plant Physiology* 52: 539–541.

Turner R. M., Alcorn S. M., Olin G., Booth J. A. (1966) The influence of shade, soil, and water on saguaro seedling establishment. *Botanical Gazette* 127: 95–102.

Warrick G. D., Krausman P. R. (1989) Barrel cacti consumption by desert bighorn sheep. *Southwestern Naturalist* 34: 483–486.

Yeaton R. I. (1978) A cyclical relationship between *Larrea tridentata* and *Opuntia leptocaulis* in the northern Chihuahuan Desert. *Journal of Ecology* 66: 651–656.

6 Ecological Genetics of Cactophilic *Drosophila*

William J. Etges, William R. Johnson, Garry A. Duncan,
Greg Huckins, and William B. Heed

This paper is dedicated to the most steadfast member of the desert *Drosophila* group, Ms. Jean S. Russell.

The large number of cactus species in the deserts and semideserts of North and South America is one of the most remarkable aspects of these arid lands. Their presence has shaped a wide variety of animal-plant associations, including that between cacti and many insect species that have become intricately dependent on either their living or dead tissues. Most cacti are well armed with spines that fend off herbivory and affect surrounding microclimatic characteristics (Nobel and Loik, chapter 5, this volume). Senescence or injury may cause the death of older tissues or, at times, the entire plant. The decaying tissues attract and contain an interesting community of saprophytic organisms that ingest the available nutrients and moisture released from the plant cells. This community consists chiefly of opportunistic bacteria and yeasts as well as immature stages of insects, including the larvae of various species of flies of the order Diptera. The larvae feed on the bacteria, the yeasts, and the plant cells, including the cell sap of the depleted tissues. These invading organisms are not among the more obvious arid land inhabitants, yet studies of *Drosophila*/microorganism/cactus relationships have provided enormous insight into their ecology and evolution. For reviews of this subject, see Barker and Starmer (1982), Heed and Mangan (1986), Fogleman and Heed (1989), and Barker et al. (1990).

For ecological geneticists who work with *Drosophila* (pomace flies) and are interested in the processes that change gene and chromosomal frequencies within and between species, the community inhabiting a "rot pocket" is a wellspring of information just waiting to be tapped. In the Sonoran Desert, two of the four species of *Drosophila* that are endemic to the desert and whose larvae and adults live in and around these fermenting tissues, have variable polytene chromosomes (inversions of gene order) that differ in frequency from one area to another. These chromosomes can be easily studied by dissecting out the salivary glands of the larvae just about to enter pupation. The tissues are then histochemically stained and squashed under a cover glass on a microscope slide to allow chromosome identification. The larvae may be obtained directly from the cactus rot pocket or adult flies may be aspirated from the rot, where they are feeding, mating, and laying eggs, and transferred to the laboratory where they continue the same activities on artificial media in glass vials.

The chief goal of this chapter is to compare the degree of the polytene chromosomal variability in *Drosophila pachea* and *Drosophila mojavensis* in relation to the host cacti used for feeding and breeding within the relevant vegetational subdivisions of the Sonoran Desert and adjacent regions as described by Gentry (1942), Shreve (1951, 1964), Brown and Lowe (1980), and Brown (1982). These subdivisions have been made with reference strictly to vegetation, which consists of the character and organization of communities (fig. 6.1). Even though the host cacti for the two *Drosophila* species traverse many of the subdivisions, it seems reasonable to infer that environmental conditions that differ somewhat between these areas could affect gene and chromosomal frequencies of the flies inhabiting their respective hosts. In fact, data on mean temperature plotted against mean precipitation for five of the six subdivisions, and including the Mojave desertscrub, has been presented as "climatographs" by Turner and Brown (1982). Even though there is a strong degree of overlap of the climatographs in certain seasons, representing shared environmental conditions, each subdivision also has its own characteristics. Variable edaphic conditions throughout the desert undoubtedly complicate any analysis of climatic factors by contracting or extending the limits of various species of cacti and/or the vegetational subdivisions themselves in certain cases, but this has not been taken into account in our analyses.

The Flies and Their Chromosomes

The polytene chromosomes of *Drosophila* have been a tool for geneticists in gene mapping and gene action since they were first discovered by

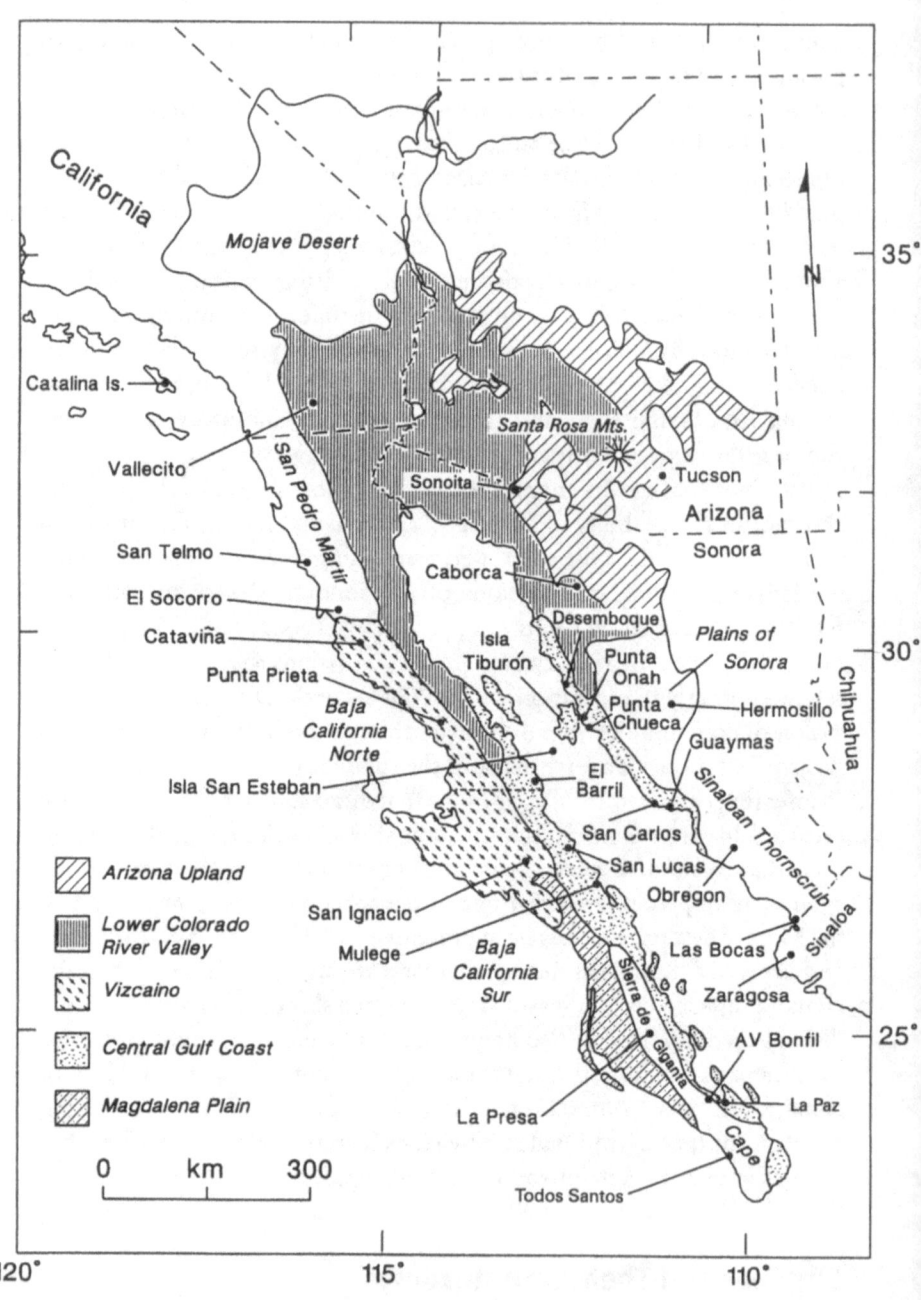

Figure 6.1 *(opposite page)* Map of the major vegetational subdivisions of the Sonoran Desert as presently envisioned (Turner and Brown 1982). Five subdivisons are listed whereas the sixth one (Plains of Sonora) is designated on the map. Adjacent vegetational regions (from Brown 1982) are indicated as the Cape region (San Lucan deciduous scrub), San Pedro Martir (Californian coastal scrub and chaparral), Mojave Desert, and Sinaloan thornscrub. Major landscape features and several place names referred to in the text are included.

Kostoff (1930) and elaborated upon by T. S. Painter (1933) at the University of Texas. The first significant analysis of salivary chromosome evolution was made by Sturtevant and Dobzhansky (1936). Since then many studies have been made of natural populations of *Drosophila* (Patterson and Stone 1952; Krimbas and Powell 1992; Powell 1997), but none of them include an in-depth analysis of host plants as possible factors that could affect inversion frequencies.

Inversion rearrangements, as well as other chromosomal mutations, can be detected in polytene chromosomes because of their large size, visible complexity, and a very defined aperiodic structure. Polytene chromosomes, which are found throughout the evolutionarily more advanced Diptera, consist of band sequences that correspond to the organization of particular genes. The bands vary in width, detail, and density of staining, the sequence of which is rarely repeated anywhere, except as short duplications. Therefore all chromosomes of an individual, as well as subunits of a single chromosome, may be separately identified. Between individuals within a species the banding sequences are often identical, while other individuals may differ by one or more inversion rearrangements. Also, closely related species usually have completely identifiable sequences that are similar or differ by one or more gene arrangements. Thus any number of genes may differ between related species, but this does not necessarily affect the visible structure of the chromosome. Therefore, the fingerprint of the polytene chromosome may be considered a living fossil record. In the simplest case, when two or more species have the same inversion fixed (homozygous), we know that their lineages trace back to a common ancestor. All inversions are thought to arise by unique mutational events (Aquadro et al. 1991; Krimbas and Powell 1992).

Another advantage of the polytene chromosome derives from its property of somatic pairing. When there are differences in gene order between homologous chromosomes, the only way they can pair is in the form of a loop. The loop represents the pairing of inverted and noninverted homologous chromosomes in the heterozygous state. In *Drosophila pachea*, there are three possible karyotypes with a single inversion in gene order: 7 +/+, 7 +/A, and 7 A/A. This is the only inversion polymorphism in the species. The Sonoran part of the distribution of inversion 7 A has been reported earlier (Ward et al. 1975). The Baja California part of the distribution, analyzed chiefly by Duncan (1979), will be presented here and a summary of the total distribution will be made. Inversions in *D. mojavensis* have been recently described by Ruiz et al. (1990). However, most of the quantitative data on inversion frequency variation in this species was presented by Johnson (1980) and is reviewed and updated in the present text.

The significance of inversion heterozygotes is that they rarely allow the recombination that occurs within them to be incorporated into the next generation, so the genes associated within the inverted sequence are usually inherited as a unit. When inversions rise in frequency in populations due to their effects on fitness, the genes contained within them are considered coadapted. Thus, inversions may have various levels of adaptive significance depending on the coadaptedness of the genes within them. Inversions have significant effects on fitness characters important in the life cycle of their carriers, such as longevity, fecundity, viability, speed of development and mating, and mating success (Dobzhansky 1955; Dobzhansky et al. 1964; Anderson and Watanabe 1974; Prakash 1967; Spiess and Spiess 1967; Ruiz et al. 1986; Etges 1989a; Salceda and Anderson 1988; Etges 1996). However, inversions are also very useful for the study of population structure, and it is in this context that we analyze the various gene orders discovered in the two desert species. The comparison of degrees of differentiation within and between local populations and larger geographical areas may be made in a hierarchical fashion with fixation indices following the pioneering work of Sewall Wright, which he summarized in volumes 2 and 4 of *Evolution and the Genetics of Populations* (Wright 1969, 1978). The fixation indices for each chromosome may then be compared in a general way to degrees of differentiation in other species in different geographical areas with different ecologies, as was done for *Drosophila pseudoobscura* in western North America (Wright 1978) and *D. subobscura* in Europe (Ferrari and Taylor 1981). In addition, historical factors certainly contribute to the variation of inversion frequency distributions. We have the added advantage in this context that the ancestral sequence of gene arrangements is known with high certainty in both *D. mojavensis* and *D. pachea*.

The Flies and Their Ecology

The phylogenetic lineages of *Drosophila mojavensis* and *D. pachea* have been separated by millions of years of evolution.[1] To make this clear, the species were assigned to different "species groups." *D. pachea* is classified as a member of the nannoptera species group with relatives in central and southern Mexico (Ward and Heed 1970; Pitnick and Heed 1994), and *D. mojavensis* is a member of the large repleta species group (Patterson and Crow 1940; Wasserman 1992). Furthermore, the host plants of the two species are very distinct. *D. pachea* breeds in the fermenting tissues of the senita cactus (*Lophocereus schottii*). This species contains unique phytosterols that are necessary for the larvae to complete development and for adults to lay fertile eggs (Heed and Kircher 1965; Fogleman et al. 1986). In addition, the presence of certain toxic alkaloids in senita excludes most other drosophilids, including *D. mojavensis*, from this host (Kircher et al. 1967; Fogleman et al. 1982). Figure 6.2 shows the distribution of *L. schottii* and its subdivision into three subspecies (Lindsay 1963). It is axiomatic that wherever senita is found in the desert, in associated thornscrub areas, or in deciduous scrub, *D. pachea* is present also. So the distribution of senita determines that of *D. pachea*. *Drosophila pachea* (strain A202) has also been reared once from *L. gatesii*, a species closely related to senita, found in allopatry in a small area of southwestern Baja California (Lindsay 1963). It is most probable that *L. gatesii*, which is not included in fig. 6.2, also contains the required sterols for larval growth and adult fecundity.

In contrast, *Drosophila mojavensis* uses a variety of different cacti, as shown in figure 6.3 (Heed and Mangan 1986). Pitahaya agria (*Stenocereus gummosus*) is the primary host for the entire peninsula of Baja California where the plant occurs, as well as for the midriff islands and other islands in the Gulf of California and the coastal area between Bahía Kino and Desemboque del Rio San Ignacio in Sonora, Mexico. A few records exist of *D. mojavensis* using organ pipe cactus (*S. thurberi*), cochal (*Myrtillocactus cochal*), and *Opuntia* in peninsular Baja California, but these records are rare. The principal host for *D. mojavensis* in mainland southern Arizona and northern Sonora is the organ pipe cactus. In southern

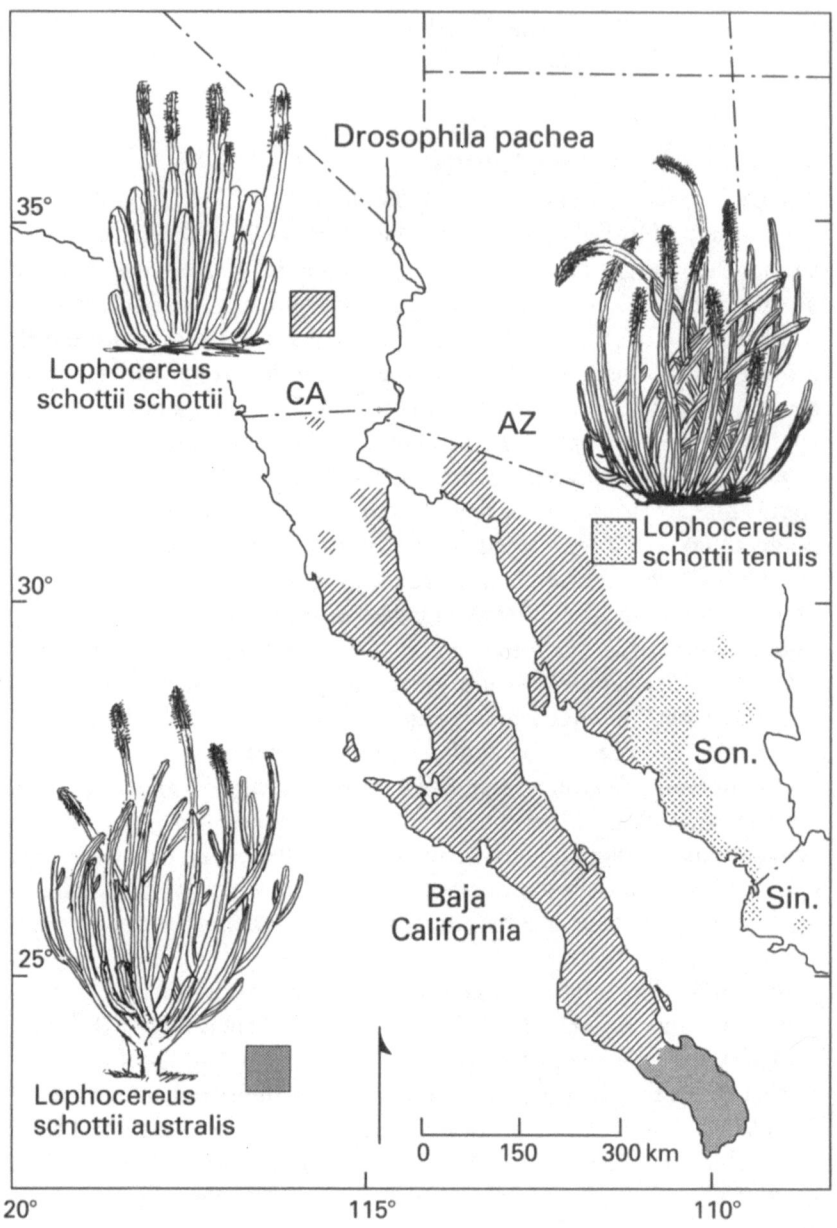

Figure 6.2 *(opposite page)* The distribution of senita cactus, the main host plant used by *Drosophila pachea*. The approximate distributions of the three subspecies in Mexico and the United States are shown, following Lindsay (1963) and Hastings et al. (1972). *Lophocereus schottii* var. *schottii* occurs in Baja California and the mainland (Arizona and Sonora), *L.s.* var. *tenuis* occurs in southern Sonora and northern Sinaloa, and *L. s.* var. *australis* occurs in the Cape region of Baja California.

Sonora and northern Sinaloa, the hosts are organ pipe and sina (*S. alamosensis*; Ruiz and Heed 1988). In the Lower Colorado subdivision of the Sonoran Desert, the Mojave Desert, and the Grand Canyon, *D. mojavensis* uses the California barrel cactus (*Ferocactus cylindraceous* [= *acanthodes*]; Taylor 1979; Turner et al. 1995). On Santa Catalina Island near Los Angeles, California, the host is *Opuntia demissa* (mostly the fruit). Thus *D. mojavensis* is oligophagous in its choice of hosts, and its distribution is neither limited by one plant nor is it limited to the Sonoran Desert (fig. 6.3). Since the senita cactus is sympatric with the pitahaya agria in the lower three quarters of Baja California and with organ pipe and sina cacti in Sonora, Mexico, we can determine whether the chromosome polymorphisms are responding similarly to environmental factors that should be common to both *Drosophila* species, such as temperature and humidity, in addition to the various host cacti to which each species is restricted[2] or whether they are the result of random forces.

Collecting Methods and the Chromosomes

Adult flies from natural populations of *Drosophila mojavensis* and *D. pachea* were obtained from 51 and 47 localities, respectively, in southern Arizona, California, and Sonora, Sinaloa, and Baja California, Mexico (appendix 6.1; appendix 6.2). Most of the collections were made from 1968 to 1980 (Johnson 1980; Duncan 1979), and in 1984 and 1985. A majority of the population samples were collected by locating rot pockets in agria, organ pipe, or senita cacti and aspirating the adult flies from the rots in the morning and evening hours. If sufficient numbers of adults were not present, intact cactus arms containing the fermenting tissues were carefully cut from the plant, wrapped in newspaper, and returned to the laboratory, where flies eclosed from the rots. Inversion frequencies

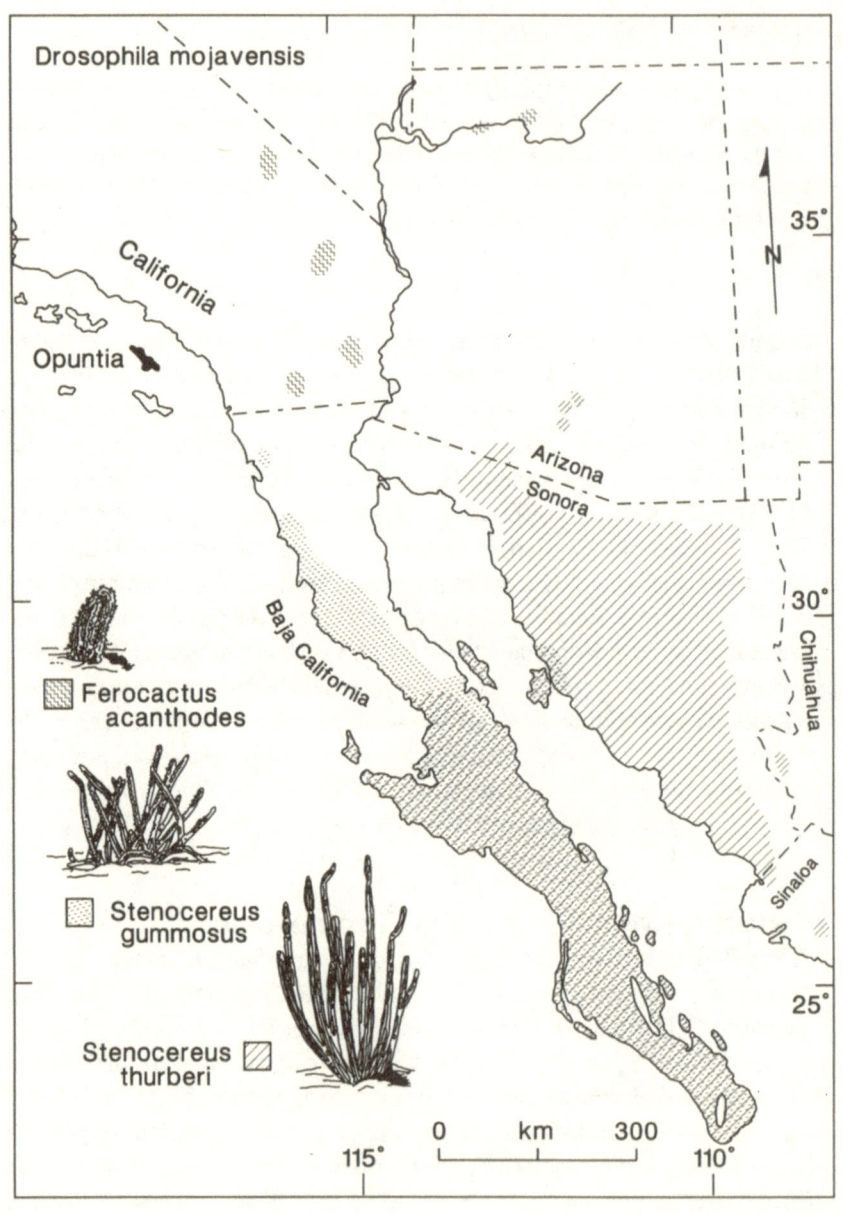

Figure 6.3 *(opposite page)* The distribution of the major host cacti of *Drosophila mojavensis* (modified from Heed and Mangan 1986, with permission). Pitahaya agria (*Stenocereus gummosus*) and organ pipe (*S. thurberi*) cacti occur together in the lower one half of Baja California and across the midriff islands to the mainland of Sonora, Mexico. Pitahaya agria extends beyond the desert in Baja California into the San Pedro Martir region in the north and into the Cape region in the south. Organ pipe cacti on the mainland extend beyond the desert eastward into the foothills of the Sierra Madre Occidental and southward into the Sinaloan thornscrub (Hastings et al. 1972). *D. mojavensis* may be found in all these areas. The distribution of the California barrel cactus, *Ferocactus cylindraceous* (labelled *acanthodes*) is shown only where collections of *D. mojavensis* have been made in association with it (Lower Colorado Valley subdivision and Mojave Desert, including the Colorado River in Arizona). *D. mojavensis* has also become established on Santa Catalina Island, where it breeds chiefly in the fruits of *Opuntia demissa*.

were calculated from the numbers of observed karyotypes for each population sample.[3]

All populations of *Drosophila pachea* were scored for chromosome seven karyotypes involving the 7 A and 7 + gene arrangements (Ward and Heed 1970). Gene arrangement designations for *D. mojavensis* for the second chromosome are ST = Standard, LP = La Paz, BA = Baja, and SL = San Lucas (Mettler 1963), which correspond to 2s, $2q^5$, $2r^5$, and $2v^7$, respectively (Ruiz et al. 1990; Wasserman 1992). ST is a single inversion removed from a hypothetical 2 abcfghqr arrangement. LP is derived from and overlaps ST. The two shorter inversions, BA and SL, are derived from and included within LP. However, BA and SL have not been recorded to be on the same homologue; therefore, the four gene arrangements may be treated statistically as though they were alleles at a single locus. On the third chromosome, ST (Standard, 3d) is the ancestral arrangement, and MU (Mulegé, 3y) is a centrally located, medium-length inversion derived from ST. Of 10 possible karyotypes expected from the four arrangements on the second chromosome (ST, LP, BA, and SL) and three possible from the third chromosome, all were identified in natural populations, except SL in the homozygous state (SL/SL). This previously undescribed short basal inversion has been found in eight localities in Baja California and on Tiburón Island and appears to be restricted to the Gulf Coast region. For *D. mojavensis*, all scoring was made simultaneously with the particular gene arrangements on both the second and third chromosomes.

Another chromosome, s1 (San Ignacio), was discovered by M. Wasserman (1976, personal communication) in a laboratory culture (A367) derived from a collection made in March 1972 near San Ignacio, Baja California. When this culture was initially analyzed for chromosome frequencies, the chromosome was not observed, but when other cultures from early collections were reanalyzed, the chromosome was found in one from San Lucas, Baja California (A422). Collections from the same area (A566, San Ignacio) also indicated the presence of the chromosome in very low frequencies (1%). Recent assays of these stocks kept in the laboratory showed that s1 was still present after eight years in culture (Etges, unpublished data). This chromosome is interesting in that it is derived via three paracentric inversions ($2s^7$, $2t^7$, and $2u^7$) from the ancestral, hypothetical chromosome (2abcfghqr), which also gave rise to the ST arrangement via one paracentric inversion. For a phylogeny of the arrangements of the second chromosome, see Ruiz et al. (1990). The s1 chromosome was not included in the analysis which follows.

Patterns of Karyotypic Variation
Drosophila pachea

Inversion polymorphism in natural populations of *Drosophila pachea* in Baja California is comparable to that described for mainland populations (Ward et. al. 1975). We recognize that some of the frequency data may be biased due to incomplete sampling of larvae for karyotyping using the PM7 and FCF7 methods (see appendix 6.2), but here we are concerned with the overall pattern of variation in relation to the vegetational subdivisions. In Baja California, the 7 + gene arrangement was not found north of Mulegé (approximately 27° N lat.), unlike in mainland populations, where 7 + has been found in low frequencies as far north as Sonoita, Sonora (approximately 31°45' N lat.). A cline is apparent in Baja California, as 7 + increases in frequency southward to its highest frequency of 0.74 north of La Paz (appendix 6.2; fig. 6.4). Thus, clinal variation in Baja California parallels that on the mainland but is shifted 3° to 4° southward on the peninsula. South of 24°15' N lat. in Baja California, the 7 + gene arrangement decreases in frequency, unlike on the mainland where 7 + continues to increase southward until all populations are karyotypically 7 +/+. Therefore, the 7 A arrangement in Baja California is not only the most common in the north, but it is also the most common gene arrangement in the far south in the Cape region. We consider the populations of *D. pachea* that inhabit southern Sonora to be ancestral for the species, because they are homozygous 7 +/+, as are all three nannoptera group species in southern Mexico (Ward and Heed 1970).

Drosophila mojavensis

Gene arrangement frequencies varied throughout the range of *Drosophila mojavensis* showing a classical central-marginal pattern of variation (Carson 1959; Heed 1981; Brussard 1984), where populations near the center of the species' range exhibited greater inversion heterozygosity than those at the periphery of the range as well as distinctive clines along peninsular Baja California (appendix 6.1; fig. 6.5; fig. 6.6). At least five categories of second chromosome inversion polymorphism are apparent: (1) Highly heterozygous populations occupy the Central Gulf Coast of Baja California. (2) Inversion LP predominated in populations occupying the Cape region south of La Paz and decreased in frequency in the Magdalena Plains where inversion ST is most common. (3) There was a gradual exchange of predominantly ST populations in the Magdalena Plains northward through heterozygous LP, ST, and BA populations in the Vizcaíno region to predominantly LP populations in the San Pedro Martir region. These more southern populations are topographically separated from the Central Gulf Coast populations by the Sierra de Giganta. (4) Homozygous LP populations occupied southern Arizona, Sonora, and Sinaloa, with the exception of several populations in the Desemboque area, including Tiburón Island. (5) Homozygous ST populations were found in the Lower Colorado River subdivision of southern California, on Santa Catalina Island, and in the Grand Canyon, Arizona (Ruiz et al. 1990). These populations are apparently discontinuous with the northern Baja populations due to high mountain ranges and the lack of suitable host cacti.

In contrast to the widespread distribution and high frequencies of LP and ST, BA and SL have more localized distributions and were sometimes quite rare. BA was most abundant (0.02 to 0.38) in the San Pedro Martir and Vizcaíno regions, decreasing in frequency southwards. BA was found again in the narrow isthmus north of the Cape region and on Isla San Jose. From these population samples, it appears that the distribution of BA is discontinuous east of the Sierra de Giganta, but it is possible that it is continuous along the Pacific Coast. BA was also present in some populations in the Desemboque region in Sonora.

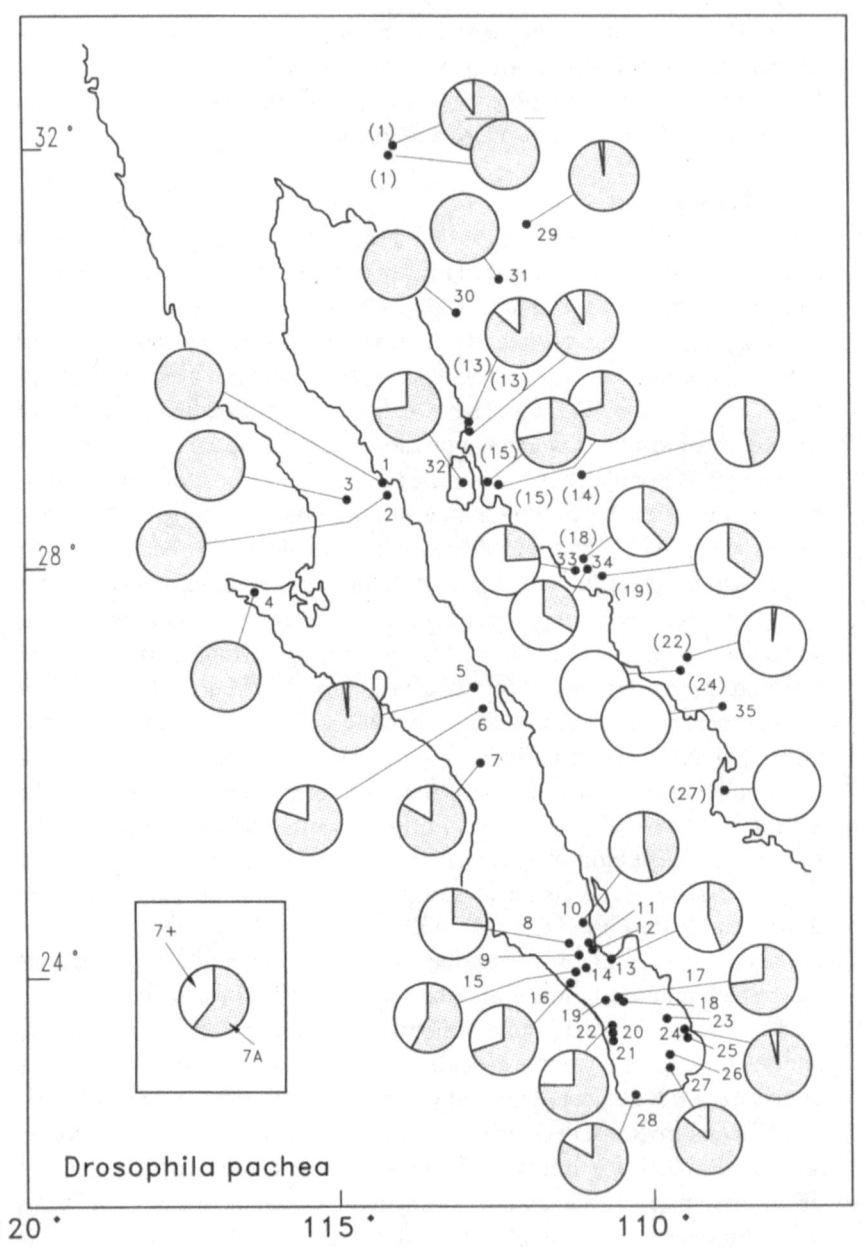

Figure 6.4 *(opposite page)* Geographical variation in chromosome 7 gene arrangement frequency in populations of *Drosophila pachea*. Population localities in parentheses are numbered according to table 1 in Ward et al. (1975). They represent 12 of the 24 localities listed in that table. The others correspond to appendix 6.2 in the present report. Not all populations are plotted for inversion frequencies, but their locations are numbered as in appendix 6.2.

SL was present in Santiago, in the Cape region, in the Central Gulf Coast region of the peninsula north of La Paz to El Barril, and on Tiburón Island. This short inversion was not discovered until the spring of 1974 even though four previous collections were made in this region (Johnson 1973).

The geographic pattern exhibited by third chromosome gene arrangements of *Drosophila mojavensis* was similar to the pattern for chromosome 2 (fig. 6.6; appendix 6.1). The important similarities are the complementary karyotypes between the Cape and the lower Magdalena Plain regions (MU decreases and ST increases in frequency), similar frequency changes with latitude in Baja California, and near fixation or fixation of ST in southern California, northern Baja California, Sonora, and southern Arizona. The important distinctions are (1) little east-west differentiation across the Sierra de Giganta, (2) replacement of karyotypes between the Vizcaíno and the San Pedro Martir regions, in contrast to the replacement occurring between the San Pedro Martir region and southern California as observed in the second chromosome, and (3) presence of heterokaryotypes in populations not using agria along coastal Sonora.

Hierarchical Analysis of Karyotypic Variation
The Method
In both *Drosophila mojavensis* and *D. pachea*, we partitioned karyotypic variation into within-population, between-population within regions (the vegetational subdivisions and adjoining regions), and between regions. We used the Analysis of Molecular Variance (AMOVA, ver. 1.53) described by Excoffier et al. (1992) to produce variance components and associated Φ (phi) statistics[4] for each level of geographic variation and Wright's F statistics (Wright 1951). We were most interested in comparing the hierarchical structures of *D. mojavensis* and *D. pachea* populations based on

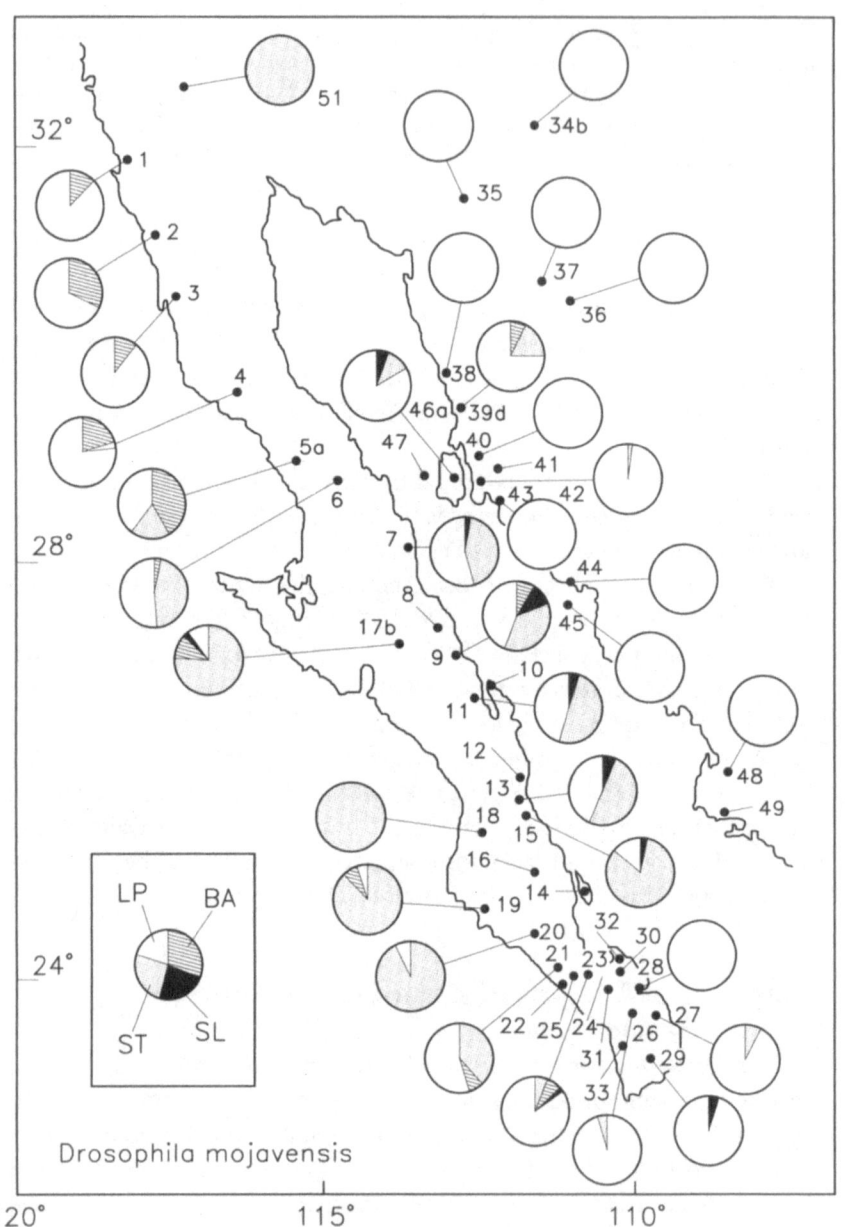

Drosophila mojavensis

Figure 6.5 *(opposite page)* Geographical variation in chromosome 2 gene arrangement frequency in the majority of the sampled populations of *Drosophila mojavensis*. Numbers refer to localities in appendix 6.1. Locality 50, Santa Catalina Island, is not included in the figure. The remaining sampled populations are found in Ruiz et al. (1990).

the nonarbitrary grouping of populations within the vegetational subdivisions, as well as comparing these results with previous analyses of population structure in *D. pseudoobscura* and *D. subobscura* (Wright 1978; Ferrari and Taylor 1981).

Populations were grouped into the vegetational subdivisions as objectively as possible, even when obvious transition zones were apparent (Johnson 1980). The Central Gulf Coast populations were separated into either peninsular (including the islands in the Gulf of California close to the peninsula) or mainland (including the islands of Tiburón and San Pedro Nolasco) regions. This was done because of the large differences between populations in these regions due to isolation caused by the Gulf of California. The Santa Catalina Island population was grouped with the southern California population because these populations share similar chromosomal constitutions. Later, however, the island was found to have closer vegetational affinities with the San Pedro Martir subdivision and was designated Coastal Sage (Axelrod 1978), even though the karyotypic affinity of *Drosophila mojavensis* on the island lies with the Vallecito sample and other collections in southern California and not with the San Pedro Martir populations (see fig. 6.1; fig. 6.5; appendix 6.1; Ruiz et al. 1990).

We also estimated the extent of population structure using host cacti as natural groups. For *Drosophila pachea*, the three subspecies of senita were used, *Lophocereus schottii schottii*, *L. s. tenuis*, and *L. s. australis*. For *D. mojavensis*, these groups were agria, organ pipe, California barrel, and *Opuntia* cacti.

Relation to the Vegetational Subdivisions

Both *Drosophila mojavensis* and *D. pachea* populations exhibited significant levels of karyotypic variation at all levels of the hierarchical analysis (table 6.1; table 6.2). Even though population sampling was carried out in different years by different investigators, and in many cases different populations were sampled within subdivisions (appendix 6.1; appendix 6.2),

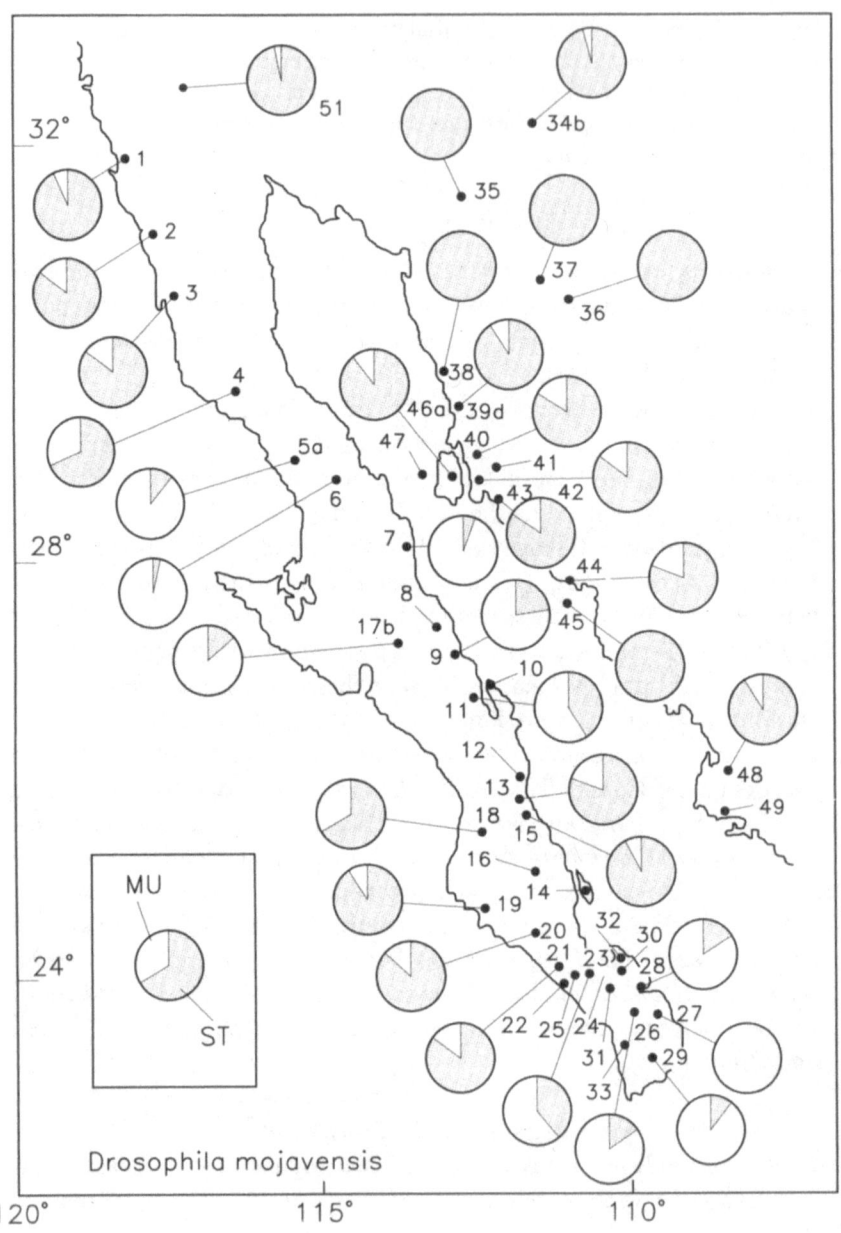

Figure 6.6 *(opposite page)* Geographical variation in chromosome 3 gene arrangement frequency in the majority of the sampled populations of *Drosophila mojavensis*. Numbers refer to localities in appendix 6.1. Locality 50, Santa Catalina Island, is not included in the figure. The remaining sampled populations are found in Ruiz et al. (1990).

both species showed significant within-population, among-population-within-subdivision, and among-subdivision variation in karyotype frequency. We were most interested in the among-subdivision levels of variation (Φ_{ct}), for at this level we can investigate the effects of the vegetational subdivisions on karyotype variation relative to other levels. Only qualitative comparisons can be made between variance components because confidence intervals around them are currently unavailable, yet these comparisons lead to some intriguing insights. When all populations were included, *D. pachea* exhibited greater genetic variance between subdivisions (61.47% of the variance, $\Phi_{ct} = 0.615$) than *D. mojavensis* (27.88% of the variance, $\Phi_{ct} = 0.279$). At first, this result suggested that populations of *D. pachea* were much more structured around the ecological pattern of the vegetational subdivisions than those of *D. mojavensis*, whether or not the Central Gulf Coast (CGC) populations were separated into Baja and mainland groups (table 6.1; table 6.2). However, after removing the Sinaloan thornscrub (STS) subdivision and reanalyzing the remaining data, estimates of Φ_{ct} and F_{rt} for *D. pachea* approached those of *D. mojavensis*. Populations of *D. pachea* in the STS subdivision are all homokaryotypic for 7 + except for three 7 +/A heterokaryotypes found near Ciudad Obregon (Ward et al. 1975). Therefore, much of the apparent intersubdivision structure was due to the influence of these STS populations.

We sequentially removed other subdivisions in a similar fashion and reanalyzed the remaining data. Removal of the Lower Colorado populations further decreased Φ_{ct} and F_{rt} and when populations from Baja California were reanalyzed by themselves, no significant difference among subdivisions was apparent (all data [STS, LC, PLS, ARIZ] analysis, table 6.2). A similar approach was used for *Drosophila mojavensis* and a negligible reduction in among-subdivision variation was detected after removing all mainland populations except the mainland CGC populations (table 6.1). Therefore, both species exhibited low, but significant, karyotypic variation between vegetational subdivisions in the Sonoran Desert. The

Table 6.1 Results of hierarchical population structure analyses in *Drosophila mojavensis* for both AMOVA analyses and Wright's F statistics for karyotypic variation within and between 51 populations.

Variance Components	df	Variance	% Total	P^a	Φ Statistic F Statistic
A. Among Subdivisions					
All data — σ^2_a	9	0.761	27.88	<0.02	$\Phi_{ct} = 0.27$
	9	0.098	22.89	<0.02	$F_{rt} = 0.22$
All data (STS, LC, ARIZ)^c — σ^2_a	6	0.744	25.39	<0.02	$\Phi_{ct} = 0.25$
	6	0.081	18.28	<0.02	$F_{rt} = 0.18$
Among Populations within Subdivisions					
All data — σ^2_b	41	0.783	28.69	<0.02	$\Phi_{sc} = 0.39$
	41	0.103	24.22	<0.02	$F_{dr} = 0.31$
All data (STS, LC, ARIZ) — σ^2_b	36	0.801	27.31	<0.02	$\Phi_{sc} = 0.25$
	36	0.102	22.87	<0.02	$F_{dr} = 0.28$
Among Individuals within Populations					
All data — σ^2_c	2289	1.185	43.44	<0.02	$\Phi_{st} = 0.56$
	2289	0.226	52.89	<0.02	$F_{id} = 0.47$
All data (STS, LC, ARIZ) — σ^2_c	1951	1.387	47.30	<0.02	$\Phi_{st} = 0.52$
	1951	0.262	58.85	<0.02	$F_{id} = 0.41$
B. Among Host Plants^d					
All data — σ^2_a	3	0.615	20.26	<0.02	$\Phi_{ct} = 0.20$
	3	0.128	25.57	<0.02	$F_{rt} = 0.25$
Among Populations within Cactus Species					
All data — σ^2_b	47	1.237	40.73	<0.02	$\Phi_{sc} = 0.51$
	47	0.145	29.13	<0.02	$F_{dr} = 0.39$
Among Individuals within Populations					
All data — σ^2_c	2289	1.185	39.01	<0.02	$\Phi_{st} = 0.61$
	2289	0.226	45.30	<0.02	$F_{id} = 0.54$

Note: The 51 populations are grouped according to (A) the vegetational subdivisions in and around the Sonoran Desert, and (B) the major host plants used by *D. mojavensis*. The data are from appendix 6.1. See text for details.

[a] The probability that the observed variance components, Φ statistics, and F statistics would be larger by chance alone. Following Excoffier et al. (1992), Φ_{ct} and σ^2_a were tested by 50 random permutations of populations across regions, Φ_{sc} and σ^2_b were tested by 50 random permutations of individuals across populations but within the same region, and Φ_{st} and σ^2_c were tested by 50 random permutations of individuals across populations without specifying populations or regions.

[b] For each level of analysis, results of the AMOVA analyses are listed in the first line of each pair, with Wright's F statistics listed in the second line. Subscripts for Φ statistics are consistent with Excoffier et al. (1992). Subscript for F statistics are consistent with Wright (1978), except for F_{id}, which is used here because these analyses are based on karyotypes rather than inversion frequencies. F_{rt} is the correlation between randomly chosen karyotypes within a group of populations relative to the entire species, F_{dr} is the correlation between randomly picked karyotypes within populations relative to that of the region, and F_{id} is the correlation of random karyotypes within populations relative to the entire species.

[c] Abbreviations in parentheses refer to the vegetational subdivisions that were excluded in each analysis. See appendix 6.1 for abbreviations.

[d] In this analysis, host cactus replaces subdivision: the host cacti are *Stenocereus gummosus*, *S. thurberi*, *Ferocactus cylindraceous*, and *Opuntia demissa*.

steep transition across the STS into the Sonoran Desert in southern Sonora almost entirely coincident with a shift in host subspecies use is the most notable aspect of karyotype variation in D. pachea.

It would not be possible to infer causes for the subdivision-related karyotype differences between *Drosophila mojavensis* and *D. pachea* if other independent genetic data were unavailable. However, Rockwood-Sluss et al. (1973) suggested that little population structure existed across the range of *D. pachea* based upon an absence of allelic variation at four protein encoding loci among eleven mainland populations. Furthermore, they found that a single population from Guaymas, Sonora (fig. 6.4, population 34) contained as much genetic variability as all the populations taken together. Such genetic uniformity among partially isolated populations is most parsimoniously explained as the result of gene flow, which homogenizes gene frequencies, particularly at enzyme encoding loci. *Drosophila pachea* adults must therefore be capable of dispersing long distances. In fact, Johnston (1974) concluded, based on intensive mark-recapture studies, that the size of mainland populations of *D. pachea* are on average smaller than those of *D. mojavensis* and that *D. pachea* individuals migrate further. The population structuring based on karyotype variability at the level of vegetational subdivisions in *D. pachea* must therefore result from forces strong enough to overcome gene flow, i.e., natural selection shaping karyotypic frequencies in response to local ecological conditions. Similar patterns exist for *D. mojavensis* based on allozyme variation (Zouros 1973), particularly at the Adh-2 locus (Starmer et al. 1977), but mainland *D. mojavensis* populations are nearly homokaryotypic for second chromosome gene arrangements, except for several that use agria (*Stenocereus gummosus*) and exhibit limited third chromosome polymorphism (fig. 6.3; fig. 6.5; fig. 6.6).

For *Drosophila mojavensis*, 43.44% of the overall variability (Φ_{st} = 0.566) was found within populations, contrasting with 22.86% (Φ_{st} = 0.771) in *D. pachea*. Φ_{st} is an estimate of the correlation between karyotypes within the same population relative to randomly chosen karyotypes across the entire species. The magnitude of Φ_{st}, as well as F_{id} (among individuals within demes), decreased somewhat and the percent of the total variance increased as mainland groups were removed, as described above. In local populations, these large Φ_{st} are expected simply on the basis of local panmixia relative to the karyotypic variation in *D. mojavensis* and *D. pachea* populations throughout their respective ranges. There is

Table 6.2 Results of hierarchical population structure analyses in *Drosophila pachea* for both AMOVA analyses and Wright's F statistics for karyotypic variation within and between 58 populations.

Variance Components	df	Variance	% Total	P^a	Φ Statistics[b] / F Statistics
A. Among Subdivisions					
All data — σ^2_a	8	0.515	61.47	< 0.02	$\Phi_{ct} = 0.615$
	8	0.144	45.09	< 0.02	$F_{rt} = 0.451$
All data (CGC combined)[c] — σ^2_a	7	0.520	60.69	< 0.02	$\Phi_{ct} = 0.607$
	7	0.146	44.92	< 0.02	$F_{rt} = 0.449$
All data (STS)[d] — σ^2_a	7	0.170	31.23	< 0.02	$\Phi_{ct} = 0.312$
	7	0.061	23.08	< 0.02	$F_{rt} = 0.231$
All data (STS, LC)[d] — σ^2_a	6	0.149	24.46	< 0.02	$\Phi_{ct} = 0.245$
	6	0.052	17.49	< 0.02	$F_{rt} = 0.175$
All data (STS, LC, PLS, ARIZ)[d] — σ^2_a	4	0.095	14.82	ns	$\Phi_{ct} = 0.148$
	4	0.027	8.66	ns	$F_{rt} = 0.087$
Among Populations within Subdivisions					
All data — σ^2_b	49	0.131	15.67	< 0.02	$\Phi_{sc} = 0.407$
	49	0.040	12.42	< 0.02	$F_{dr} = 0.226$
All data (CGC combined) — σ^2_b	50	0.145	16.97	< 0.02	$\Phi_{sc} = 0.432$
	50	0.043	13.31	< 0.02	$F_{dr} = 0.242$
All data (STS) — σ^2_b	44	0.152	27.82	< 0.02	$\Phi_{sc} = 0.405$
	44	0.047	17.27	< 0.02	$F_{dr} = 0.225$
All data (STS, LC) — σ^2_b	35	0.191	31.28	< 0.02	$\Phi_{sc} = 0.414$
	35	0.057	19.08	< 0.02	$F_{dr} = 0.231$
All data (STS, LC, PLS, ARIZ) — σ^2_b	28	0.230	35.82	< 0.02	$\Phi_{sc} = 0.421$
	28	0.069	21.67	< 0.02	$F_{dr} = 0.237$
Among Individuals within Populations					
All data — σ^2_c	2227	0.192	22.86	< 0.02	$\Phi_{st} = 0.771$
	2227	0.135	42.50	< 0.02	$F_{id} = 0.575$
All data (CGC combined) — σ^2_c	2227	0.192	22.34	< 0.02	$\Phi_{st} = 0.777$
	2227	0.135	41.77	< 0.02	$F_{id} = 0.582$
All data (STS) — σ^2_c	1893	0.224	40.95	< 0.02	$\Phi_{st} = 0.591$
	1893	0.158	59.65	< 0.02	$F_{id} = 0.404$
All data (STS, LC) — σ^2_c	1522	0.270	44.26	< 0.02	$\Phi_{st} = 0.557$
	1522	0.188	63.42	< 0.02	$F_{id} = 0.366$
All data (STS, LC, PLS, ARIZ) — σ^2_c	1255	0.317	49.35	< 0.02	$\Phi_{st} = 0.506$
	1255	0.221	69.66	< 0.02	$F_{id} = 0.303$
B. Among Senita Cactus Subspecies[e]					
All data — σ^2_a	2	0.742	72.74	< 0.02	$\Phi_{ct} = 0.727$
	2	0.181	50.25	< 0.02	$F_{rt} = 0.502$
Among Populations within Cactus Subspecies					
All data — σ^2_b	55	0.087	8.49	< 0.02	$\Phi_{sc} = 0.311$
	55	0.044	12.21	< 0.02	$F_{dr} = 0.245$
Among Individuals within Populations					
All data — σ^2_c	2227	0.192	18.77	< 0.02	$\Phi_{st} = 0.812$
	2227	0.135	37.54	< 0.02	$F_{id} = 0.625$

Table 6.2 continued

Note: The 58 populations are grouped according to (A) the vegetational subdivisions in and around the Sonoran Desert, and (B) the three subspecies of senita cactus. The data from appendix 6.2 were combined with all published (Ward et al. 1975) and unpublished *D. pachea* karyotype data. See text for details.

^a The probability that the observed variance components, Φ statistics, and F statistics would be larger by chance alone. Following Excoffier et al. (1992), Φ_{ct} and σ^2_a were tested by 50 random permutations of populations across regions, Φ_{sc} and σ^2_b were tested by 50 random permutations of individuals across populations but within the same region, and Φ_{st} and σ^2_c were tested by 50 random permutations of individuals across populations without specifying populations or regions.

^b For each level of analysis, results of the AMOVA analyses are listed in the first line of each pair, with Wright's F statistics listed in the second line. Subscripts for Φ statistics are consistent with Excoffier et al. (1992). Subscripts for F statistics are consistent with Wright (1978), except for F_{id}, which is used here because these analyses are based on karyotypes rather than inversion frequencies. F_{rt} is the correlation between randomly chosen karyotypes within a group of populations relative to the entire species, F_{dr} is the correlation between randomly picked karyotypes within populations relative to that of the region, and F_{id} is the correlation of random karyotypes within populations relative to the entire species.

^c All Central Gulf Coast populations from both sides of the Gulf of California combined into one group.

^d Abbreviations in parentheses refer to the vegetational subdivisions that were excluded in each analysis. See appendix 6.2 for abbreviations.

^e In this analysis, cactus subspecies replace subdivisions: the subspecies are *Lophocereus schottii* var. *schottii*, *L. s.* var. *tenuis*, and *L. s.* var. *australis*.

no commonly used F statistic at this level because our analyses are based on karyotype frequencies, not inversion frequencies.

Estimates of Φ_{sc} and F_{dr}, reflecting karyotypic correlations among populations within subdivisions relative to the correlation between randomly picked karyotypes, were similar in both species. At this level of population structuring, overall karyotypic variation within vegetational subdivisions was not markedly different between species.

Comparisons with Other Species

Comparisons of population structure with other *Drosophila* species are potentially difficult to interpret because of differences in absolute sizes of species' ranges and the lack of breeding site data. Other studies involve large-scale differentiation over continent-wide ranges in North American *D. pseudoobscura* (Wright 1978) and European and northwest African *D. subobscura* (Ferrari and Taylor 1981) that were collected by trapping; i.e., breeding sites for these species are not known except in a few cases (Carson 1951; Begon 1976). There has been no analysis of how breeding site variation (as opposed to meteorological variation) in these species might influence inversion polymorphism or gene flow in natural populations.

Trapping (or baiting) can significantly distort estimates of dispersal and, thus, population structure, because movement is strongly correlated with the distance between traps (Johnston and Heed 1975). Also, grouping into regions in such large-scale studies has been somewhat subjective, based on geographic proximity (Wright 1978) and "proximity and the probable paleolithic distribution of the species" (Ferrari and Taylor 1981).

Nevertheless, extensive population structuring has been reported across the ranges of *Drosophila pseudoobscura* and *D. subobscura* based on inversion frequencies. We can compare only two levels of population structure given the AMOVA results: F_{rt}, variation among regions within a species, and F_{dr}, variation among populations within regions. For *D. pseudoobscura*, $F_{rt} = 0.322$ and $F_{dr} = 0.075$, based on 87 populations grouped into 15 regions. For *D. subobscura*, $F_{rt} = 0.208$ and $F_{dr} = 0.149$, based on 38 populations grouped into 10 regions. Both studies also included groupings of regions in the hierarchy that we did not. Judging by the reported variation across the ranges of each species, the vegetational subdivisions in and around the Sonoran Desert are most similar to the regions, in terms of scale, used by Wright (1978) and Ferrari and Taylor (1981). *Drosophila pachea* showed the highest degree of variation among regions, but we have already demonstrated that this is due to the large effect of the STS populations. With this region removed from the analysis, F_{rt} declined from 0.451 to 0.231. It is striking to us that this single region has such a large effect on population structure in *D. pachea*, lowering the magnitude of the among-region variation to below that calculated for *D. pseudoobscura*, which included populations from British Columbia and northwestern California to Guatemala and Colombia. When Wright (1978) considered only populations north of Mexico, F_{rt} decreased to 0.213. Thus, genetic differentiation of populations of *D. mojavensis* and *D. pachea* among the smaller-scaled vegetational subdivisions of the desert and adjacent areas is similar to that of all European populations of *D. subobscura* and United States and Canadian populations of *D. pseudoobscura*.

The degree of interdemic variation within regions, F_{dr}, was lower in both widespread species, particularly *Drosophila pseudoobscura*. F_{dr} was approximately twice the value in the desert species than in the others, showing a much finer scale of variation even among populations within subdivisions. This is probably the result of the method of measurement, i.e., *D. pseudoobscura* and *D. subobscura* were grouped according to regional similarity of inversion frequencies while among-population-within-subdivision structuring in *D. mojavensis* and *D. pachea* was measured independently of karyotypic similarity and in accordance with the vegetational subdivisions. There is also the potential that host plant effects have played a major role in the level of local population structuring in the desert species.

Variation Correlated with Substrate
Drosophila pachea

Since *Drosophila pachea* can essentially use only senita cactus as a host plant, the three subspecies were partitioned to determine any effect they may have on karyotype frequencies. Results of this AMOVA analysis for populations grouped by subspecies revealed large Φ_{ct} and F_{rt}, indicating strong karyotypic differences associated with them (table 6.2B).

The average frequency of the 7 + chromosome among *Drosophila pachea* populations associated with the three cactus subspecies was 0.060 for *Lophocereus schottii schottii*, 0.258 for *L. s. australis*, and 0.804 for *L. s. tenuis*. There were no significant karyotypic differences between populations using *L. s. schottii* in Baja California and those from Sonora, so the two areas have been treated as a unit. An analysis of variance testing for the differences among populations that use the different host plant subspecies was also highly significant (F = 99.99, $P <$ .001). Host plant use accounted for 80.64% of the total variance in inversion frequencies among *D. pachea* populations. These large differences are a reflection of the high values of Φ_{ct} = 0.727 and F_{rt} = 0.502 mentioned above and found in table 6.2B.

On the mainland, the steep transition in inversion frequency from the midpart of the Central Gulf Coast subdivision in the vicinity of Bahía Kino into the Sinaloan thornscrub is coincident in the Guaymas region with the replacement of *Lophocereus schottii schottii* with *L. s. tenuis* (fig. 6.2; fig. 6.4). This apparent coincidence is a problem for disentangling the effects of host plants and subdivisions on inversion frequencies. However, the Guaymas–San Carlos–Empalme area lies within the Central Gulf Coast subdivision and is inhabited by the westward extension of *L. s. tenuis* from the Sinaloan thornscrub (Lindsay 1963). This subspecies also extends directly north into the Plains of Sonora as far as Hermosillo (Lindsay 1963).

A total of 236 karyotypes analyzed from the Hermosillo and Guaymas areas (appendix 6.2) exhibited the highest heterozygosity, 0.453, of chromosome 7 in *Drosophila pachea* on the mainland. The average frequency of the 7 + arrangement was 0.669 in these populations. It is the abrupt step from this frequency to an average frequency of 0.996 ($n = 170$ individuals) in the Sinaloan thornscrub that has probably given the STS its large effect in the analyses of population structure (table 6.2).

In summary, the shift from *Lophocereus schottii schottii* to *L. s. tenuis* in the Guaymas region was accompanied by high heterozygosity for chromosome 7 in *Drosophila pachea* populations (see also Ward et al. 1975). Similarly, *L. s. schotti* merges with *L. s. australis* near La Paz in and around the Cape region in Baja California (Lindsay 1963), and these populations are characterized by the highest heterozygosities for *D. pachea* in the peninsula (0.489, $n = 91$ individuals). The average frequency of the 7 + gene arrangement in these populations (nos. 8 to 16, appendix 6.2) was 0.505. Notice that San Agustin and La Aguja are listed in the Magdalena Plain subdivision while E. La Paz is located within the Central Gulf Coast subdivision. Thus *L. s. australis* extends beyond the Cape Region, analogous to the extension of *L. s. tenuis* in the Guaymas area on the mainland. Shreve (1937) characterized the transition from the desert to the Cape as an interdigitation of the two regions over rugged and varied country rather than a gradual transition over many kilometers, as occurs with the merging of the desert with the thornscrub in Sonora.

South of La Paz the samples in appendix 6.2 may be readily pooled into groups according to similarity in gene arrangement frequencies within similar geographic areas. A cline in frequencies of the 7 + gene arrangement was also present west of the Sierra La Laguna from the San Pedro area (frequency of 7 + = 0.186), directly south to the tip of the peninsula at Cabo San Lucas and then north on the east side of the sierras (frequency of 7 + = 0.138) to San Bartolo (frequency of 7 + = 0.037). Other interpretations for the Cape region must include a varied and patchy distribution of the 7 + gene arrangement. Even so, the difference in frequencies between the latter two areas mentioned is significant ($G = 12.674$, $P < .005$) and they are only 25 km apart.

Collection records north of La Paz indicate a gradual increase of the 7 + gene arrangement until ca. 27° N lat. where it decreases to zero. Up to this point the frequencies north and south of La Paz are mirror images, except the distances involved are much less in the Cape region.

The overall characteristics of the two major tropical subspecies, *Lophocereus schottii tenuis* and *L. s. australis*, are the presence of more and thinner stems with a correspondingly higher number of ribs (Borg 1937; Lindsay 1963). In addition, *L. s. australis* has a noticeable trunk, which makes it more treelike than the others (Lindsay 1963; Cody 1984; fig. 6.2). Cody (1984) also recorded a significantly greater amount of stem branching in *L. s. australis* than in *L. s. schottii* in Baja California.

That changes in the diameter of the stems are gradual, at least on the mainland, has been observed by Felger and Lowe (1967) and Nobel (1980). Felger and Lowe (1967) reported a cline in surface-volume ratio of the stems. The ratio increased gradually southward from Sonoita to Bahía Kino and then more abruptly from there and from Hermosillo to the Playa Cochorit-Potam area south of Guaymas. The correlation of the transformed clinal frequency of the 7 + gene arrangement of *Drosophila pachea* to this ratio, and also the increasing rib number, was shown to be significant in both cases (Ward et al. 1975).

Nobel (1980) measured midstem diameter of senita populations along approximately the same transect as Felger and Lowe (1967). His data illustrate that plants in the three northern localities have significantly greater mean stem diameters, in both mature and immature stems, than the plants in the three more southern localities (mature stems: $t = 14.41$; $P < .01$; $n_1 = n_2 \cong 50$; immature stems: $t = 12.51$; $P < .01$; $n_1 = n_2 \cong 50$). Furthermore, the rate of change in relation to latitude increased from Hermosillo to San Carlos and then to Obregon, the southern localities, as did the rate of change in the frequency of the 7 + gene arrangement.

From these two sets of data it appears that *Lophocereus schottii tenuis* responds to a change in latitude to a greater degree than *L. s. schottii*. The important characteristic of senita cacti in regard to the larvae of *Drosophila pachea* is probably the diameter of the stem, because it could indirectly influence larval development time. The fermenting tissues in thicker stems should ultimately allow for slower developmental rates since these rots, on average, should be longer lived.

This scenario is in agreement with the data for the mainland, yet it is only partially in agreement in the Cape region where there is a decrease in frequency of the 7 + gene arrangement southeast from La Paz. The Cape region exhibits a cline in the reverse direction compared to the mainland as previously discussed. Duncan (1979) also found a reverse cline in rib number for the Cape and that the correlation with the 7 + arrangement was positive and significant. Duncan recorded the lowest rib number on all plants scored with a mean of nine ribs in the La Paz region, which gradually declined to a mean of six ribs near Cabo San Lucas at the tip of the peninsula. However, since senita stem diameter has not been directly measured in the Cape region, to our knowledge, this correlation only suggests senita stems increase in width to the southeast. A possible regulating factor in this context is the presence of an isolated section of the Central Gulf Coast subdivision of the desert in the southeast corner of the Cape north of 23° N lat. Shreve (1964) noted the similarity in aspect and composition of this area to the region around Bahía Concepción between 26° and 27° N lat. The frequencies of the 7 + gene arrangement are similar near both areas.

Drosophila mojavensis

The host plants and their distribution have been described earlier in this chapter for *Drosophila mojavensis*. It is clear that karyotypic variation is host plant related; without exception, all heterozygosity on the second chromosome has been found in pitahaya agria–inhabiting *D. mojavensis* populations. This association is epitomized by two samples separated by 50 km of uninterrupted desert in coastal Sonora. One collection taken from pitahaya agria (locality 39d, appendix 6.1) was polymorphic with the frequencies of LP = 0.78, ST = 0.18, and BA = 0.04, while the sample from an organ pipe population (locality 40) was homozygous for LP. ST/ST homokaryotypes are found throughout southern California and the Grand Canyon in barrel cactus–breeding populations and in Santa Catalina Island populations that use *Opuntia*. LP is fixed throughout the organ pipe–breeding and sina-breeding populations in southern Arizona, Sonora, and Sinaloa. Genetic differentiation mediated by the cacti has also been proposed by Richardson et al. (1977) for allozyme frequencies. This substrate correlation does not hold true, however, for heterozygosity on the third chromosome of *D. mojavensis* because of the presence of MU in many of the populations along coastal Sonora where agria is not present.

The AMOVA results confirm this host-karyotype association (table 6.1B) as well as significant variation between populations using the same host, and a large within-population correlation among karyotypes as seen when the populations were grouped according to the vegetational subdivisions. Significant Φ_{sc} and F_{dr}, indicative of significant variation among populations using particular hosts, were due to the degree of polymorphism among populations using the same hosts. Several Baja California populations of *Drosophila mojavensis* are homokaryotypic for both second and third chromosome gene arrangements, and some organ pipe–breeding populations are polymorphic for the third chromosome while others are not (fig. 6.5; fig. 6.6). Thus, other factors, particularly plant density and host preference behavior, as well as direct climatic influences have been included as factors controlling inversion polymorphism.

A large portion of southern Baja California, most of the islands in the Gulf, and the Desemboque region on the mainland are occupied by both agria and organ pipe cacti. However, our rearing records from these areas, which extend beyond 30 years, show very few instances of the use of organ pipe in this region of overlap. Adult *Drosophila mojavensis* also tend to prefer the volatile profiles of agria over organ pipe cacti in choice tests (Downing 1985; Fogleman and Heed 1989; Newby and Etges 1998). *D. mojavensis* probably uses agria to a greater degree because it is consistently a more abundant and uniform trophic resource. Pitahaya agria plants frequently grow in large thickets of interconnected stems that reproduce themselves mainly by vegetative growth. Organ pipe cacti rely entirely on fruiting and animal dispersion of seeds and therefore grow as individual plants. The density of agria stems can exceed that of organ pipe stems by seven times, yet agria stems are on average half the diameter of organ pipe stems. Most significantly, rot densities can be approximately 40 times higher in agria than in organ pipe cacti in certain localities (Mangan 1982). Therefore, agria is a more predictable resource for ovipositing adults, but organ pipe rots last longer when they occur, due to larger stem diameters (Heed 1981; Heed and Mangan 1986; Etges and Heed 1987).

Rot density, or trophic predictability, is probably the most significant factor influencing population structure, as it will determine levels of rot-to-rot dispersal and, therefore, gene flow among demes (Endler 1979). Studies of dispersal, similar to Johnston (1974), have yet to be undertaken in Baja, where inter-rot migration in agria-inhabiting populations of *Drosophila mojavensis* is suspected to be lower than in organ pipe–inhabiting populations because of the higher abundance of agria rots. Higher density and aggregation of rots in agria patches vs. organ pipe patches represents a major shift in trophic predictability and has influenced the evolution of life history differences between Baja and mainland *D. mojavensis* (Etges and Heed 1987). Baja California populations have shorter egg-to-adult development times and show greater genetic homeostasis (more constant expression of life history traits), with increasing larval densities, than mainland populations. This is associated with higher rates of tissue fermentation in agria than in organ pipe cacti (Etges 1989b). Longer development times, on the other hand, usually engender larger adult body size and reproductive potential in *Drosophila;* this is also the case in *D. mojavensis* (Etges and Heed 1992). Therefore, the longer egg-to-adult development times of mainland *D. mojavensis* are probably an adaptation resulting from the greater longevity of breeding sites that produce larger adults better able to disperse between the comparatively rarer, but larger, organ pipe rots (Etges 1993).

In summary, the switch to organ pipe from pitahaya agria has caused a number of shifts in the life history, physiology, and behavior of *Drosophila mojavensis* with a concomitant loss of significant heterozygosity for chromosomal inversions. For these reasons, *D. mojavensis* is considered to consist of at least two host races (Etges 1990).

Variation Correlated with Climate

A direct assessment was made of covariation of inversion heterozygosity and annual temperature fluctuation for populations of *Drosophila mojavensis* and *D. pachea*. The rationale for this approach is as follows: heterozygosity is a convenient measure of karyotypic variability within populations of *Drosophila* that frequently results from balancing selection (Lewontin et al. 1981). This approach does not assume that there is selection for heterozygosity per se, only that polymorphism is retained because of variable selection (see Orzack 1985; Etges 1989b for further details). Because precipitation and average temperature fluctuation through the year are among the principal causes for the formation of the vegetational subdivisions (Hastings and Turner 1965; Turner and Brown 1982), we calculated average annual temperature range, TMR, by taking the difference between the highest average monthly temperature and lowest average monthly temperature recorded at the nearest weather station (Hastings and Humphrey 1969a, 1969b). Inversion heterozygosity was calculated using $h_x = 1 - \Sigma x_i^2$, where x_i is the frequency of the ith gene arrangement (appendix 6.1; appendix 6.2). Correlations were calculated between arcsine transformed heterozygosity values and TMR, degrees latitude, degrees longitude, and elevation above sea level of the weather station nearest to each population. Populations were separated into mainland and Baja groups prior to analysis.

Inversion heterozygosity was negatively correlated with TMR and elevation in *Drosophila pachea* on the mainland and marginally so on the peninsula (table 6.3). Levels of chromosome polymorphism in *D. pachea* were therefore lowest in populations experiencing the greatest annual temperature fluctuations, such as in the Lower Colorado and Arizona Upland subdivisions. Heterozygosity increased along Baja California to the south near La Paz and then decreased in the southern Cape region as the frequency of 7 A increased (fig. 6.4). Thus the possibility of finding a basis for variable selection due to temperature variation has not been realized. In fact, the response of average heterozygosity (HBAR) on the mainland compared to Baja was quite different. Significant correlations, all negative, were found for latitude and longitude in Baja California but with TMR and elevation on the mainland. TMR was strongly correlated with latitude, longitude, and elevation on the mainland, but less so on the peninsula due to the prevailing maritime conditions and the lack of weather stations at consistently higher elevations.

Duncan (1979) suggested that the pattern of heterozygosity in Baja California resulted from secondary contact of isolated populations after the Cape region was rejoined to the peninsula during the Pleistocene, prior to which it was an island. However, the area of highest heterozygosity on the mainland is localized in the Guaymas region and thus resembles the region around La Paz with respect to placement on the southern edge of the desert. It is this phenomenon which could be most significant. In sum, inversion polymorphism in *Drosophila pachea* seems unrelated to local temperature fluctuations that could cause fluctuating selection, suggesting that host plant use is a major determinant of the degree of inversion polymorphism in this species (but see Ward et al. 1975).

Because of the disparity in the distribution of polymorphism between mainland and Baja populations of *Drosophila mojavensis*, most of the correlations with heterozygosity were found for Baja populations. TMR was uncorrelated with heterozygosity of either chromosome. Both latitude and longitude were positively associated with heterozygosity on the second chromosome due to the number of populations close to fixation in southern Baja and the Cape region. Throughout the range of *D. mojavensis*, levels of second and third chromosome heterozygosity were uncorrelated (table 6.3), suggesting these polymorphisms are influenced by other factors.

Table 6.3 Pearson product-moment correlations between observed inversion heterozygosity and relevant environmental variables in *Drosophila pachea* and *D. mojavensis*.

D. pachea

	HBAR	TMR	LAT	LONG	ELEV
HBAR	—	-0.453*** (50)	-0.223 (49)	0.067 (44)	-0.416** (44)
TMR	-0.360+ (28)	—	0.756*** (81)	0.686*** (44)	0.434*** (76)
LAT	-0.508** (28)	0.409** (60)	—	0.832*** (76)	0.742*** (76)
LONG	-0.450* (28)	0.218+ (60)	0.966*** (60)	—	0.333** (76)
ELEV	-0.247 (28)	0.281* (59)	0.303* (59)	0.264* (59)	—

D. mojavensis

	H2BAR	H3BAR	HBAR	TMR	LAT	LONG	ELEV
H2BAR	—	0.208 (21)	0.776*** (21)	0.198 (19)	-0.010 (21)	0.025 (21)	0.241 (15)
H3BAR	0.0516 (37)	—	0.764*** (21)	0.093 (19)	-0.199 (21)	-0.073 (21)	-0.481+ (15)
HBAR	0.813*** (37)	0.622*** (37)	—	0.219 (19)	-0.115 (21)	-0.021 (21)	-0.476+ (15)
TMR	0.186 (37)	0.150 (37)	0.228 (37)	—	0.244+ (61)	-0.032 (61)	0.022 (55)
LAT	0.576*** (37)	-0.102 (37)	0.395* (37)	0.230+ (71)	—	-0.128 (63)	0.602*** (55)
LONG	0.433** (37)	-0.092 (37)	0.290+ (37)	0.042 (71)	0.949*** (71)	—	0.107 (55)
ELEV	-0.043 (28)	-0.039 (28)	-0.056 (28)	0.176 (62)	0.195 (62)	0.173 (62)	—

Note: TMR is defined in Appendix 6.1, LAT is degrees latitude, LONG is degrees longitude, and ELEV is distance above sea level. HBAR is average heterozygosity. For *D. mojavensis*, H2BAR is observed second chromosome heterozygosity, and H3BAR is observed third chromosome heterozygosity. Values above the diagonal refer to mainland populations, and those below the diagonal to Baja California populations. Values in parentheses are sample sizes. See Appendices 6.1 and 6.2 for those populations included in the analyses. Additional meteorological data were taken from weather stations listed in Hastings and Humphrey (1969a, 1969b). Some data were missing.

Inversion Phylogeny and the History of the Sonoran Desert

Distance trees were constructed based on the inversion frequencies in *Drosophila mojavensis* using PHYLIP (ver. 3.51c; Felsenstein 1993) to further scrutinize population structuring. We were most interested in assessing the shape of the tree with respect to the clustering of populations in the phytogeographic subdivisions. Inversions were considered "alleles" of the second and third chromosomes, while the latter were considered "loci" for a subset of the 51 populations (appendix 6.1). We used the neighbor-joining method (Saitou and Nei 1987) to explore the sensitivity of this technique with our rather restricted data set, which consisted of the frequencies of two polymorphic chromosomes in 43 populations. The analyses were used to order populations based on genetic distances and to form heuristic population clusters.

The inversion phylogeny for *Drosophila mojavensis* and its relatives is well known (Ruiz et al. 1990; Wasserman 1992). The ST and SI second chromosomes are derived from a hypothetical ancestor that gave rise to *D. mojavensis* and its closest relative *D. arizonae*. All of the other second chromosome inversions arose from the ST chromosome. Therefore, the genetic distance tree based on inversion frequency similarity among populations has an evolutionary basis (fig. 6.7). This analysis revealed a different degree of structuring than that explained by the presence of the vegetational subdivisions. This tree also provides more evidence for the proposed history of divergence of *D. mojavensis* throughout the Sonoran Desert (Heed 1982), despite the difficulties of using gene arrangements in this instance that are usually not selectively neutral characters. Of the 51 populations, eight were deleted because they were all homokaryotypic populations from the mainland and added no information to the pattern of relationships (Felsenstein 1993). These are grouped under a single label, Arizona/Sonora. We decided to use a presumed ancestral population from Vallecito in southern California (Ruiz et al. 1990) as an outgroup even though the neighbor-joining method does not produce a rooted tree.

Population clusters in the genetic distance tree relate only marginally to the vegetational subdivisions discussed above (fig. 6.7; fig. 6.8). Whereas the vegetational subdivisions are predominantly an east-west division on the peninsula, the genetic distance clusters are entirely subdivided along the axis of the peninsula in a northwest-southeast direction. The result is a mosaic of disjunctions that are difficult to interpret without

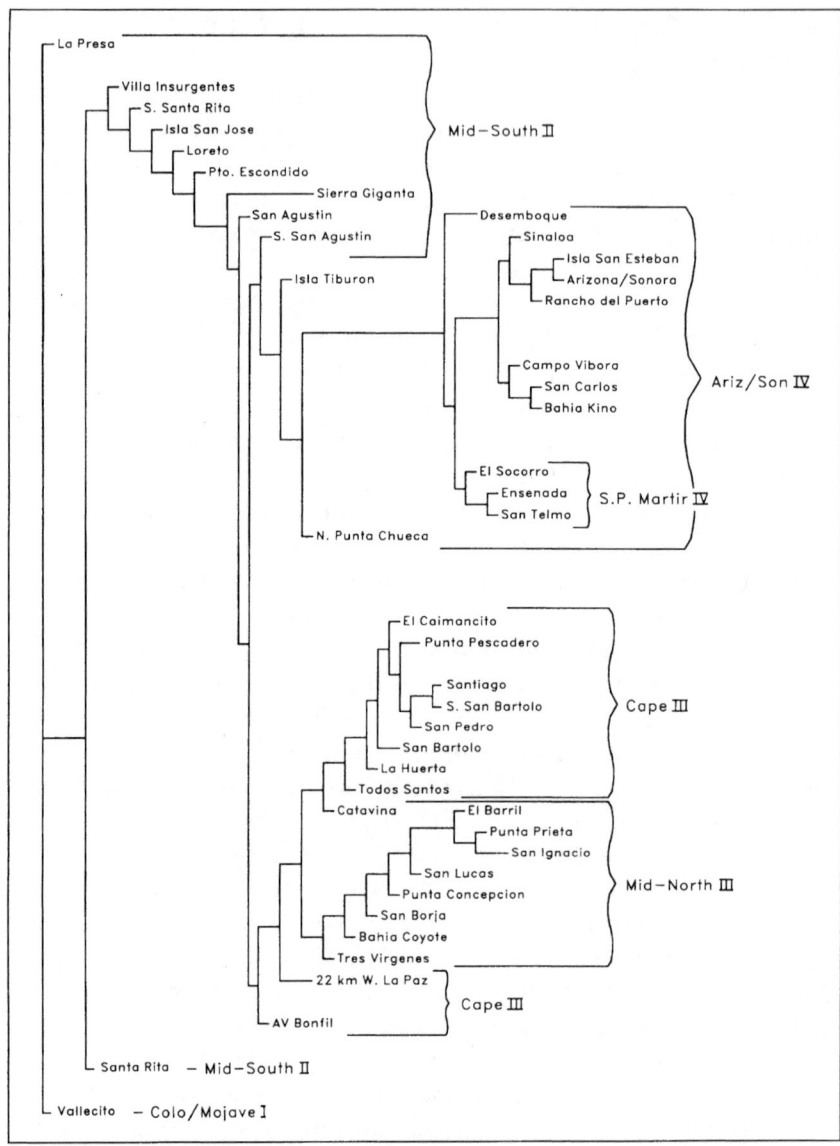

Figure 6.7 Populations of *Drosophila mojavensis* are clustered by their genetic distances using the neighbor-joining method (Saitou and Nei 1987). The ancestral Colorado/Mojave cluster, represented by the Vallecito population, was used as the "outgroup." Clusters are labelled according to their approximate geographic region.

either invoking convergence in inversion frequencies due to natural selection or a series of migrations (dispersals) and extinctions in the history of the species. In any event, there is evidence in the tree that an ancestral condition for *Drosophila mojavensis* exists in Baja California. The midsouth cluster in Baja California is positioned as the ancestral group that gave rise to the diversification of all Baja and mainland Mexico and Arizona populations.

The history of diversification of *Drosophila mojavensis* into its present range has also been inferred from other comparative data. Johnson (1980) suggested that the central-marginal pattern of population structure in *D. mojavensis* resulted from the divergence of Baja populations into mainland Mexico by dispersing across the midriff islands in the Gulf of California at about 29° N lat. Cody et al. (1983) suggested that the spread of plants across these islands, mostly from west to east, including pitahaya agria, occurred about 15 000 years ago when aridity became the dominant climatological factor in the region. Johnson (1980) also observed that all of the present-day gene arrangements can be found in Baja, including the rare s1 chromosome found only in San Ignacio and San Lucas, Baja California Sur. These localities are located in the southern part of the Mid-North group (fig. 6.7; fig. 6.8). Since s1 contains three inversions and is derived from the same hypothetical ancestral chromosome as st, it may be considered a relictual survivor of events described above. All of the host plants used by *D. mojavensis* are present in Baja, except sina, yet agria is the most widespread and it appears to be the most important host plant in the region. Divergence onto the mainland eventually resulted in a host plant switch to organ pipe and sina cacti. Mainland populations have lost most of the inversion heterozygosity (fig. 6.5; fig. 6.6), have shifted in allele frequencies at several enzyme loci (Zouros 1973; Heed 1978), and have undergone evolution in life history, morphological and physiological traits associated with adaptation to organ pipe cacti (Starmer et al. 1977; Etges and Heed 1987; Etges 1989c; Etges and Klassen 1989; Etges 1990; Etges 1993).

The origin of *Drosophila mojavensis* is thought to have occurred in Baja California as the peninsula formed. The movement north of what is now Baja California from mainland Mexico resulted from tectonic drift starting four to six million years ago (Gastil et al. 1975). This split the proto-*D. mojavensis* from its mainland ancestors that evolved into *D. arizonae* and provided the isolation required for *D. mojavensis* to speciate.

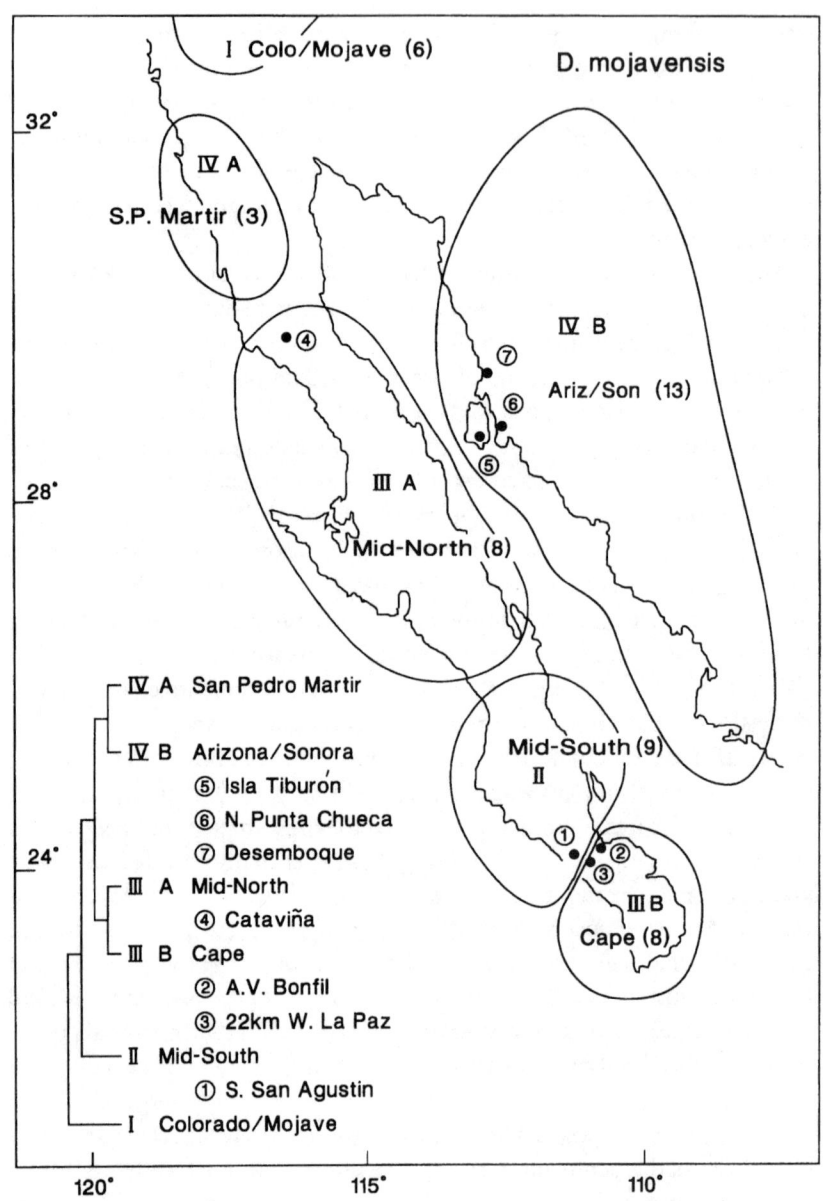

Figure 6.8 *(opposite page)* A summary of the genetic distance diagram in figure 6.7 is presented and the chief clusters are mapped on to the Sonoran Desert and adjacent regions. Four main clusters are considered (two of them are subdivided [A and B]). Several geographic disjunctions in relatedness are noticeable among the clusters and among several localities within them. The Colorado/Mojave cluster (Cluster I with 6 populations) is 900 km removed from the Mid-South portion of the Baja California penninsula (Cluster II with 9 populations). Cluster III is divided by Cluster II into a Mid-North region (8 populations) and the Cape region (8 populations). Cluster IV is centered in the Arizona/Sonora region (13 populations) with an element in the San Pedro Martir region (3 populations).

The samples from localities 1 to 7 have inversion frequencies most similar to other regions or are nebulously placed in a pre-cluster position. This is probably due to their ecotonal nature (situated between vegetational subdivisions). For instance, the S. San Agustin population is in the midst of a cline in inversion frequencies from the Magdalena plain into the Cape region and by chance is more similar to the Isla Tiburón population than those of the Mid-South cluster (see fig. 6.7). Similar reasoning may be applied to populations from localities 2, 3, and 4. Populations represented by localities 5, 6, and 7 are remnants of an earlier west-to-east dispersal across the midriff islands similar to the probable dispersal of their most favored host cactus, the pitahaya agria.

Conclusions

The present distribution of inversion polymorphisms in both *Drosophila mojavensis* and *D. pachea* has been influenced by the same climatic factors that have shaped the vegetational subdivisions in and around the Sonoran Desert. The same forces that have produced the vegetational subdivisions and influenced the distribution of the host cacti found within them have secondarily molded the karyotypic variation in both species. These results imply that the ecological mosaic of the vegetational subdivisions of the Sonoran Desert surely influence genetic variation in many other insect species that use desert plants to carry out their life cycles. Surprisingly, within the relatively small range of the Sonoran Desert and adjacent provinces, the extent of genetic variation among regions (F_{rt}) for both species was comparable to the whole of western North America for *D. pseudoobscura* and of Europe and North Africa for *D. subobscura*. In this light, the vegetational subdivisions are considered a potent source of ecological variability that have structured the organization of karyotypic diversity in populations of *D. mojavensis* and *D. pachea*.

The AMOVA analyses have shown host plant effects to be equally pronounced in *Drosophila mojavensis* and even more so in *D. pachea* than the effects of the vegetational subdivisions on chromosomal polymorphisms. This is not surprising in light of the monophagy exhibited by *D. pachea* on senita cacti and the distinct polytypic monophagy exhibited by *D. mojavensis* on pitahaya agria, organ pipe, sina, and barrel cacti in their respective geographic areas.

The major part of the polymorphism for *Drosophila pachea* resides in the transition zones between *Lophocereus schottii schottii* and *L. s. tenuis* on the mainland in the Guaymas area, and between its nominate race and *L. s. australis* in the peninsular La Paz area. Changes in gene arrangement frequency across these zones account for the localization of 45% and 50% heterozygosities in each area, respectively.

In *Drosophila mojavensis*, all polymorphism on the second chromosome is restricted to pitahaya agria–using populations. The mean heterozygosity for the entire peninsula for this chromosome is 27% (range, 0% to 63%). The peninsular frequencies drop rapidly as one progresses from west to east across the midriff islands to the Desemboque area on the mainland and then reach zero in the remainder of the species distribution. The density of breeding sites is significantly greater in Baja California than the other two major areas, and this may have influenced the formation and retention of second chromosome polymorphism.

There is also a concomitant change in stem volume of the host plants of *Drosophila pachea* and *D. mojavensis* in relation to their degree of chromosomal heterozygosity. It appears that stem volume differences have influenced genetic differences in larval longevity by natural selection, which in turn has caused evolution in other components of the life history of these insects. Extensive studies by Etges have shown that major changes between agria- and organ pipe–using populations of *D. mojavensis* have resulted in the evolution of host races.

In regard to climate, tests for the effects (correlations) of mean ranges of temperatures, longitude, latitude, and elevation on chromosomal heterozygosity were not in accord across the entire geographic distributions of either *Drosophila pachea* or *D. mojavensis* in the Sonoran Desert and adjacent regions. Thus responses to climatic effects are difficult to detect directly by the use of the degree of heterozygosity as a variable.

With respect to evolutionary history, it may be seen that the ancestral karyotype for *Drosophila pachea* is found in southern Sonora, while the ancestral karyotype for *D. mojavensis* is found in southern California, northern Arizona, and Santa Catalina Island. Thus the history of the two species must be quite different. We have attempted to reconstruct the history of *D. mojavensis* by use of distance trees based on karyotype frequencies. The results suggest that a number of extinctions and recolonizations must have taken place along the length of Baja California, which suggests a dynamic sequence of events leading up to the present geographic pattern of the various gene arrangements. This highlights that the commonality in karyotypic response in the two species analyzed in the present study is much more than a chance phenomenon.

The interaction between hosts and karyotypes in these species within the context of the biogeography, history, and present-day vegetational subdivisions of the Sonoran Desert has indeed provided a wellspring of understanding into the maintenance of inversion polymorphisms in natural populations of *Drosophila* not heretofore possible to obtain in other species. The study of insect-host plant interactions in these arid lands has provided a unique insight into the history of insect diversification and the maintenance of genetic polymorphism.

Acknowledgments

We thank all of the interested students of the desert *Drosophila* system, past and present, especially H. Alonso-Pimentel, J. A. Endler, R. S. Felger, D. P. Fellows, J. C. Fogleman, A. Fontdevila, P. F. Ganter, A. Gathman, A. C. Gibson, J. S. Grove, M. C. Jefferson, J. S. Johnston, D. E. Jurgenson, H. W. Kircher, R. L. Mangan, T. A. Markow, L. E. Mettler, R. H. Richardson, E. S. Rockwood-Sluss, A. Ruiz, A. Russell, J. S. Russell, F. de M. Sene, W. T. Starmer, R. H. Thomas, D. C. Vacek, B. Ward, M. Wasserman, and E. Zouros. We thank the Tohono O'odham Nation for allowing us to collect on their reservation, the Seri Indians for permitting us to collect on their lands, the staff at Anza Borrego State Park for allowing us to collect flies there, the Catalina Island Reserve for helping us carry out fieldwork there, and the government of Mexico for allowing us to collect both cacti and flies. P. E. Smouse provided an early copy of the AMOVA program and advice, patience, and immeasurable assistance to W.J.E. in implementing the AMOVA. We couldn't have done it without him. L. Excoffier and J. B. Walsh provided much insight into the interpretation of the AMOVA results. W. T. Starmer made his computer available to us for using PHYLIP, and S. A. Cameron helped interpret the trees. J. S. F. Barker,

R. Robichaux, A. Ruiz, and several anonymous reviewers provided helpful comments on the manuscript. Most of this research was funded by grants from the National Science Foundation to W.B.H.

Notes

1. It is significant that each of the four species of *Drosophila* endemic to the Sonoran Desert and adjacent areas traces its lineage independently to central and southern Mexico (Heed 1982). The endemics *D. nigrospiracula* and *D. mettleri* are not discussed in the present text because the former species has not been found to have any inversion polymorphism (Cooper 1964) and the latter species has not been analyzed. The host plant for *D. nigrospiracula* is the saguaro (*Carnegiea gigantea*) in Arizona and Sonora, and the cardón (*Pachycereus pringlei*) in coastal Sonora and Baja California. Whether or not these cacti have an immediate or even indirect effect on blocking incipient inversion polymorphism can only be conjectured at the present time.

2. Manuel del Barco, a Jesuit priest who lived in Lower California from 1738 to 1768, gives a wonderful description of organ pipe, pitahaya agria, and senita cacti in *The Natural History of Baja California,* Dawson's Book Shop, Los Angeles, 1980.

3. Adult female *Drosophila mojavensis* collected in the field were isolated into separate vials in the laboratory containing 15 to 20 ml of banana-agar-yeast-malt food supplemented with dried agria powder. Adults reared from cactus rots were randomly pair-mated and their progeny were reared under the same conditions. One third-instar larva was chosen at random from each vial and the salivary glands were dissected out, fixed, and stained in 1% natural orcein in lactoacetic acid solution for 5 to 10 minutes and squashed directly on a slide. Karyotypes of *D. pachea* larvae were obtained in five different ways: 1) offspring of wild caught females; 2) F1s of wild caught males mated to females of a laboratory homokaryotypic stock; 3) pair matings between rot-reared adults; 4) matings between rot-reared virgin adults and adults of a laboratory homokaryotypic stock; 5) rot larvae. All polytene chromosomes were viewed with a standard light microscope with oil immersion.

4. The AMOVA analyses produce variance components and associated statistics, as well as F statistics for estimating hierarchical population structure. Significance values of the variance components were calculated by obtaining null distributions of individuals in populations and populations in regions and testing for significance after 50 random permutations. Unlike F statistics, the AMOVA requires an input matrix of all pairwise, inter-karyotype distances for use in calculating statistics. This matrix was formed by calculating the inter-karyotype distance between all possible karyotypes weighted by global inversion frequencies for

those gene arrangements (see Excoffier et al. 1992). For *Drosophila pachea*, there were three observed karyotypes, and therefore a 3 × 3 distance matrix was used. For *D. mojavensis* with four second-chromosome gene arrangements and two third-chromosome gene arrangements, there were 30 possible karyotypes, and therefore a 30 × 30 distance matrix was formed. Frequency data for each population were pooled if sampled more than once. Temporal variation was small relative to the overall pattern of geographic variation in each species (appendix 6.1; appendix 6.2).

Literature Cited

Anderson W. W., Watanabe T. K. (1974) Selection by fertility in *Drosophila pseudoobscura*. *Genetics* 77: 559–564.

Aquadro C. F., Weaver A. L., Schaeffer S. W., Anderson W. W. (1991) Molecular evolution of inversions in *Drosophila pseudoobscura*: The amylase gene region. *Proceedings of the National Academy of Sciences USA.* 88: 305–309.

Axelrod D. I. (1978) The origin of coastal sage vegetation. *American Journal of Botany* 65: 1117–1131.

Barker J. S. F., Starmer W. T. (eds) (1982) *Ecological Genetics and Evolution: The Cactus-Yeast-Drosophila Model System.* Academic Press, Sydney.

Barker J. S. F., Starmer W. T., MacIntyre R. J. (eds) (1990) *Ecological and Evolutionary Genetics of Drosophila.* Plenum, New York.

Begon M. (1976) Dispersal density and microdistribution in *Drosophila subobscura* Collin. *Journal of Animal Ecology* 45: 441–456.

Borg J. (1937) *Cacti.* Macmillan, London.

Brown D. E. (ed) (1982) *Biotic Communities of the American Southwest: United States and Mexico. Desert Plants* 4(1–4): 3–341.

Brown D. E., Lowe C. H. (1980) *Biotic Communities of the Southwest.* USDA Forest Service General Technical Report RM-78. Rocky Mountain Forest and Range Experiment Station, Fort Collins, Colo.

Brussard P. F. (1984) Geographical patterns and environmental gradients: the central-marginal model in *Drosophila* revisited. *Annual Review of Ecology and Systematics* 16: 25–64.

Carson H. L. (1951) Natural breeding sites of *Drosophila pseudoobscura* and *Drosophila persimilis* in the transition zone of the Sierra Nevada. *Evolution* 5: 91–96.

Carson H. L. (1959) Genetic conditions which promote or retard the formation of species. *Cold Spring Harbor Symposium on Quantitative Biology* 24: 87–105.

Cody M. L. (1984) Branching patterns in columnar cacti. In: Margaris N. S., Aridnoustou-Faraggitaki M., Oechel W. C. (eds) *Being Alive On Land,* pp 201–236. W. Junk, The Hague.

Cody M. L., Moran R., Thompson H. (1983) The plants. In: Case T. J., Cody M. L. (eds) *Island Biogeography in the Sea of Cortez*, pp 49–97. University of California Press, Los Angeles.
Cooper J. W. (1964) Genetic and cytological studies of *Drosophila nigrospiracula* in the Sonoran Desert. Master's thesis, University of Arizona, Tucson.
del Barco M. (1980) *The Natural History of Baja California*. Dawson's Book Shop, Los Angeles.
Dobzhansky T. (1955) A review of some fundamental concepts and problems of populations genetics. *Cold Spring Harbor Symposium on Quantitative Biology*. 20: 1–13.
Dobzhansky T., Lewontin R. C., Pavlovsky O. (1964) The capacity for increase in chromosomally polymorphic and monomorphic populations of *Drosophila pseudoobscura*. *Heredity* 19: 597–614.
Downing R. J. (1985) The chemical basis for host plant selection in *Drosophila mojavensis*. Master's thesis, University of Denver, Denver.
Duncan G. A. (1979) Chromosomal polymorphism in *Drosophila pachea*. Ph.D. dissertation, University of Arizona, Tucson.
Endler J. A. (1979) Gene flow and life history patterns. *Genetics* 93: 263–284.
Etges W. J. (1989a) Chromosomal influences on life history variation along an altitudinal transect in *Drosophila robusta*. *American Naturalist* 133: 83–110.
Etges W. J. (1989b) Divergence in cactophilic *Drosophila*: the evolutionary significance of adult ethanol metabolism. *Evolution* 43: 1316–1319.
Etges W. J. (1989c) Evolution of developmental homeostasis in *Drosophila mojavensis*. *Evolutionary Ecology* 3: 189–201.
Etges W. J. (1990) Direction of life history evolution in *Drosophila mojavensis*. In: Barker J. S. F., Starmer W. T., MacIntyre R. J. (eds) *Ecological and Evolutionary Genetics of Drosophila*, pp 37–56. Plenum, New York.
Etges W. J. (1993) Genetics of host-cactus response and life history evolution among ancestral and derived populations of *Drosophila mojavensis*. *Evolution* 47: 750–767.
Etges W. J. (1996) Sexual selection operating in a natural population of *Drosophila robusta*. *Evolution* 50: 2095–2100.
Etges W. J., Heed W. B. (1987) Sensitivity to larval density in populations of *Drosophila mojavensis*: influences of host plant variation on components of fitness. *Oecologia* 71: 375–381.
Etges W. J., Heed W. B. (1992) Remating effects on the genetic structure of female life histories in populations of *Drosophila mojavensis*. *Heredity* 68: 515–528.
Etges W. J., Klassen C. S. (1989) Influences of atmospheric ethanol on adult *Drosophila mojavensis*: altered metabolic rates and increases in fitness among populations. *Physiological Zoology* 62(1): 170–193.
Excoffier L., Smouse P. E., Quattro J. M. (1992) Analysis of molecular variance

inferred from metric distances among DNA haplotypes: application to human mitochondrial DNA restriction data. *Genetics* 131: 479–491.

Felger R. S., Lowe C. H. (1967) Clinal variation in the surface-volume relationships of the columnar cactus *Lophocereus schottii* in northwestern Mexico. *Ecology* 48: 530–536.

Felsenstein J. (1993) PHYLIP vers. 3.51c, University of Washington, Seattle.

Ferrari J. A., Taylor C. E. (1981) Hierarchical patterns of chromosome variation in *Drosophila subobscura*. *Evolution* 35: 391–394.

Fogleman J. C., Duperret S. M., Kircher H. W. (1986) The role of phytosterols in host plant utilization by cactophilic *Drosophila*. *Lipids* 21: 92–96.

Fogleman J. C., Heed W. B. (1989) Columnar cacti and desert *Drosophila*: the chemistry of host plant specificity. In: Schmidt J. O. (ed) *Interaction among Plants and Animals in the Western Deserts*, pp 1–24. University of New Mexico Press, Albuquerque.

Fogleman J. C., Heed W. B., Kircher H. W. (1982) *Drosophila mettleri* and senita cactus alkaloids: fitness measurements and their ecological significance. *Comparative Biochemistry and Physiology* 71A: 413–417.

Gastil R. G., Phillips R. P., Allison E. C. (1975) Reconnaissance geology of the state of Baja California. In: *Geological Society of America, Inc. Memoir 140*, pp 139–143. Boulder, Colo.

Gentry H. S. (1942) *Rio Mayo Plants*. Carnegie Institute of Washington Publication no. 527, Washington, D.C.

Hastings J. R., Humphrey R. R. (1969a) Climatological data and statistics for Baja California. University of Arizona Institute for Atmospheric Physics, Technical Report 18: 1–96.

Hastings J. R., Humphrey R. R. (1969b) Climatological data and statistics for Sonora and Northern Sinaloa. University of Arizona Institute for Atmospheric Physics, Technical Report 19: 1–96.

Hastings J. R., Turner R. M. (1965) Seasonal precipitation regimes in Baja California, Mexico. *Geografiska Annaler* 47: 204–233.

Hastings J. R., Turner R. M., Warren D. K. (1972) An atlas of some plant distributions in the Sonoran Desert. University of Arizona Institute of Atmospheric Physics, Technical Report 21: 1–255.

Heed W. B. (1978) Ecology and genetics of Sonoran Desert *Drosophila*. In: Brussard P. F. (ed) *Ecological Genetics: The Interface*, pp 109–126. Springer-Verlag, New York.

Heed W. B. (1981) Central and marginal populations revisited. *Drosophila Information Service* 56: 60–61.

Heed W. B. (1982) The origin of *Drosophila* in the Sonoran Desert. In: Barker J. S. F., Starmer W. T. (eds) *Ecological Genetics and Evolution: The Cactus-Yeast-Drosophila Model System*, pp 65–80. Academic Press, Sydney.

Heed W. B., Kircher H. W. (1965) Unique sterol in the ecology and nutrition of *Drosophila pachea*. *Science* 149: 758–761.

Heed W. B., Mangan R. L. (1986) Community ecology of the Sonoran Desert *Drosophila*. In: Ashburner M., Carson H. L., Thompson J. N. Jr (eds) *The Genetics and Biology of Drosophila*, Vol. 3e, pp 311–345. Academic Press, New York.

Johnson W. R. (1973) Chromosome variation in natural populations of *Drosophila mojavensis*. Master's thesis, University of Arizona, Tucson.

Johnson W. R. (1980) Chromosomal polymorphism in natural populations of the desert adapted species, *Drosophila mojavensis*. Ph.D. dissertation, University of Arizona, Tucson.

Johnston J. S. (1974) Dispersal in natural populations of cactiphilic *Drosophila pachea* and *D. mojavensis*. *Genetics* 77: s32–s33.

Johnston J. S., Heed W. B. (1975) Dispersal of *Drosophila*: the effect of baiting on the behavior and distribution of natural populations. *American Naturalist* 109: 207–216.

Kircher H. W., Heed W. B., Russell J. S., Grove J. (1967) Senita cactus alkaloids: their significance to Sonoran Desert *Drosophila* ecology. *Journal of Insect Physiology* 13: 1969–1974.

Kostoff D. (1930) Discoid structure of the spireme. *Journal of Heredity* 21: 323–324.

Krimbas C. B., Powell J. R. (1992) *Drosophila Inversion Polymorphism*. CRC Press, Boca Raton, Fla.

Lewontin R. C., Moore J. A., Provine W. B., Wallace B. (eds) (1981) *Dobzhansky's Genetics of Natural Populations, I–XLIII*. Columbia University Press, New York.

Lindsay G. (1963) The genus *Lophocereus*. *Cactus and Succulent Journal* 35: 176–192.

Mangan R. L. (1982) Adaptations to competition in cactus breeding *Drosophila*. In: Barker J. S. F., Starmer W. T. (eds) *Ecological Genetics and Evolution: The Cactus-Yeast-Drosophila Model System*, pp 257–272. Academic Press, Sydney.

Mettler L. E. (1963) *Drosophila mojavensis baja*, a new form in the mulleri complex. *Drosophila Information Service* 38: 57.

Newby B. D., Etges W. J. (1998) Host preference among populations of *Drosophila mojavensis* that use different host cacti. *Journal of Insect Behavior* 11: 691–712.

Nobel P. S. (1980) Morphology, surface temperatures, and northern limits of columnar cacti in the Sonoran Desert. *Ecology* 61: 1–7.

Orzack S. H. (1985) Population dynamics in variable environments. V. The genetics of homeostasis revisited. *American Naturalist* 125: 550–572.

Painter T. S. (1933) A new method for the study of chromosome rearrangements and the plotting of chromosome maps. *Science* 78: 585–586.
Patterson J. T., Crow J. F. (1940) Hybridization in the mulleri group of *Drosophila*. *University of Texas Publications* 4032: 251–256.
Patterson J. T., Stone W. S. (1952) *Evolution in the Genus Drosophila*. Macmillan, New York.
Pitnick S., Heed W. B. (1994) New species of cactus-breeding *Drosophila* (Diptera: Drosophilidae) in the nannoptera species group. *Annals of the Entomological Society of America* 87: 307–310.
Powell, J. R. (1997) *Progress and Prospects in Evolutionary Biology: the Drosophila Model*. Oxford University Press, New York.
Prakash S. (1967) Association between mating speed and fertility in *Drosophila robusta*. *Genetics* 57: 655–663.
Richardson R. H., Smouse P. E., Richardson M. E. (1977) Patterns of molecular variation. I. Associations of electrophoretic mobility and larval substrate within species of the *Drosophila mulleri* complex. *Genetics* 85: 141–154.
Rockwood-Sluss E. S., Johnston J. S., Heed W. B. (1973) Allozyme genotype-environment relationships. I. Variation in natural populations of *Drosophila pachea*. *Genetics* 73: 135–146.
Ruiz A., Fontdevila A., Santos M., Torroja E. (1986) The evolutionary history of *Drosophila buzzatii*. VII. Evidence for endocyclic selection acting on the inversion polymorphism in a natural population. *Evolution* 40: 740–755.
Ruiz A., Heed W. B. (1988) Host plant specificity in the cactophilic *Drosophila mulleri* species complex. *Journal of Animal Ecology* 57: 237–249.
Ruiz A., Heed W. B., Wasserman M. (1990) Evolution of the *mojavensis* cluster of cactophilic *Drosophila* with descriptions of two new species. *Journal of Heredity* 81: 30–42.
Saitou N., Nei M. (1987) The neighbor-joining method: a new method for reconstructing phylogenetic trees. *Molecular Biology and Evolution* 4: 406–425.
Salceda V. M., Anderson W. W. (1988) Rare male mating advantage in a natural population of *Drosophila pseudoobscura*. *Proceedings of the Natural Academy of Sciences USA* 85: 9870–9874.
Shreve F. (1937) The vegetation of the Cape region of Baja California. *Madroño* 4: 105–113.
Shreve F. (1951) *Vegetation of the Sonoran Desert*. Carnegie Institution of Washington Publication no. 591, Washington, D.C.
Shreve F. (1964) *Vegetation of the Sonoran Desert*. In: Shreve F., Wiggins I. L. *Vegetation and Flora of the Sonoran Desert*. Stanford University Press, Stanford.
Spiess E. B., Spiess L. D. (1967) Mating propensity, chromosomal polymorphism, and dependent conditions in *Drosophila persimilis*. *Evolution* 21: 672–678.

Starmer W. T. (1982) Analysis of the community structure of yeasts associated with the decaying stems of cactus. I. *Stenocereus gummosus*. *Microbial Ecology* 8: 71–81.

Starmer W. T., Heed W. B., Rockwood-Sluss E. S. (1977) Extension of longevity in *Drosophila mojavensis* by environmental ethanol: Differences between subraces. *Proceedings of the National Academy of Sciences USA*. 74: 387–391.

Sturtevant A. H., Dobzhansky T. (1936) Inversions in the third chromosome of wild races of *Drosophila pseudoobscura*, and their use in the study of the history of the species. *Proceedings of the National Academy of Sciences USA*. 22: 448–450.

Taylor N. P. (1979) Notes on *Ferocactus* B & R. *Cactus and Succulent Journal (Great Britain)* 41: 88–94.

Turner R. M., Bowers J. E., Burgess T. L. (1995) *Sonoran Desert Plants: An Ecological Atlas*, University of Arizona Press, Tucson.

Turner R. M., Brown D. E. (1982) Sonoran desertscrub. *Desert Plants* 4: 181–221.

Ward B. L., Heed W. B. (1970) Chromosomal phylogeny of *Drosophila pachea* and related species. *Journal of Heredity* 61: 248–258.

Ward B. L., Starmer W. T., Russell J. S., Heed W. B. (1975) The correlation of climate and host plant morphology with a geographic gradient of an inversion polymorphism in *Drosophila pachea*. *Evolution* 28: 565–575.

Wasserman M. (1992) Cytological evolution of the *Drosophila repleta* species group. In: Krimbas C. B., Powell J. R. (eds) *Drosophila Inversion Polymorphism*, pp 455–552. CRC Press, Boca Raton, Fla.

Wright S. (1951) The genetical structure of populations. *Annals of Eugenics* 15: 323–354.

Wright S. (1969) *Evolution and the Genetics of Populations*. Vol. 2, *The Theory of Gene Frequencies*. University of Chicago Press, Chicago.

Wright S. (1978) *Evolution and the Genetics of Populations*. Vol. 4, *Variability Within and Among Populations*. University of Chicago Press, Chicago.

Zouros E. (1973) Genic differentiation associated with the early stages of speciation in the *mulleri* subgroup of *Drosophila*. *Evolution* 27: 601–621.

Appendix 6.1 Collection Data and Chromosome Frequencies of *Drosophila mojavensis*.

Locality		Collection			Second Chromosome				Third Chromosome						
Name[a]	Region[b]	No.[c]	Date	N[d]	LP	ST	BA	SL	ST	MU	h_2	b_3	H	TMR[e]
Baja California Norte														
1. Ensenada	SPM	A756	12-79	14	0.86	0.00	0.14	0.00	0.93	0.07	0.24	0.13	0.19	8.6
2. San Telmo	SPM	A758	12-79	106	0.70	0.00	0.30	0.00	0.86	0.14	0.42	0.24	0.33	10.8
3. El Socorro	SPM	A519	7-74	132	0.93	0.00	0.07	0.00	0.89	0.11	0.13	0.20	0.17	5.9
4. Cataviña	VIZ	A761	12-79	100	0.77	0.02	0.21	0.00	0.30	0.70	0.36	0.42	0.39	13.5
5. Punta Prieta a	VIZ	A420	2-74	106	0.44	0.18	0.38	0.00	0.04	0.96	0.63	0.08	0.36	11.4
Punta Prieta b	VIZ	A896	5-85	230	0.43	0.31	0.25	0.01	0.06	0.94				
6. San Borja	VIZ	A350*	11-71	56	0.52	0.46	0.02	0.00	0.30	0.70	0.52	0.42	0.47	11.1
7. El Barril	CGC	A570	3-75	54	0.52	0.46	0.00	0.02	0.02	0.98	0.52	0.04	0.28	14.3
Baja California Sur														
8. Tres Vírgenes	CGC	A567*	11-74	100	0.61	0.29	0.08	0.02	0.42	0.58	0.54	0.49	0.51	15.3
9. San Lucas	CGC	A422+	2-74	132	0.35	0.49	0.07	0.09	0.20	0.80	0.62	0.32	0.47	16.3
10. Punta Concepción	CGC	A352*	11-71	94	0.39	0.60	0.01	0.00	0.38	0.62	0.49	0.47	0.48	16.2
11. Bahía Coyote	CGC	A427	4-74	64	0.58	0.38	0.00	0.04	0.37	0.63	0.52	0.47	0.50	16.2
12. Loreto	CGC	A385*	3-72	106	0.24	0.76	0.00	0.00	0.95	0.05	0.36	0.09	0.22	14.1
13. Puerto Escondido	CGC	A428	4-74	68	0.40	0.57	0.00	0.03	0.78	0.22	0.51	0.34	0.42	14.1
14. Isla San Jose	CGC	A593*	3-76	54	0.09	0.83	0.07	0.00	0.96	0.04	0.30	0.08	0.19	12.1
15. Sierra Giganta	GIG	A429*	4-74	70	0.11	0.88	0.00	0.01	0.96	0.04	0.21	0.08	0.14	14.9
16. La Presa	GIG	A376*	3-72	104	0.00	1.00	0.00	0.00	0.98	0.02	0.00	0.04	0.02	14.9
17. San Ignacio a	MAG	A367*	4-72	54	0.24	0.67	0.09	0.00	0.02	0.98	0.42	0.12	0.27	12.4
San Ignacio b	MAG	A421	2-74	126	0.08	0.74	0.16	0.02	0.10	0.90				
San Ignacio c	MAG	A566+	11-74	82	0.12	0.73	0.14	0.00	0.70	0.30				
18. V. Insurgentes	MAG	A430	4-74	60	0.00	1.00	0.00	0.00	0.70	0.30	0.00	0.42	0.21	12.4
19. Santa Rita	MAG	A564	11-74	64	0.03	0.94	0.03	0.00	0.96	0.04	0.11	0.08	0.10	11.6

Appendix 6.1 continued

Locality		Collection			Second Chromosome				Third Chromosome						
Name[a]	Region[b]	No.[c]	Date	N[d]	LP	ST	BA	SL	ST	MU	h_2	b_3	H	TMR[e]
20. 50 km S. S. Rita	MAG	A431	4-74	80	0.05	0.95	0.00	0.00	0.91	0.09	0.09	0.16	0.13	12.0
21. San Agustin	MAG	A563	11-74	82	0.61	0.33	0.06	0.00	0.88	0.12	0.52	0.21	0.37	7.8
22. 10 km S. San Ag.	MAG	A587	3-76	70	0.89	0.11	0.00	0.00	0.76	0.24	0.20	0.36	0.28	7.8
23. 22 km W. La Paz	CAPE	A560	11-74	86	0.92	0.03	0.03	0.02	0.40	0.60	0.15	0.48	0.32	11.8
24. AV Bonfil	CAPE	A768	3-80	100	0.84	0.14	0.02	0.00	0.69	0.31	0.27	0.43	0.35	11.8
25. Virgin Maria	MAG	A770	3-80	100	0.94	0.05	0.00	0.01	0.30	0.70	0.11	0.42	0.27	12.7
26. San Bartolo	CAPE	A561	11-74	56	0.96	0.04	0.00	0.00	0.16	0.84	0.08	0.27	0.18	11.9
27. 10 km S. San Bart.	CAPE	A433	4-74	80	0.95	0.05	0.00	0.00	0.00	1.00	0.09	0.00	0.05	11.9
28. Punta Pescadero	CGC	A374*	3-72	78	1.00	0.00	0.00	0.00	0.11	0.89	0.00	0.20	0.10	12.3
29. Santiago	CAPE	A597	3-76	82	0.98	0.00	0.00	0.02	0.06	0.94	0.04	0.11	0.08	13.5
30. La Huerta	CGC	A772	3-80	40	0.95	0.05	0.00	0.00	0.20	0.80	0.10	0.32	0.21	13.0
31. San Pedro	CAPE	A767	3-80	74	0.97	0.03	0.00	0.00	0.08	0.92	0.06	0.15	0.10	11.0
32. El Caimancito	CGC	A769	3-80	80	0.96	0.04	0.00	0.00	0.11	0.89	0.08	0.20	0.14	11.8
33. Todos Santos	CAPE	A771	3-80	76	0.82	0.08	0.10	0.00	0.23	0.77	0.31	0.35	0.33	9.3
Arizona														
34. S. Rosa Mtns.[f] a	ARIZ	A572	2-75	72	1.00	0.00	0.00	0.00	1.00	0.00	0.00	0.00	0.00	23.1
S. Rosa Mtns. b	ARIZ	A900	3-85	214	1.00	0.00	0.00	0.00	0.99	0.01	0.00	0.00	0.00	
35. Organ Pipe NM[f]	ARIZ	A345*	9-71	30	1.00	0.00	0.00	0.00	1.00	0.00	0.00	0.00	0.00	20.8
Sonora														
36. Altar Valley[f]	ARIZ	A319	4-71	58	1.00	0.00	0.00	0.00	1.00	0.00	0.00	0.00	0.00	18.9
37. Caborca[f]	LC	A316	4-71	44	1.00	0.00	0.00	0.00	1.00	0.00	0.00	0.00	0.00	20.8
38. Punta Libertad[f]	CGC	A514*	4-74	56	1.00	0.00	0.00	0.00	1.00	0.00	0.00	0.00	0.00	17.3
39. Desemboque[g] a	CGC	A361*	12-71	56	0.96	0.04	0.00	0.00	0.88	0.12	0.08	0.21	0.15	17.9
Desemboque b	CGC	A366*	3-72	102	0.99	0.01	0.00	0.00	0.95	0.00				
Desemboque c	CGC	A388	3-72	100	0.98	0.02	0.00	0.00	0.98	0.02				

Locality	Subdiv.	Stock	Date	N	2a	2b	2c	2d	3a	3b	3c	h_2	h_3	H	TMR
Desemboque d	CGC	A509	3-74	100	0.78	0.18	0.04	0.00	0.94	0.06	0.00				
Desemboque e	CGC	A856	2-84	40	0.98	0.02	0.00	0.00	0.95	0.05	0.00				
40. Campo Vibora[g]	CGC	A510	3-74	100	1.00	0.00	0.00	0.00	0.91	0.09	0.00	0.00	0.16	0.08	17.8
41. Ran. del Puerto[g]	CGC	A580	7-75	84	1.00	0.00	0.00	0.00	0.97	0.03	0.00	0.00	0.06	0.03	17.8
42. 25 km N. Chueca[g]	CGC	A581	7-75	42	0.93	0.07	0.00	0.00	0.95	0.05	0.13	0.13	0.10	0.12	17.8
43. Bahía Kino[f]	CGC	A511	3-74	46	1.00	0.00	0.00	0.00	0.87	0.13	0.00	0.00	0.23	0.12	17.7
44. San Carlos[f]	CGC	A512	3-74	100	1.00	0.00	0.00	0.00	0.82	0.18	0.00	0.00	0.30	0.15	13.1
45. I. S. Ped. Nolas.[f]	CGC	A513	3-74	100	1.00	0.00	0.00	0.00	1.00	0.00	0.00	0.00	0.00	0.00	—
46. Isla Tiburón a	CGC	A506*	11-74	80	0.87	0.12	0.00	0.01	0.93	0.07	0.07	0.23	0.13	0.18	17.8
Isla Tiburón b	CGC	A732	5-78	100	0.97	0.03	0.00	0.00	1.00	0.00	0.00				
47. Isla San Esteban	CGC	A731	5-78	100	0.99	0.01	0.00	0.00	1.00	0.00	0.02	0.02	0.00	0.01	—
Sinaloa															
48. Los Mochis[f]	STS	A337	11-71	110	1.00	0.00	0.00	0.00	0.96	0.04	0.00	0.00	0.08	0.04	13.4
49. Topolobampo[f]	STS	A559	11-74	4	1.00	0.00	0.00	0.00	1.00	0.00	0.00	0.00	0.00	0.00	11.2
California															
50. S. Catalina Is.[h]		A826	10-81	60	0.00	1.00	0.00	0.00	1.00	0.00	0.00	0.00	0.00	0.00	—
51. Vallecito[i]	LC	A753	12-79	100	0.00	1.00	0.00	0.00	0.99	0.01	0.01	0.00	0.00	0.00	—

Note: Inversion heterozygosity estimates of second (h_2) and third (h_3) chromosomes and their mean (H) for *D. mojavensis* are presented where $h_x = 1 - \Sigma x_i^2$, with x_i being the frequency of the ith inversion. For populations that were sampled more than once, heterozygosities were calculated for pooled data. Temperature data were taken from Hastings and Humphrey (1969a, 1969b).

[a] Site numbers refer to the collecting sites as numbered in figure 6.5.
[b] Vegetational subdivisions: ARIZ = Arizona Upland; CAPE = Cape region in Baja California Sur; CGC = Central Gulf Coast; GIG = Sierra Giganta; LC = Lower Colorado Valley; MAG = Magdalena; SPM = San Pedro Martir; STS = Sinaloan thornscrub forest; VIZ = Vizcaíno.
[c] Cultures and chromosome analyses were derived from field-captured isofemales, except those indicated by *, which were derived from flies reared from cactus rots. + SI gene arrangement in the second chromosome is also present.
[d] N = number of chromosomes observed per collection.
[e] TMR is the difference between the highest average monthly temperature and the lowest average monthly temperature at the closest weather station.
[f] Mainland and island localities where organ pipe is present and pitahaya agria is not.
[g] Mainland and island localities where pitahaya agria is present.
[h] Locality where *Opuntia* is present and other hosts are not.
[i] Locality where the California barrel cactus, *Ferocactus cylindraceous*, is present and other hosts are not.

Appendix 6.2 Collection Records and Karyotypes of *Drosophila pachea* from Arizona, Baja California, and Sonora, Mexico.

Locality		Region[c]	Collection			Type of Collection[d]	No. Observed Karyotypes[a]				Frequency (+)	h_7	TMR
Name[b]			Cactus	No.	Date		+/+	+/A	A/A				
Baja California Norte													
1. Bahia de Los Angeles[e]		CGC	*	A191	3-68		0	0	17		0.00	0.00	15.4
2. Bahia de Los Angeles		CGC	*	A606	3-76	PM7	0	0	13		0.00	0.00	16.2
3. San Borja		VIZ	*	A605	3-76	PM7	0	0	16		0.00	0.00	11.1
Baja California Sur													
4. Bahia Tortugas		VIZ	*	A608	3-76	RL	0	0	35		0.00	0.00	8.3
5. Mulege		CGC	*	A603	3-76	PM7	0	1	31		0.02	0.23	16.2
6. Mulege		CGC	*	A607	3-76	PM1	1	19	33		0.20	0.00	
7. Comondu		MAG	*	A622	3-62		1	6	16		0.17	0.28	11.7
8. San Agustin		MAG	†	A720	1-78	FCF7, PM1	9	7	1		0.74	0.49	7.8
9. San Agustin		MAG	†	A719	1-78	PM1	3	6	2		0.55		
10. La Paz		CAPE	†	A596	3-76	PM7	7	12	5		0.54	0.50	11.8
11. La Paz		CAPE	†	A679	3-77	PM1, PM7	2	5	2		0.50		
12. La Paz		CAPE	†	A677	3-77	PM1, PM7, MF, FCF1	4	13	8		0.42		
13. E. La Paz		CGC	†	A595	3-76	PM7	13	26	7		0.56	0.51	12.5
14. La Aguja		MAG	†	A674	3-77	PM1, PM7, MM	3	4	5		0.42	0.54	9.8
15. La Aguja		MAG	†	A672	3-77	PM1, PM7, MM, MF	2	7	4		0.42		
16. La Aguja[e]		MAG	†	A202	4-68		3	9	13		0.30		
17. San Pedro		CAPE	†	A721	1-78	FCF7, PM1	5	24	43		0.27	0.32	12.0
18. San Pedro		CAPE	†	A686	3-77	FCF1, PM1, MM	0	14	32		0.15		
19. El Carrizal		CAPE	†	A691	3-77	MF, MM	0	4	8		0.17	0.28	12.0
20. Todos Santos		CAPE	†	A684	3-77	FCF1, MM, MF	0	2	9		0.09	0.46	9.3

21. Todos Santos	CAPE	†	A693	3-77	MF, MM	1	2	7	0.20	0.46	9.3
22. Todos Santos	CAPE	†	A725	1-78	PM7	0	6	6	0.25	0.09	11.9
23. San Bartolo	CAPE	†	A688	3-77	FCF1, PM1	0	4	41	0.04	0.08	12.7
24. La Ribera	CAPE	†	A689	3-77	FCF1, PM1	0	4	48	0.04	0.00	12.7
25. Santiago	CAPE	†	A694	3-77	FCF1, FCM	0	0	12	0.00	0.21	13.0
26. Caduaño	CAPE	†	A696	3-77	PM1, MM	1	3	17	0.12	0.21	13.0
27. S. Jose d. Cabo	CAPE	†	A723	1-78	FCF1, PM1	1	10	32	0.14	0.24	13.0
28. Cabo San Lucas	CAPE	†	A724	1-78	FCF7, PM1	1	2	9	0.17	0.28	9.9
Sonora											
29. Sasabe[f]	ARIZ	*	A650	11-76	PM	0	1	22	0.02	0.04	16.3
Sonoita[g]	ARIZ	*			RL	0	4	68	0.03	0.05	20.8
Magdalena[g]	ARIZ	*			FCF1	0	1	49	0.01	0.02	17.2
Santa Ana[g]	ARIZ	*			FCF1, RL	0	1	64	0.01	0.02	17.4
Altar Valley[g]	LC	*			RL	0	2	118	0.01	0.02	18.9
P. Peñasco[g]	LC	*			RRF3	0	0	21	0.00	0.02	18.9
Desemboque[g]	CGC	*			RL	1	29	111	0.10	0.20	17.9
Kino[g]	CGC	*			RRF2	3	21	23	0.29	0.41	17.7
Empalme[g]	CGC	•			RRF1	45	41	5	0.72	0.40	12.2
Guaymas[g]	CGC	•			RRF2	73	87	24	0.63	0.46	13.1
30. Cerro Colorado	LC	*	A652	11-76	PM1	0	0	25	0.00	0.00	19.5
31. Caborca[f]	LC	*	A651	11-76	RL, PM1	0	0	47	0.00	0.07	20.8
32. Isla Tiburón	CGC	•	A663	11-76	PM7	1	18	23	0.27	0.39	17.8
33. N. Guaymas	CGC	•	A660	11-76	FCFF2	23	21	0	0.76	0.42	—
34. N. Guaymas	CGC	•	A665	1-77	FCF1	63	54	18	0.67		
35. Navojoa[f]	STS	•	A658	11-76	FCFF2	29	0	0	1.00	0.00	14.6
Vicam[g]	STS	•			RRF1	16	0	0	1.00	0.00	17.4
Cd. Obregón[g]	STS	•			RL	110	3	0	0.99	0.03	15.1
Zaragosa[g] (Sinaloa)	STS	•			RL	90	0	0	1.00	0.00	14.6
Hermosillo[g]	PLS	•			RRF1	4	11	3	0.53	0.50	15.2

Appendix 6.2 continued

Note: Localities are usually the closest weather station. Inversion heterozygosity estimates of the seventh chromosome (b_7) are presented along with TMR data. Chromosome data from populations sampled more than once were pooled. The subspecific status of senita cactus for each collection is designated as follows: *Lophocereus schottii schottii* (*), *L. s. australis* (†), and *L. s. tenuis* (•). The temperature data were taken from Hastings and Humphrey (1969a, 1969b). For collection records for populations from Sonora and Sinaloa, see Ward et al. (1975).

[a] All collections were found to be in Hardy-Weinberg equilibrium (X^2 analyses, all p values $> .05$).

[b] Site numbers refer to the collecting sites as numbered in figure 6.4.

[c] Vegetational subdivisions: ARIZ = Arizona Upland; CAPE = Cape region in Baja California Sur; CGC = Central Gulf Coast; LC = Lower Colorado Valley; MAG = Magdalena Plain; PLS = Plains of Sonora; SPM = San Pedro Martir; STS = Sinaloan thornscrub forest; VIZ = Vizcaino.

[d] FCF1 = field-caught isofemales, 1 larva squashed per female; FCF7 = field-caught isofemales, 7 larvae squashed per female; FCM = field-caught males mated to Zaragosa females, 7 progeny squashed per male to determine his karyotype; PM1 = pair matings between rot-reared adults, 1 larva squashed per pair; PM7 = pair matings between rot-reared adults, 7 larvae squashed per pair to determine karyotypes of the parents; MM = matings between rot-reared males and Zaragosa females, 7 progeny squashed per female; MF = matings between rot-reared females and Zaragosa males, 7 larvae squashed per female; RL = rot larvae squashed and scored; FCFF2 = F$_2$ progeny of field-caught isofemales; RRF1 = F$_1$ larvae of rot-reared flies; RRF2 = F$_2$ larvae of rot-reared flies; RRF3 = F$_3$ larvae of rot-reared flies.

[e] From the unpublished records of W. B. Heed.

[f] Inversion data were combined with those from Ward et al. (1975) for heterozygosity estimates.

[g] Inversion data from Ward et al. (1975). For multiple samples from a single location, data were pooled.

7 Ecological Consequences of Agricultural Development in a Sonoran Desert Valley

Laura L. Jackson and Patricia W. Comus

If one were to construct a generalized land-use map of the Sonoran Desert, showing cropland in black and other uses in white, and place it next to a similar map of the midwestern prairie/corn belt, the two would look like photographic negatives of one another. While agriculture in the Midwest occupies virtually every square meter of upland, Sonoran Desert agriculture is found nowhere but on the valley floors. The availability of water for irrigation, and fine-textured lowland soils capable of retaining that water, are prerequisites for desert agriculture. Both of these requirements are found only on valley floors.

This landscape pattern, while excusing a great deal of desert habitat from intensive use, nevertheless singles out lowland vegetation types, moisture regimes, and soils, and the lowland landscape element itself for extensive disturbance. The effects go beyond fields: reduced stream flow caused by surface and groundwater pumping, increased silt load and dissolved salts in the streams due to return of used irrigation water, and changes in river hydroperiod caused by dams all damage downstream reaches as effectively as does cultivation.

The changes wrought by agriculture on lowland deserts have ranged from slight to profound, depending on the type of agriculture. Floodwater farming by the Tohono O'odham Indians involved the careful selection of land at the mouth of an ephemeral wash, where floodwaters after winter and summer rains would collect and then spread overland. Farmers used low earthen berms, brush dams, and rock alignments to slow the water down and train it into their fields, where it dropped its load of nutrient-rich debris and soaked into the soil. One or two floods might suffice to grow a crop. This method involved only temporary, subtle changes in land form and, although an estimated 10 000 acres were farmed this way in 1913, disturbance of vegetation was limited to small patches (Nabhan 1986).

In contrast, the earlier Hohokam people (1700 to 550 years before present [BP]) diverted river water through several hundred miles of canals along the Salt and Gila Rivers. Native vegetation was removed and soils may have been salinized over a large area. Water consumption of the crops must have reduced downstream flow and changed flooding regimes. The demise of the Hohokam civilization has been attributed to various combinations of social upheaval, climate change, soil degradation, and the effects of destructive floods bracketed by long, dry periods (Crown and Judge 1991; Gumerman 1991).

Two technological developments in the twentieth century, the big dam and the deep pump, transformed our ability to modify desert lowlands. The Roosevelt Dam on the Salt River, completed in 1911, was the first federally funded, large flood and irrigation project in the country (Worster 1985). (The Salt River Project first fed agricultural development, but later was used for urban development as farms gave way to houses in the Phoenix metropolitan area.) Dams protected farmers from floods and simultaneously assured a better water supply. They also made it possible to consume a greater proportion of the river's annual flow by storing water during the rainy season and releasing it during dry months. Complete appropriation of river waters reduced their flow rate to a trickle in wet seasons and dampened the flooding regime. This threatened riparian species like cottonwood (*Populus fremontii*), which requires high groundwater and new, flood-deposited sediments for recruitment (Stromberg et al. 1991).

The second innovation was the improvement of the centrifugal pump in the 1920s–1930s, coincident with expanding rural electricity and natural gas service. Deep well pumping expanded the province of agriculture beyond the reach of surface water. In California, this process began in 1910–1920; in Arizona and the High Plains of Texas, in the 1930s–1940s; and in Mexico, the dry, western coastal plains were cleared in the 1950s–1960s (Worster 1985; Green 1973; Bowden 1977; Shapiro 1989). Several hundred thousand hectares of Sonoran Desert lowlands around the Willcox Playa, the lower Gila, the coastal plains of Sonora, Mexico, and the broad lower Santa Cruz Valley were swept up in the flurry to exploit finite groundwater reserves.

Within 10 to 30 years of initial groundwater development, fields in the pumplands were being abandoned. The inherent productivity of the sunny climate could not offset the high price of pumping water from hundreds of feet to water a crop with a fluctuating market value. About half of the fields in the lower Santa Cruz Valley, Rainbow Valley, the Willcox Playa, the Gila Valley west of Gila Bend, and the Hermosillo coastal plain no longer grow crops. Rusting pumps, crumbling ditches, abandoned cotton gins, and vacant farmsteads in these valleys attest to the tremendous capital invested in farming during its heyday.

The purpose of this study was to learn how the largest of the pumplands in Arizona, the lower Santa Cruz Valley, was changed by agricultural exploitation. Without a historical model, we do not know whether successional processes are returning the valley to what it once was or to some new state. If changes to the old ecosystem are understood, ecological restoration techniques might be used to help repair the damage. The first step towards this goal is a rigorously documented and independently corroborated timetable of human-induced changes, placed in the context of constantly changing climate, soils, and biotic communities.

Several sources of data were used to document agricultural land-use change and its effects on the valley ecosystem. To learn the time and extent of clearing for agriculture, and to quantify the pattern of fragmented habitat and abandoned farmland, we used topographic maps and aerial photos, combined with field visits. U.S. agricultural census figures were used to independently assess the extent of abandoned farmland and other changes in land use since 1944. Historic aerial photos, maps, soil surveys, and personal interviews were used to uncover surface hydrologic changes since 1885.

We also sought to understand the characteristics of the native vegetation prior to intensive modern agriculture. Aerial photos, historical accounts, and earlier studies and descriptions of the vegetation of the area were supplemented by field data in those remnants of valley vegetation that have not been plowed or otherwise severely impacted. We collected data on plant species composition, density of perennials, both winter and summer ephemerals, and textural, structural, and chemical characteristics of soils in the various major plant associations (Comus, unpublished data).

Methods and Approaches

The lower Santa Cruz Valley (Pinal County, Arizona, 32°30′–33°00′ N, 111°22′30″–112°00′ W; elevation 347 to 610 m) is a basin surrounded by mountains that has filled with alluvium from the Gila River, the Santa Cruz River, and tributary drainages. Mean annual precipitation is 150 to 250 mm; 40% falls in locally intense summer thunderstorms from July to September, and the balance in more widespread, predictable winter rains from December to March. Mean maximum temperature in July is 40.5°C (105°F), mean annual temperature is 20°C, and the frost-free season is 240 to 325 days (Sellers and Hill 1974). The original vegetation is described by Brown (1982) as Sonoran saltbush desertscrub, dominated by *Atriplex* spp.

We wanted to know how much land had been cleared, farmed, and abandoned, and the spatial pattern of land use in the valley. No data on the extent or distribution of natural desert remnants or abandoned farms were available in any federal, state, or county office. Therefore, we had to piece together the history and geography of the region from a variety of sources. To estimate the rate of desert conversion and field abandonment, we consulted Pinal County agricultural production statistics (Arizona Crop and Livestock Reporting Service 1966, 1981; Arizona Agricultural Statistics Service 1944, 1952, 1972–1974, 1976–1992) and land-use estimates by county from the U.S. Census of Agriculture from 1944 to the latest published census in 1987 (U.S. Bureau of the Census 1956, 1961, 1966, 1972, 1977, 1981, 1984, 1989). The lower Santa Cruz Valley is completely contained within Pinal County (fig. 7.1) and comprises 90% of the farming area, so countywide statistics accurately characterize the valley region.

We wished to obtain a more direct estimate of abandoned farmland and determine its spatial distribution. Therefore, land-use patterns were interpreted from 1:24000 scale, panchromatic, unrectified, 1983 aerial photos of the eastern half of the Pinal County farming region and its bordering lands, comprising 102 400 ha (32°30–33°00′N, 111°22′30″–111°45′W).

In order to interpret aerial photos accurately, we studied a complete set of historical aerial photos (1936, 1949, 1954, 1964, 1970, 1979, and 1983) and 7.5′ topographic maps (USGS 1961–1983) for two 2330 ha regions consisting of actively farmed, abandoned, and (apparently) never-cleared land. Each region was completely surveyed by car or on foot, and land-use history determined by presence or absence of irrigation pumps,

Figure 7.1 Map of Pinal County, Arizona, showing the Gila River, the approximate path of the Santa Cruz River, and irrigated regions in 1963. Irrigated regions coincide with the limits of the lower Santa Cruz Valley outside of Indian Reservations. Redrawn from a 1963 U.S. Bureau of Reclamation map, "Irrigated areas in Arizona," produced in cooperation with the University of Arizona Departments of Agricultural Economics and Agricultural Engineering.

ditches, furrows, and other signs of disturbance. These field surveys were compared to recent and historical aerial photos and topographic maps to "calibrate" our interpretations.

Topographic contours (obtained from USGS 7.5′ topographic maps) were an important clue to land-use history, since completely straight contours could only be the result of field leveling. All abandoned fields and desert remnants recognizable on the ground were visible in aerial photos, and two additional old fields were recognized from the aerial photos only. Abandoned fields visible on older (1936, 1949) aerial photos were still recognizable in 1983 photos, so these methods appeared sufficient to detect fields abandoned by or before 1936. In addition to the systematic surveys, field visits were made to dozens of native desert remnants and abandoned fields throughout the valley, and their current vegetation compared to 1936 and 1983 aerial photos. We judged these methods adequate to identify fields abandoned within this century. One area was initially identified as undisturbed desert using both aerial photography and field surveys; however, we later found Hohokam pottery and two earthen irrigation ditches suggesting it had been cultivated in prehistoric times (1200 to 1300 BP).

After creating the land-use maps, we drove two 40 km transects through the valley and recorded land use every half mile. Using binoculars, we were able to verify the photo interpretations up to 0.8 km from the road; 6400 ha, or 7.8% of the photo survey, were verified in this way.

Information about the Santa Cruz River and the former courses of smaller ephemeral streams was derived from soil surveys carried out in 1917 (Eckman et al. 1920) and 1935 (Poulson et al. 1941), supplemented by Forbes' 1911 assessment of water resources in Arizona. Four maps, depicting natural waterways, reservoirs, and canals in 1885, 1930, 1980, and 1990 were redrawn from information in the historical soils maps (Poulson et al. 1941; Eckman et al. 1920), aerial photos, USGS 7.5' topographic sheets, and current county maps using a variable zoom photocopier to equalize map scales.

Untilled native remnants contained several more or less intergrading communities, especially with respect to dominant shrubby perennials. These dominant species include narrow-leaf saltbush (*Atriplex linearis*), desert saltbush (*Atriplex polycarpa*), big galleta (*Pleuraphis rigida*), creosote bush (*Larrea tridentata*), *Lycium californicum,* and seepweed (*Suaeda moquinii*). For each of these dominant species, three or four sites (in which that species constitutes more than 50% of the long-lived woody perennials) were selected. Four sites in which cactus species predominate were also chosen, with care taken to choose only such sites as have a significant proportion of cacti that reproduce sexually (to avoid the possibility that such communities might result purely from the accidental introduction of vegetative propagules).

Perennial species density and cover within these sites were determined using McAuliffe's log series sampling method (1991b) or, where individual shrubs could not be distinguished from a distance of more than a few meters, by measuring canopy interception in 1 m wide, stratified random transects.

Winter and summer annuals were surveyed at each site, with species presence noted, and total percent cover and percent cover of dominant species estimated visually. Representative herbarium specimens of perennial and annual species were collected with data for documentations.

The Santa Cruz Valley Ecosystem Before Pioneer Agriculture
Pre-settlement Hydrology

Figure 7.2 (About 1885) shows the surface hydrology of the lower Santa Cruz Valley as it may have looked before any modification by European settlers. The main channel of the Santa Cruz carried floodwaters from the upper Santa Cruz River near Tucson. The channel disappeared south of Eloy, then fanned out into the "Santa Cruz Flats," a broad, flat region lacking obvious drainage. The dispersed waters flowed around Casa Grande Mountain through various shallow, ephemeral washes, and converged again 31 km from the channel's disappearance, to meet the Gila River. McClellan Wash, according to Eckman et al. (1920), was not a well-defined, separate drainage. It sometimes carried overflow from the Santa Cruz River but in addition received runoff from the Picacho Reservoir area, draining the northeastern side of the valley. A soil series was named after McClellan Wash in the oldest soils map (Eckman et al. 1920), indicating the wash's regional importance at the time. McClellan Wash has virtually disappeared from modern maps. Poulson et al.'s 1941 soils map shows two distinct McClellan's washes, one south and one north of the Florence–Casa Grande Canal, that do not appear to be integrated (fig. 7.2).

The far southwestern drainage, called Greene's Wash since 1910, apparently was fed by runoff from the Sawtooth and Silver Reef Mountains to the south (fig. 7.2). Two permanent Tohono O'odham villages, Shopishk and Chuichu, were located along its length. Aerial photos from 1936 reveal several buildings and active fields in these villages, positioned to receive floodwater both from Greene's Wash and from surrounding mountains. Their presence suggests that the wash was an important source of seasonal surface water. It is not clear whether this wash received floodwaters from the Santa Cruz before they were connected by the construction of Greene's Canal (fig. 7.2, About 1930). The Santa Cruz River and Greene's Wash met on the Ak Chin Indian Reservation and were later met by Vekol Wash (not shown) at the site of the village. "Ak Chin" means "mouth of the wash," and these people successfully practiced floodwater

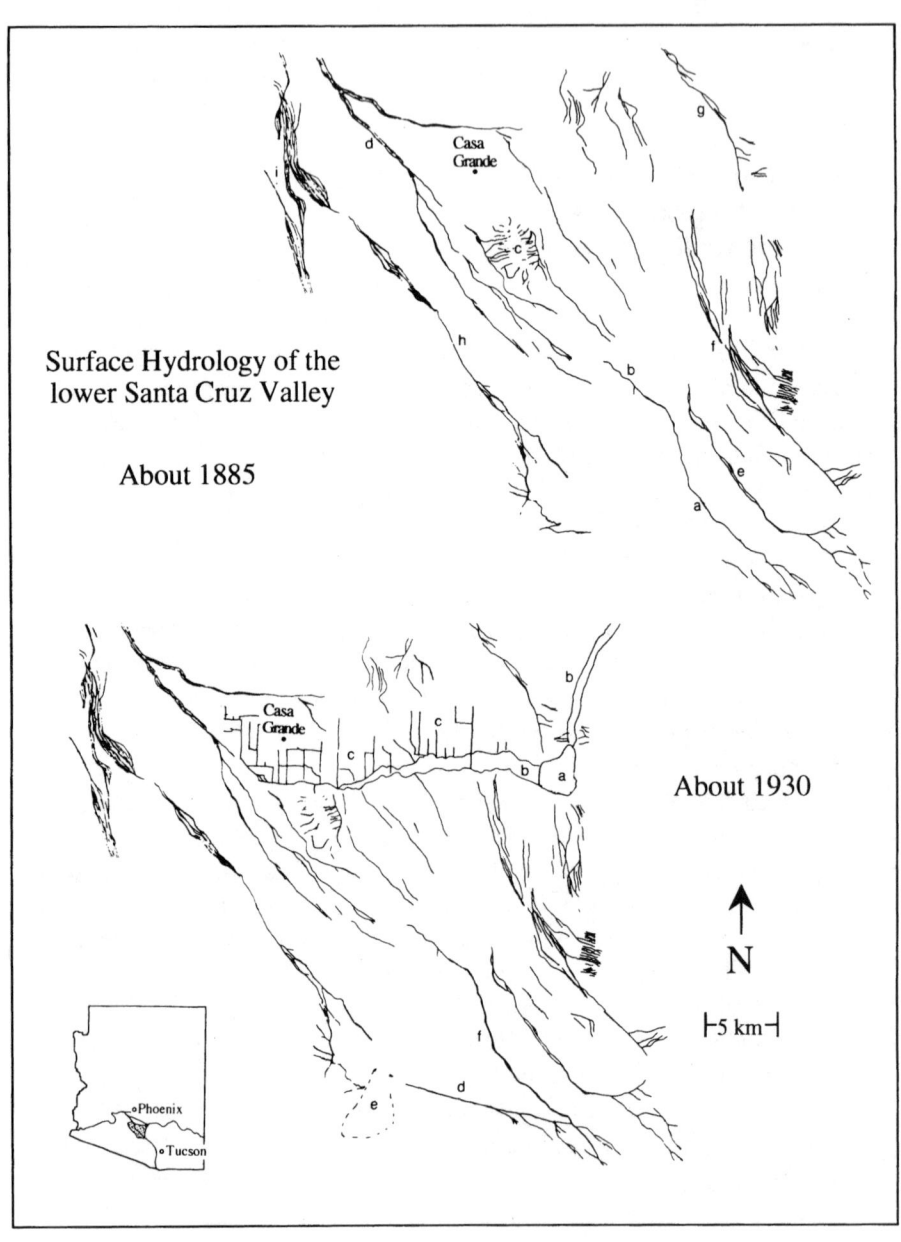

Figure 7.2 *(opposite page)* Surface hydrology of the lower Santa Cruz Valley, including all known natural waterways, constructed waterways, reservoirs, and canals in 1885 and 1930.
About 1885 The location of ephemeral streams was obtained from the earliest complete county soils map, which was based in part on fieldwork and aerial photos from the mid-1930s (Poulson et al. 1941) and checked against 1936 aerial photos. Water control systems constructed after 1885 were omitted. a. Southern channel of the Santa Cruz. b. The Santa Cruz Flats, an active alluvial fan north of the Santa Cruz channel without obvious drainageways. Several areas of the valley are referred to as "flats" on modern maps. c. Casa Grande Mountain, showing ephemeral streams coming off the mountain. d. Reemergence of the Santa Cruz River as a distinct channel west of the town of Casa Grande. e. Red Rock Wash. f. The southern McClellan Wash. g. The northern McClellan Wash. h. Greene's Wash. i. Convergence of the Santa Rosa and Greene's Washes, near the current town of Stanfield.
About 1930 Information comes from 1963 USGS 7.5' topographic maps and 1936 aerial photos in combination with the 1941 county soils map (Poulson et al. 1941). Before groundwater pumping, most farming was limited to a 2 to 4 km band north (downstream) of the Florence–Casa Grande Canal and Picacho Reservoir system (Smith 1940). New developments since 1885: a. Picacho Reservoir. b. The linked Florence–Casa Grande Canals originate to the northeast from the Gila River, pass around the east side of Picacho Reservoir, and run west. c. Feeder canals run north to irrigate fields. d. Greene's Canal, extending from the junction of the Santa Cruz (west) and Robles (east) Washes at Sasco Flats. Constructed 1908–1910. e. The ruins of Greene's Reservoir, constructed 1908–1910. f. Soon-to-be-abandoned streambed of the Santa Cruz River (c. 1939–1940) due to new channel formation in Greene's Canal.

(also known as "ak chin") farming before surrounding agricultural practices dried up their water supply (Nabhan 1986; White 1990).

Settlers who ventured south from the long-cultivated floodplains of the Gila River into the Santa Cruz Valley in the late 1880s probably met people who had been grazing cattle and sheep since the 1860s. Water from washes draining the Picacho Mountains, stored in a reservoir at their base, would have sustained ranchers, their livestock, and probably occasional crops. This may have been supplemented by windmills or steam-driven pumps tapped into shallow groundwater. European-style agriculture began when canals were constructed beginning in 1886, with wells drilled on homesteads after 1910. The availability of irrigation water fluctuated widely from year to year, so much of this land was temporarily abandoned until the 1930s, when Coolidge Dam became operational and pumping became more feasible (Poulson et al. 1941).

In these early years, the agriculture of both the pioneers and indigenous peoples were governed by the limitations of soil and rainfall and the unpredictable nature of the Santa Cruz River and its associated washes. An understanding of the frequency and nature of flooding and its relationship to groundwater, soils, and vegetation would have required long-term residence, since intervals between floods varied from months to years. Incomplete appreciation for the nature of the valley's hydrologic system by newcomers may have been the norm. In 1891, when the Southern Pacific Railroad first went through the valley, engineers aligned its route to avoid crossing the (historical) northwest flow of the Santa Cruz. However, the tracks washed out that same year (Dobyns 1981) and railroad service was spotty through at least 1917 (Eckman et al. 1920). Further evidence of early engineers' lack of appreciation of the power of flooding comes from the failure of Greene's Reservoir, built in 1911, to hold back even one storm's floodwaters (F. M. Barrios, undated, unpublished manuscript; discussed in greater detail below).

The Santa Cruz Valley was essentially an inland delta, receiving water and silts from the surrounding mountains and only rarely delivering any water to the Gila River. Groundwater was reported to be high—less than 50 feet throughout the valley (Forbes 1911). Drainage was dispersed and indistinct, and these conditions—usually dry but sometimes inundated for days or weeks—created barren, high sodium "slick spots" where nothing would grow. According to Poulson et al.:

> Floodwaters of the Santa Cruz River cross the area in a northwesterly direction and join the Gila River through numerous shallow ill-defined channels and elongated playalike depressions....
>
> [T]he chief activity of the Santa Cruz River is deposition of materials during floods. Though water does not remain long on the surface of the land after floods, the soil materials are charged to various degrees with salts left by the evaporating water. . . . at present the water table is nearly everywhere more than 25 feet below the surface, and it is 150 or more below the surface in the higher parts of the valleys. *It seems probable that the water table has been lowered by the cutting of the channel of the Gila River [1890's].* Many of the intermittent streams, or desert washes, do not reach the main stream but spread out in sheets, sorting and depositing the materials. (1941, 4–5; our emphasis)

At the time of early settlement, the water table may already have been dropping. Eckman et al. (1920), whose fieldwork for their soils map was conducted 20 years before the Poulson soil survey, emphasized the imperfect drainage much more than did Poulson's group, and it is possible that they saw a wetter valley: "Drainage is also poorly developed locally in the shallow and imperfect drainage ways of the desert. Here the gradient is low, and the runoff and percolation much retarded. . . . Little attempt has been made to provide drainage for the broad flats of the desert, cultivation being confined to the better drained soils" (32).

Also supporting this view is the earliest settlers' avoidance of fine-textured soils, with easier access to irrigation and floodwaters, but poor drainage. This practice changed with the advent of deep plowing techniques (B. Hooper 1992, personal communication).

Pre-settlement Vegetation

The earliest published reports describe the valley as a shrubland composed primarily of desert saltbush, creosote bush, velvet mesquite (*Prosopis velutina*), and wolfberry (*Lycium* spp.) (Eckman et al. 1920; Shantz and Piemeisel 1924 [field observations made 1914–1917]; Poulson et al. 1941; fig. 7.3A). Shantz and Piemeisel, working for the USDA, investigated the flora of the Gila River Valley, especially between the Gila and Salt Rivers, and (less intensively) the land in the vicinity of Casa Grande, in an effort to discover which plant associations indicated soils with the best potential for successful agriculture (1924). They identified desert saltbush as the dominant species in a plant association that had formerly covered more land in these valleys than had any other type of vegetation. They noted that desert saltbush mixed, alternated, and interdigitated with the creosote bush association, with creosote bush occupying coarser-grained, better-drained soils more common higher on the valley slopes. Similarly, desert saltbush intermixed with or gave way to narrow-leaf saltbush and/or seepweed where soil included an argillic horizon beginning within the top 0.5 m of the surface or contained appreciable concentrations of salts. They described the distribution of mesquite thickets along washes and, especially, in "adobe" soils of lowlands, where soils of high water-holding capacity retained moisture throughout hot months. They also noted the scattered growth of individual mesquite trees or shrubs in the lighter soils of the desert saltbush associations. Another saltbush species, *Atriplex canescens*, was noted to occupy the infrequent sandy ridges of the area, where it sometimes coexisted with burroweed (*Isocoma acradenia*)

Figure 7.3 (A, top) Native desert remnant dominated by desert saltbush, with individuals of wolfberry, barrel cactus, and mesquite. The Sacaton Mountains are in the background. (B, bottom) Abandoned farmland dominated by burroweed in the same region.

or more rarely with desert broom (*Baccharis sarothroides*). Burroweed and desert broom were described as "less frequent or rare" in the desert saltbush association (fig. 7.3B).

The most important ephemeral species of both the creosote bush and desert saltbush associations was Indian wheat (*Plantago patagonica*). Annual fescue (*Vulpia octoflora*) and goldfields (*Lasthenia coulteri*) were the second most important ephemerals in creosote bush and desert saltbush associations, respectively.

We do not know how accurately species composition reported by Shantz and Piemeisel (1924) might have reflected that of the plant communities that existed prior to nineteenth century grazing. Pioneers in the 1880s reported cranesbill (*Erodium* sp.; species not clear) and grasses "as high as a horse's belly" (R. Edmond 1994, personal communication).

There is little evidence that the valley shares the history of widespread conversion of grassland to shrubland that was so well documented in southeastern Arizona (Hastings and Turner 1965). First, there are no comparable accounts of rapid vegetation change near the turn of the century. Second, the climate is less conducive to grass: the elevation is lower, temperatures are hotter, and rainfall is 200 to 250 mm (as opposed to 300 to 350 mm in southeastern Arizona).

Nevertheless, two kinds of terrain that are significantly wetter than most of the valley due to landscape position and soils have been reported as grassland. The Sasco Flats, a 15 to 20 km stretch where the Robles and Santa Cruz Washes converge in the extreme southeast portion of the valley (fig. 7.2), supported dense stands of Johnsongrass (*Sorghum halepense*) intermixed with velvet mesquite up through the 1930s (Smith 1940). Smith indicated that Johnsongrass, a European weed, became abundant after the flood of 1916; originally this area may have contained giant sacaton (*Sporobolis airoides* var. *wrightii*; D. James 1992, personal communication). A second type of grassland still persists in small dune fields in the western part of the valley where sparse populations of big galleta are growing. Topographic maps show old ranch sites strategically located next to at least two of these grassland areas.

Early European Agriculture, 1885–1935

Conversion of desert for cultivation was uncommon in the early 1900s due to lack of surface water. The Pinal County soils map of Eckman et al.

(1920; fieldwork completed before 1917) reports that "only small patches are cultivated in the desert section," meaning the Santa Cruz Valley.

The first two (European) modifications of the valley's hydrology appear to have been the Florence–Casa Grande canals (completed in 1886) and Picacho Reservoir (completed in 1908) (fig. 7.2, About 1930). According to Forbes (1911) and Eckman et al. (1920), they were constructed in 1885–1886 for irrigation purposes. However, it is possible that both the canals and the reservoir were there before 1886. Picacho Reservoir is an ideal location for passive storage of seasonal floodwaters from the Picacho Mountains (USGS 7.5' topographic series, Picacho Quadrangle), which may have been exploited and modified by Native American farmers. An elderly resident of the valley told us that he once knew a person who began cattle ranching after the Civil War, 10 miles west of Picacho Reservoir (J. A. Roberts 1994, personal communication). This suggests the availability of water before 1885. Roberts believed that before Picacho Reservoir was linked to the Gila River in 1886, the reservoir functioned as a local catch basin.

The Florence–Casa Grande canal (actually two parallel canals; fig. 7.2) originates to the northeast from the Gila River, goes around Picacho Reservoir, and then snakes west toward the town of Casa Grande. Its complex curves along this reach (as opposed to the straight 1911 canal dug further south by the Santa Cruz Canal Company) suggest engineers who took full advantage of local microtopography and soils to make the job easier or more efficient. The canals would have originally cut through thick layers of rocklike caliche, also calling for a careful choice of route (Hall 1991). Would newcomers to the area, virtually ignorant of the nature of this desert floodplain (as Dobyns [1981] has argued), have been able to ascertain this route and quickly build the canal? Given the sophisticated irrigation systems of Native American desert farmers and their historical importance in the area (Crown and Judge 1991), there is no reason to assume that the Florence–Casa Grande Canal Company designed these canals. Like most of the modern canals around Phoenix, the Florence–Casa Grande Canal may have been built on Hohokam ruins.

The Florence–Casa Grande feeder canals served farms to their north (fig. 7.2) only sporadically: when the Gila was flooding or when Picacho Reservoir received enough runoff to overflow. The Coolidge Dam on the Gila River, completed in 1929, was supposed to have provided settlers with a much more dependable water supply, but water yields were poorer than expected (Smith 1940). Aerial photos from 1936 show fields limited to a 2 to 5 km band around the canals, interspersed with many areas of uncleared desert.

In 1908 an ambitious group of investors called the Santa Cruz Reservoir Company began an irrigation enterprise that would profoundly affect the future of virtually the entire lower Santa Cruz Valley. The company, partly financed by Colonel William Greene, sought to draw the waters of the Santa Cruz River out of its northern course and bring them 25 km west, to a reservoir constructed against the slopes of the Sawtooth Mountains, on the south end of the valley (fig. 7.2). The 2440 m earthen dike, finished in 1911, was designed to hold 30 000 acre-feet of water in "Greene's Reservoir" and irrigate 4050 ha of cropland, but the company failed and the dike was quickly washed out by the first few runoff events (F. Barrios, unpublished manuscript). The canal, merely a 2.5 m wide ditch crossing a remote and unfarmed part of the valley, remained functional.

Almost immediately, Greene's Canal developed a headcut, gradually carving a deeper, wider channel that moved upstream (east) until it reached the Santa Cruz channel. In 1931, 90% of the Santa Cruz River flowed in its original channel (fig. 7.2), spreading out over the Santa Cruz Flats, and passing through Eloy (J. A. Roberts 1992, personal communication; Smith 1940). Significant floodwaters were still reaching Eloy in 1939. After 1939 Greene's Canal redirected the waters of the Santa Cruz River into the western part of the valley (F. M. Barrios, unpublished manuscript; J. A. Roberts 1992, personal communication). The Eloy region, once the beneficiary of floodwaters and rich black silts used to grow vegetables, was robbed.

This new channel also affected the Sasco Flat upstream, an area of 1300 to 1600 ha of velvet mesquite trees and Johnsongrass between the Santa Cruz River and Robles Wash. Smith (1940) warned that:

> The overflow into the [Greene Canal] channel has started many new gullies and some of them have grown backward nearly a mile. If the cutting continues it may reach through the flat to the main channels of the Robles Wash and the Santa Cruz River. . . . The intricate control problem should be resolved as soon as possible. If the water spreading is continued, the utilization of the flat for growing feed as at present can be continued, and the spreading undoubtedly slows down the floods and flattens the flood crests materially, a result of much benefit, since farther downstream the new land use has left no definite place for the floodwaters to go.

Smith's predictions were correct (personal observations).

Pumplands Developed, 1935–1954

By the late 1930s, pumping technology had improved and electric and gas service became widely available (Shapiro 1989). For the first time, the great, dry floodplain south of the Florence–Casa Grande Canal was open for agriculture. Unlike surface water, which required a legal water right plus membership in a cooperative irrigation association or company, groundwater was unregulated, and its exploitation was individualistic. This must have been a great attraction to investors. Smith's account (1940) of the development of the Eloy district illustrates the character and speed of changes throughout the valley:

> The first drilled well in the region of Eloy was drilled in 1916 ... but was only 110 feet in depth. The first deep well was drilled 2 years later for the promoters of Cotton City (now Eloy) and was 320 feet in depth.
>
> About 1924 it was discovered that the overflow and other recently deposited alluvial lands were adapted to winter vegetables, the "black lands." By 1930 about a dozen wells of large yields were in operation in the Eloy district. The pumping lift was over 100 feet but the high value of the lettuce crop justified the high cost of water. In 1934 the growing of lettuce was abandoned due to the spread of a destructive fungus in the district. Peas, asparagus, carrots, and some other crops were continued. The total area cultivated in 1930, however, did not exceed 4,000 acres.
>
> The big development occurred in 1936 and 1937 and was the result of four factors. The price of cotton advanced in 1935 and 1936; notable improvements had been made in the design and efficiency of deep-well turbine pumps; pumping plants were offered on a credit basis; and electric power rates were reduced.
>
> About 4,000 acres of newly cleared land were planted in 1936 and 13,500 acres additional in 1937. Forty-four new wells were drilled, all of them 20-inch diameter and nearly all of them to depths of 400 to 600 feet. ... By 1939 there were 90 wells in operation.

The Arizona Crop and Livestock Association's estimates of area harvested in Pinal County show a sharp increase beginning in the mid-1930s, accompanied by an equally large increase in water pumped (fig. 7.4). The yearly expansion of crops harvested in Pinal County ceased in 1953, when the end of the Korean War brought lower cotton prices, and the federal government reimposed acreage limits (Shapiro 1989). Barr (1955), summing up the previous seven years of agricultural development for the Agricultural Experiment Station, reports that about 78 900 ha of desert

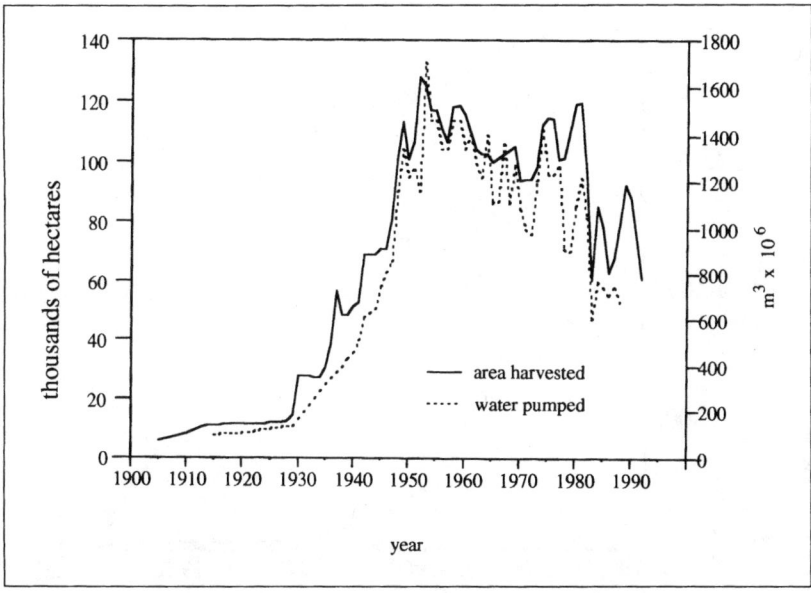

Figure 7.4 The change in agricultural activity in Pinal County, 1905–1992, in terms of area harvested and water pumped. Data on area harvested were obtained for the period 1905–1987 from the Arizona Agricultural Statistics Service (various years), the Arizona Crop and Livestock Reporting Service (1966, 1981), and from annual reports of the Arizona Agricultural Experiment Station from 1949–1954 (Barr 1950, 1951, 1952, 1955). Pumpage data are from Hammett (1992).

were cleared in Pinal County between 1947 and 1954. This is a rate of 30 ha per day, or 6% of the total arable land each year (fig. 7.4).

Agricultural Extensification, 1935–Present

In 1940 the University of Arizona hydrologist G.E.P. Smith calculated that by 1936, pumping operations near Eloy were twice the sustainable water yield, and that by 1937 they were three times the rate of groundwater recharge. The regions near Eloy with groundwater less than 18 to 27 m from the surface in 1917 were designated a critical groundwater area in 1949. By 1977, the groundwater level had declined by 90 to 150 m (Smith 1940; USDA 1977).

The clearest biological signal of groundwater decline has been the

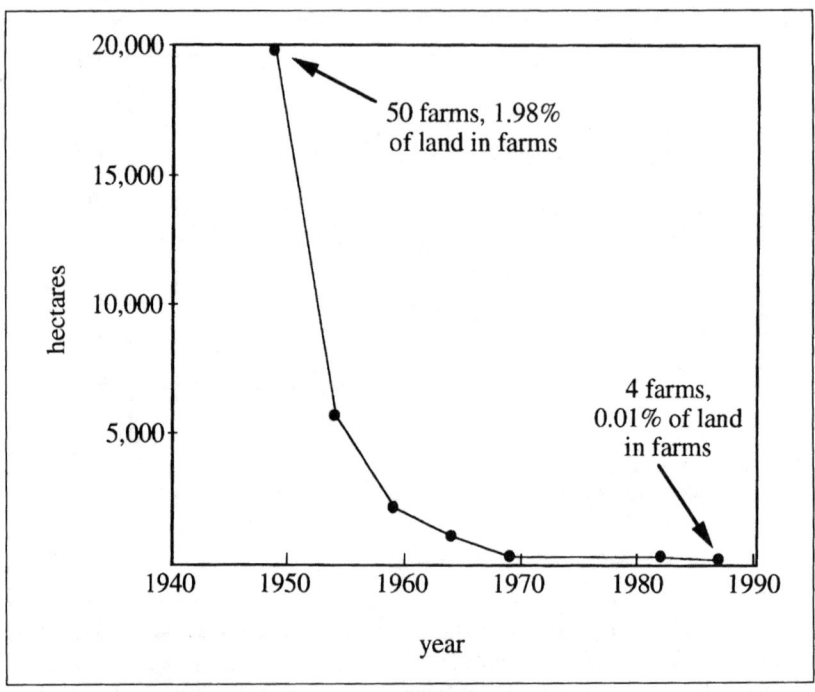

Figure 7.5 Loss of woodland in Pinal county, from 1949 to 1987 (U.S. Bureau of the Census, various years). "Woodland" is defined as "all wood lots and timber tracts and cutover land with young trees which have or will have value as wood or timber." Since the only commercially valuable timber existed on the valley floors or at high elevations outside the county, "woodland" was a good indicator of the extent of velvet mesquite (*Prosopis velutina*) forests. In 1949 there were a total of 50 farms reporting some woodland, and the area constituted almost 2% of all land in farms in the county. In 1987, only 4 farms reported any woodland, and the area accounted for only 0.01% of land in farms in the county.

death of large mesquites, a phreatophyte that once studded the valley and formed extensive forests in some areas. We have observed large (up to 75 cm diameter), many-branched mesquite stumps throughout the valley near former waterways. As late as 1949, farmers reported owning 20 000 ha of commercially valuable "woodland" composed mostly of mesquite, but in 1987 only four farms reported owning a total of 50 ha of mesquite woodland (fig. 7.5). Judd (1971) documented mesquite death along the Gila River after upstream channelization and groundwater pumping caused the water table to drop from 13 m below the surface in 1923, to 30 m in 1952, to 46 m in 1960. Prior to extensive groundwater pumping, the roots of young mesquite trees were able to reach the groundwater table. Today these conditions no longer exist, and while small ($<$ 15 cm diameter) mesquite trees are frequent throughout the valley, they are dependent on surface moisture, instead of becoming large phreatophytic trees.

At the same time that woodlands were being lost, other forms of partially wild land were being converted to economic purposes. "Wasteland" on farms was defined by the U.S. Agricultural Census as houses, lots, ponds, roads, and undefined "waste" land. This did not include cropland, idle cropland, woodland, pasture, or rangeland. During the period of rapid clearing, the amount of wasteland reported by farmers in Pinal County decreased 47%, from 45 562 ha in 1950 to 24 214 ha in 1954. After new clearing ceased, the area reported as wasteland declined another 49%, to 12 247 ha. Since it is not likely that the number or size of houses, ponds, roads, and other infrastructure had declined greatly, these data suggest that since 1954, scraps of land not used for farming, such as those along unchannelized washes, had been converted to cropland or some other intensive land use.

In addition to groundwater decline, clearing resulted in major changes in the valley's topography. In order to irrigate efficiently, farmers found it necessary to level the fields to an even, gently sloping grade. While some fields cleared and abandoned before 1949 contain significant microtopography (personal observations), fields farmed since then lacked significant microtopography other than that produced by the last tillage equipment. As a consequence of leveling, the original landscape pattern of dendritic ephemeral washes was converted to a grid of ditches. Figure 7.6 shows a portion of the eastern edge of the valley in which leveled and unleveled land lie adjacent to one another. Farmed land is easily identified by the

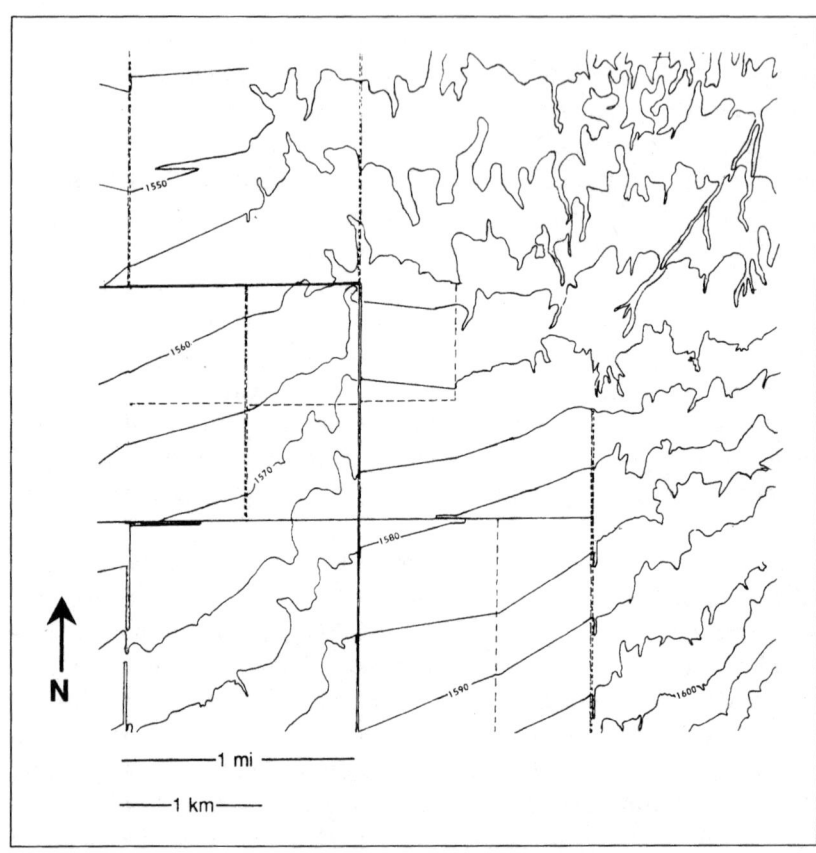

Figure 7.6 Land leveling for irrigation is evident in 5 ft (1.5 m) contour intervals (USGS 7.5″ topographic map, "Picacho Reservoir," 1981). The large-scale modification of topography for irrigation has serious implications for any future restoration.

Figure 7.7 Aerial photo of section just north of the Casa Grande Canal, near present-day 11-Mile-Corner, taken in 1936. (a, lower right) Minor wash crossing fields and native vegetation; (b, center left) Modification of existing northwest-trending wash to irrigate crops. Note dense vegetation below (north of) the fields due to excess irrigation runoff.

straight contour lines. In addition to channeling water movement, the grid of irrigation ditches, elevated roads, and leveled fields changed local runoff, creating ponding where it never existed before.

Channelization and control of washes did not begin immediately after clearing for agriculture. Aerial photos from 1936 show washes running through many cultivated fields (fig. 7.7). Some fields used and abandoned between 1936 and 1949 were arranged specifically to take advantage of floodwaters. The natural waterways were turned into northwest-trending irrigation ditches, from which waters were turned out downhill on both sides of the ditch (Jackson et al. 1991). Tailwater from this practice

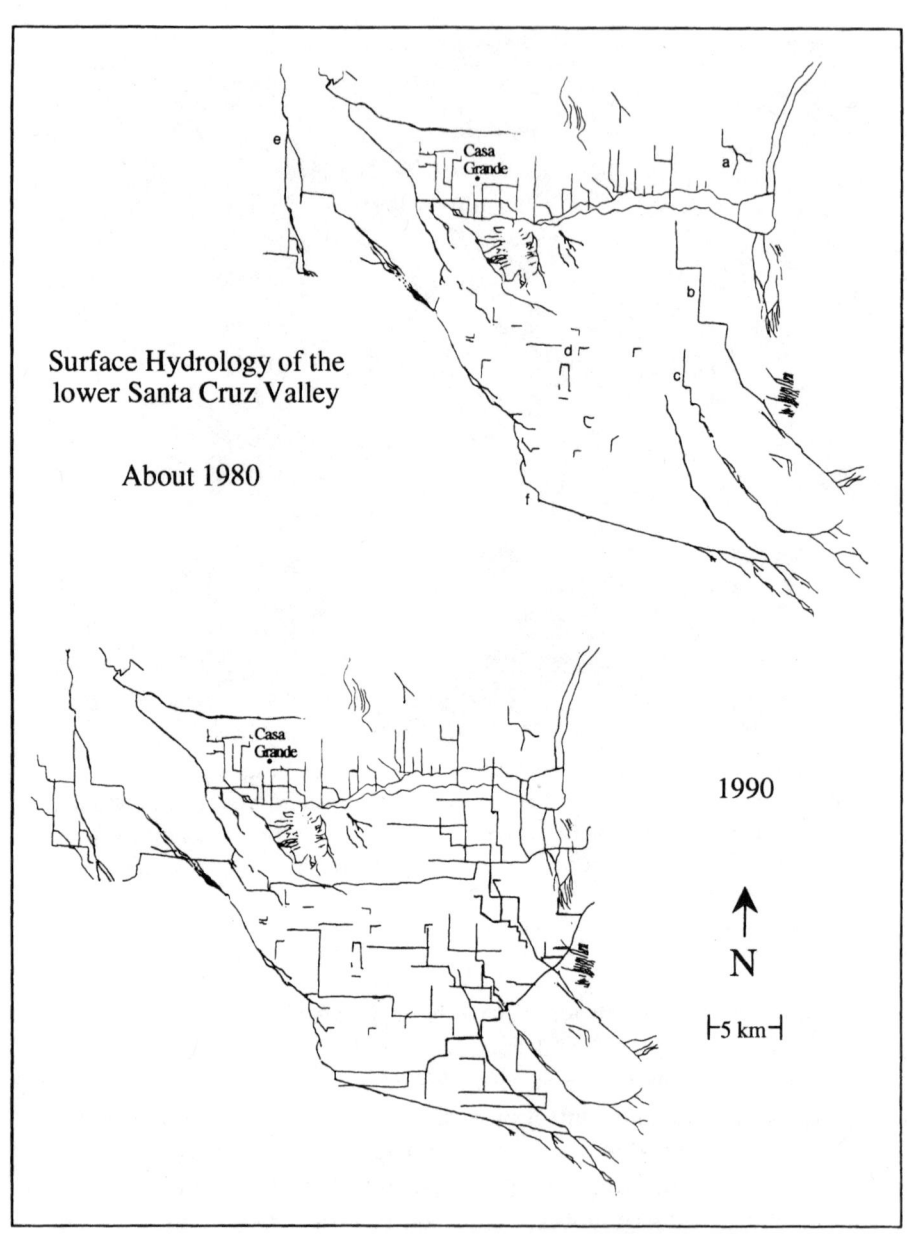

Figure 7.8 *(opposite page)* Surface hydrology of the lower Santa Cruz Valley, including all known natural waterways, constructed waterways, reservoirs, and canals in 1980 and 1990.
About 1980 Channelized washes were determined from 1963–1981 USGS topographic maps (more precise dates for channelization of particular streams are unknown). a. Remaining fragment of the northern McClellan's Wash; the rest has been leveled and farmed. b. Channelized southern McClellan's Wash. c. Channelized Red Rock Wash. d. Elimination of multiple drainageways of the Santa Cruz River flow south of Casa Grande Mountain. e. Channelized convergence of the Greene's and Santa Rosa Washes. f. Former location of Greene's Reservoir, now used for fields.
1990 Natural and constructed waterways of the lower Santa Cruz Valley, 1990. Central Arizona Project canals, carrying water from the Colorado River, run west and south from the main aqueduct on the east side of the valley (not shown).

gathered at the low end of the field and was released back onto the desert, enhancing nearby vegetation (fig. 7.7).

By the 1940s the practice of using natural washes had ceased, possibly due to the unreliable and diminished flow from the Santa Cruz River once it was captured by Greene's Canal. Between 1940 and 1983, all major washes were modified so that they would not flood fields (fig. 7.8). This became particularly important in the 1970s, as farmers switched from unlined dirt irrigation ditches to the more expensive concrete ditches. While the new concrete ditches conserved water, at a cost of $18/m to build, they were too expensive to risk being destroyed by floods. Wash channelization became even more important. The dense flush of tailwater-irrigated native vegetation visible on the low edges of fields in 1936 and 1954 aerial photos is not evident in later photos.

The most striking change was the disappearance of the northern McClellan Wash, which ran northwest from Picacho Reservoir into the Gila River (fig. 7.8). The wash was treated as a major landscape feature in the Eckman et al. (1920) soils map. Smith (1940) shows it as a wash that regularly received floodwaters and nearly formed a channel at its northern extreme; this wash was subsequently depicted in a 1963 University of Arizona map as a canal, but was absent from the 1969 USGS 7.5' topographic series. The southern McClellan Wash, Red Rock Wash, and the Santa Rosa Wash at its convergence with Greene's Wash were straightened and channelized. Other, smaller systems now follow county road ditches (fig. 7.8).

The 1992–1993 completion of the Central Arizona Project, a network of canals that link virtually all of the Santa Cruz Valley to the Colorado River (fig. 7.8), created another series of physical barriers to the movement of animals and floodwaters. The canals range from the size of a large irrigation ditch (2 m), to a major aqueduct with a 7 m wide canal and 30 m of cleared and graded levies. In low areas near former wash systems, tall (15 m) dikes protect the canal from overland flow. Local runoff water backs up behind the dikes instead of following its historic northwest course.

Abandoned and Idle Cropland, 1954–1987

While agricultural practices demanded more and more extensive modification of the lowland desert ecosystem, farms were failing. Between 1952 and the mid-1980s, the area harvested fluctuated within 70% to 94% of the record high (fig. 7.4). By 1987, total cropland as reported by the U.S. Agricultural Census was down 26% from 1954 (table 7.1). The minimum amount of farmland either abandoned completely or transformed to nonagricultural uses between 1954 and 1987 was 47 779 hectares. This number is conservative; it does not take into account the land abandoned before 1954 (probably less than 10% of the total, Karpiscak 1980) or any new cropland cleared after 1954.

The rate of abandonment in Pinal County was disproportionately greater than that of Arizona as a whole (table 7.1). Pinal County accounted for 28% of the state's agricultural land in 1954, but only 20% in 1987 (U.S. Agricultural Census 1956, 1989). Not only was cropland lost from the survey, but more cropland in the survey was unused. In 1987 "idle" cropland accounted for 48 575 ha, or 35% of all cropland, compared with 22 487 ha (12%) in the early 1950s (table 7.1). According to the Arizona Department of Water Resources (R. Edmond 1993, personal communication), farmers now set aside a minimum of 20% of active farmland in any given year. While most of this idle land is in fact cultivated every few years, a portion of it is probably no longer in use at all, although it could be brought back into production given the right economic conditions.

These results differ substantially from the estimates of Cox et al. (1983), who used the decennial U.S. Census data in a similar manner to estimate abandoned farmland. Cox and colleagues reported a maximum of 352 768 ha in production in 1950, and in 1980 a minimum of 89 684 ha, for a loss of 263 084 ha in Pinal County. The 1950 figure is much greater than the maximum number of arable hectares in the valley (Van Cleve Associ-

Table 7.1 U.S. agricultural census data for hectares of "total" cropland and "other" (1954) or "idle" (1987) cropland in Arizona and Pinal County.

	1954	1987	Change (%)	Total Idle (1987)
Arizona				
Total cropland (ha)	653 524	588 366	65 159 (−10%)	
Other/idle cropland (ha)	83 301	135 070	51 770 (+62%)	116 929
(% idle)	(12.7%)	(23.0%)		
Pinal County				
Total cropland (ha)	185 012	137 233	47 779 (−26%)	
Other/idle cropland (ha)	22 487	48 575	26 088 (+116%)	73 867
(% idle)	(12.1%)	(35.4%)		

Source: U.S. Bureau of the Census (1956, 1989).
Note: The difference between total cropland in 1957 and 1983 is a conservative estimate of total abandoned farmland. The "total cropland" census category in both surveys is the sum of land in cash or cover crops, orchards, and nurseries; pasture (as distinguished from unirrigated rangeland); land on which all crops have failed; commodity set-aside land; cultivated fallow land; and "idle" land. "Idle" land refers to other cleared land that was not currently in use.

"Other" cropland in 1954 and "idle" cropland in 1987 were estimates of the amount of total cropland not in production that year. Thus, the difference between "other" cropland in 1954 and "idle" cropland in 1987 is a conservative estimate of the change in idle land within farms. The sum of the change in total cropland plus the change in other/idle cropland is an estimate of the total amount of cleared land not in production.

ates 1963; Turner 1974a, 1974b), the maximum number of harvested hectares according to the Arizona Crop and Livestock Reporting Service, and the maximum area of cropland, including idle land, based on the agricultural census (185 012 ha in 1954; U. S. Bureau of the Census 1956.)

Distribution of Remnant Desert Vegetation and Abandoned Land

We used aerial photo interpretation to independently assess the amount of abandoned farmland for a portion of the valley and to learn the geographic distribution of abandoned and still-cultivated cropland. The eastern half (102 400 ha) of the valley was included in the study. Six land-use classes were recognizable from aerial photos and topographic maps: native, cultivated, abandoned, developed, cultivated-developed, and native-developed (table 7.2). This measurement, based on 1983 aerial photos for the eastern portion of the valley, was similar to the U.S. Agricultural Census results (1956, 1989). Of 72 158 ha of land in the study area that

Table 7.2 Land-use classes recognized from 1983 aerial photos and topographic maps of Pinal County, as a percentage of the total area measured.

Land Use	Description	% Mapped Area	% Core Area
Native	Never cleared for agriculture. Rough textured vegetation, no sign of ditches or pumps, variable topography	23.6	8.4
Cultivated	Currently in cultivation or fallow and recently tilled. Dense, even vegetation or smooth bare soil surface, actively maintained irrigation ditches and roads, and leveled topography	51.0	61.0
Abandoned	Once cleared and cultivated but no longer actively maintained as a field. Uneven and sparse distribution of plants, ranging from completely herbaceous to some small trees; presence of ditches and furrows; and leveled topography	16.9	21.6
Developed	Completely covered with buildings, roads, canals, mining operations, or other major human-made features	3.0	4.5
Cultivated-Developed	Farmed, and then converted to some other use. This primarily consisted of housing developments that contained significant areas of old field vegetation	2.5	3.0
Native-Developed	Native desert converted to another use. This primarily consisted of housing developments with scattered native vegetation and no sign of previous cultivation	3.0	1.5

Note: Percentages are of the entire mapped region (102 400 ha) and the core agricultural area in the valley's center (44 993 ha).

were once cultivated, 17 318 ha (24%) were abandoned, and 2598 ha (3.6%) were "developed" into housing districts, many of which failed to attract residents. This result (27.6%) is in close agreement with the U.S. Agricultural Census estimate (1956, 1989) of 26% loss in total farmland from 1954–1987.

Desert Remnants

Figure 7.9 shows the distribution of desert remnants in the eastern half of the Santa Cruz Valley. While the reason for each desert remnant's preservation is unique, a few common patterns are evident. Land on the edges of the valley, where soils are thinner and consist of coarser materials and depth to groundwater is greatest, were not cleared. The large size of remnants in Figure 7.9 is only indicative of the arbitrary borders of the mapped area. In some cases, the border between suitable and unsuitable land is distinct, such as where the Florence–Casa Grande Canal defines the valley's northeastern edge, and Greene's Canal defines the southern edge. In other cases, the distinction is less clear, such as in the northwest corner where native and cultivated land are intermixed. In the interior of the valley, land along large washes may have been spared because it was often subject to flooding. Railroad rights-of-way, acquired before most desert clearing, harbor desert habitat in several locations throughout the valley, just as they preserve tall grass prairie in the Midwest. A string of mostly small remnants marks the location of the Florence–Casa Grande Canal west of the Picacho Reservoir. The canal created odd field shapes, which may have been difficult to work. Although soil limitations may have preserved desert remnants in many areas, only two remnant areas are clearly a matter of unsuitable soils. First, sandy ridges in the southwest corner of the valley would not hold sufficient irrigation water to farm. Second, an area near the study's center has abundant "slick spots" where virtually nothing grows (as described by Poulson et al. 1941). These are indicative of highly sodic soils (fig. 7.9).

The size and distance between habitat patches in agricultural landscapes can have important implications for wildlife use and persistence of species (Bunce and Howard 1990; Vos and Opdam 1993). In order to learn more about the islands of native habitat that were left in the center of the valley, the three 7.5′ quadrangles running north-south down the center of the map were selected for further study (fig. 7.10). The size distribution of native patches was skewed, with 54 of the 77 patches under 100 ha and over half of those under 20 ha. These islands followed two patterns of dispersion. Nine patches adjoined a neighboring patch at one corner and thus had nearest-neighbor distances of zero. The rest of the native patches were larger and more isolated; the mean nearest neighbor distance among nonadjacent habitat islands was 524 m ($\pm$ 1 SE, 417 to 660 m).

Landscape Pattern of Abandoned Fields

Figure 7.11 shows land use, including cultivated, abandoned, and developed land in 1983. The greatest concentration of abandoned farmland is in the southern portion of the county, where development of farmland was most recent and was entirely dependent on groundwater. However, the detailed pattern of abandonment does not closely follow the local extent

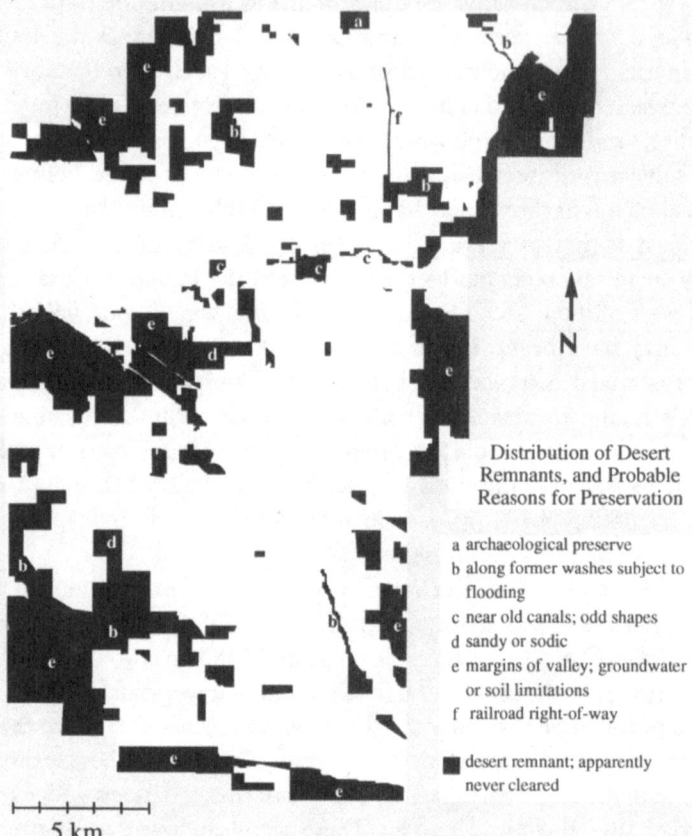

Figure 7.9 *(opposite page)* Distribution of remnant desert vegetation in the eastern half of the lower Santa Cruz Valley, showing probable reason for preservation of selected remnants. White areas on the map consist of all other land-use categories. Land use was determined section by section (one square mile or 259 ha on 1983 aerial photos [1:24000 scale]) and recorded on 1:24000 U.S. Geological Survey topographic maps. To correct for areal distortions caused by parallax, section estimates were proportionally adjusted to sum to 259 ha. This method was preferable to using rectified photos because their resolution was too poor to judge land use. Data were digitized using AutoCAD 12.0 (Autodesk Inc. 1992). Initial drawings were checked against 1991 water rights maps supplied by the Arizona Department of Water Resources, Phoenix.

of groundwater decline. Areas with high (60 to 90 m) groundwater decline in 1970 were no more likely to be abandoned than those areas with moderate (40 to 60 m) decline (U.S. Department of Agriculture 1977; Hammet 1992). This reflects the dominant role of chance events in a farm's history, including the financial vicissitudes of individual owners.

We hypothesized that fields near native remnants — presumably unsuitable for farming — would be more likely to be abandoned, but this was not true. Abandoned fields were negatively associated with desert remnants. The mean distance between random points located in cultivated land ($N = 89$) to the nearest native desert patch was 860 m, while the mean distance from abandoned land ($N = 46$) to native was 1291 m ($T = 3.008$, $p = .003$; distances square-root transformed for analysis).

These distances are interesting when compared to the ability of desert species to disperse into old fields following abandonment. While wind-dispersed shrubs and trees are abundant in virtually all old fields (depending on the region, either burroweed, desert broom, brittlebush [*Encelia farinosa*], or tamarisk [*Tamarix pentandra*]), the presence or absence of heavy-seeded perennial shrubs and trees seems to be correlated with the position of the old field in the landscape. Fields next to desert remnants generally have been colonized by at least a few species from the neighboring desert (although this might mean one or two individuals each), while isolated fields are more species-poor.

Creosote bush, a former codominant shrub, has a heavy seed moved by small mammals (Valentine and Gerard 1968). In a study of 16 old fields bordered by desert remnants containing creosote bush, we found that colonization occurs from the field edges and follows an exponential decay

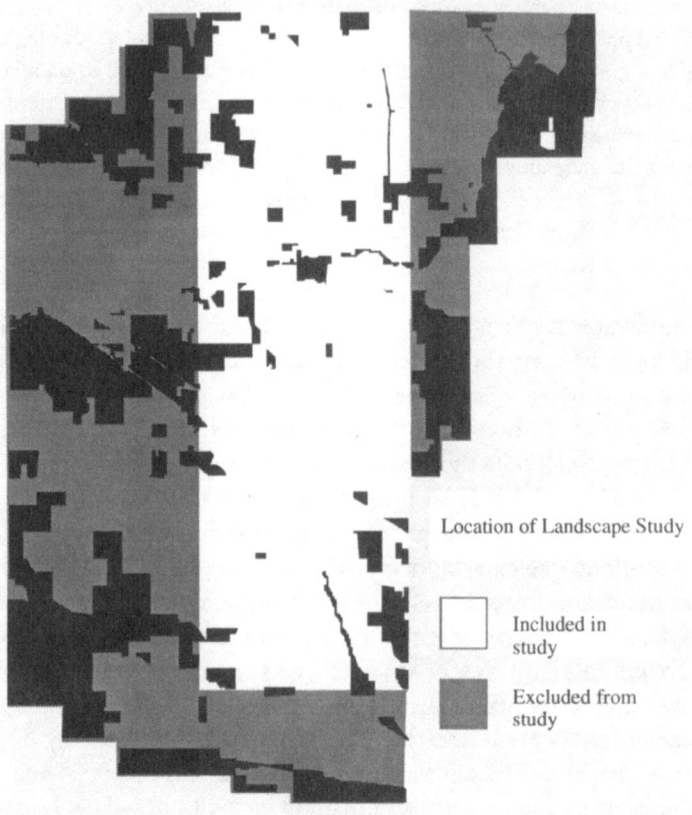

Figure 7.10 To describe spatial patterns of native and abandoned land in completely agricultural areas, an 11.5 × 41.1 km band was selected in the valley center, comprising three USGS 7.5′ quadrangles covering 44 993 ha. The perimeter and area of each native stand was measured by AutoCAD 12.0 (Autodesk Inc. 1992). Grid lines were established with a ground distance of 100 m, and random sets of coordinates generated to measure the distance from open (cultivated or abandoned) points to the nearest desert remnant.

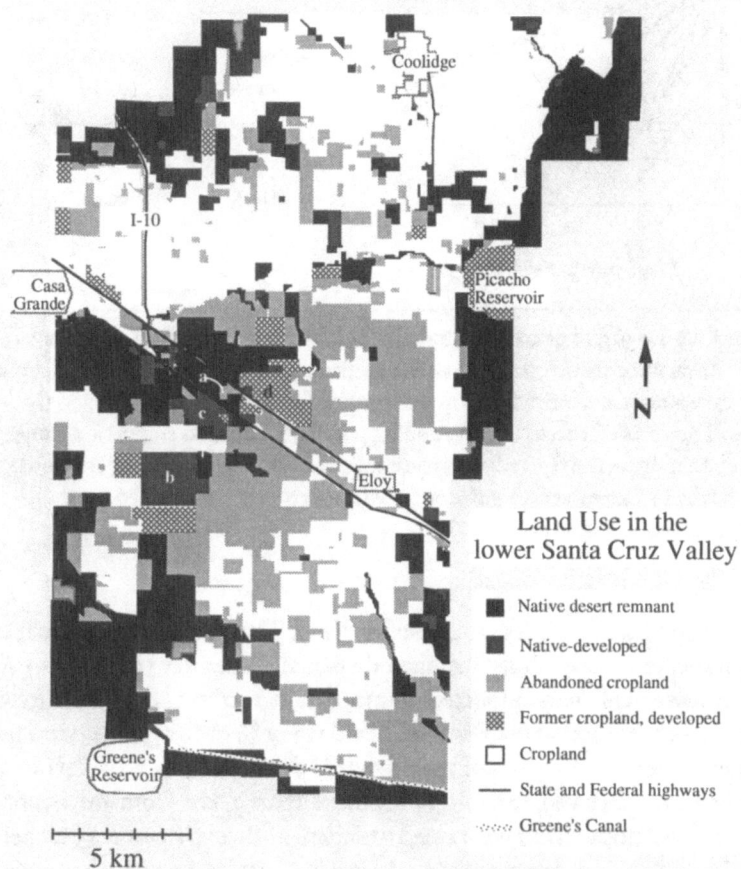

Figure 7.11 Land use in 1983, based on aerial photo interpretation. See table 7.2 for percent of land in each land use category. (a) Location of old field, surrounded by native desert, which has substantially recovered vegetative cover and composition since farming was abandoned. (b) "Arizona City," showing contiguous native-developed and developed former cropland. (c) Second area of native-developed land cleared for houses. (d) "Toltec" development; composed of former cropland.

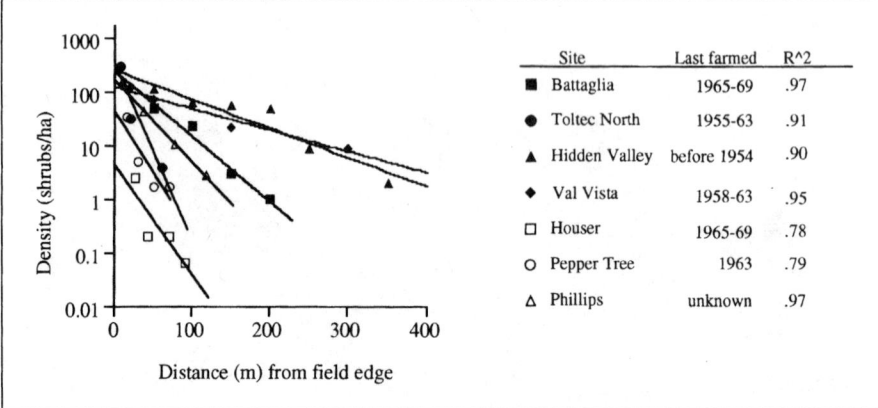

Figure 7.12 Dispersal of creosote bush into old fields as a function of field age. Six old fields were chosen that had native creosote bush vegetation on at least one edge. Time since last cultivation was determined from repeat historical aerial photos. Transects of 200 to 800 m running parallel to the field edge at a distance of 5 to 360 m from the field edge were established, and the number of creosote bush shrubs in a 20 m wide band along the transect were counted. Note results are presented on a log scale.

function ($p < .01$ in all regressions; fig. 7.12). The shapes of the distributions indicate mostly short distance dispersal. After 30 years, creosote bush in these fields had colonized no more than 100 to 360 m. Density of shrubs in the interior of a 30-year-old field was 1% of their density on the field edges and 0.1% or less of their original density (fig. 7.12). Nevertheless, we have observed two old fields more than 1 km from the nearest creosote bush population with small numbers of shrubs (in one 65 ha field last farmed in 1949, one plant; in another field of similar size abandoned in the 1960s, five plants).

With the average point in an abandoned field averaging nearly 1300 m from the nearest desert remnant in the central part of the valley, it may be over 100 years before many areas are recolonized. Roads, old irrigation ditches, and other obstructions to seed movement may further slow the recolonization process. However, rates of colonization could be greatly hastened by unusual events, such as the movement of seeds by dust devils, and the creation of "safe sites" for seed germination and seedling establishment.

Karpiscak (1980) studied rates of vegetation recovery of old fields in the Avra and Santa Cruz Valleys. He found that reestablishment of the formerly dominant saltbush (*Atriplex* spp.) or creosote bush was highly variable. In some fields, virtually no plant cover existed 25 years following abandonment. In a small number of fields farmed over 30 years ago, fields had nearly the same species composition as the surrounding desert habitat. In revisiting the fields studied by Karpiscak that had recovered well, we observed that each was in an unusual landscape position — next to desert vegetation. The best-recovered old field that we examined (based on a qualitative comparison of species lists and plant density of the old field and neighboring desert vegetation) was surrounded by native desert. We found this to be a highly unusual situation (fig. 7.11).

Two other land-use categories, native-developed and cultivated-developed, made up the balance of land not devoted to towns, highways, or gravel mines (fig. 7.11). The unincorporated settlement of Arizona City includes 1100 ha of former desert and 840 ha of former cropland, all with paved roads arranged suburb style. According to U. S. Geological Survey 7′30″ topographic maps, only one section (259 ha) has densely spaced, occupied houses; less than 5% of the house lots in the remaining area contain buildings. Native vegetation is sparse in these areas because it was bladed for construction of an artificial lake, roads, and homesites. A 555 ha area south of I-10 highway is also in the native-developed category. Although paved roads were built through more than half of it, large, mature native vegetation is present. County tax rolls showed that in 1991, all lots were owned by individual investors from out of state who purchased them beginning in the 1950s. Since it is unlikely they will ever move to this hot, waterless portion of desert, the individual ownership of small plots tends to protect the area from future, more destructive development. A third area of former cropland, known as "Toltec," consists of a few houses clustered within 970 ha of "developed" but unoccupied land, complete with streets and street signs. All but 90 ha of this is abandoned farmland.

Existing Vegetation in Native Remnants and Abandoned Farmland

Perennial species composition in the native remnants of desert shrubland is similar today to that described by Shantz and Piemeisel (1924), except for the previously discussed elimination of most large and medium-sized mesquite trees (fig. 7.3A). Burroweed and desert broom are infrequent, occurring only at wash edges where, perhaps, perturbation from flooding gives these weedy species a competitive advantage.

In native remnants where narrow-leaf saltbush grows, data from soil studies indicate that this species grows in soils with calcium carbonate at the surface and dominates often in heavy clay soils or in soils that have a reddish argillic (clay) horizon within the top 22 cm. Seepweed typically inhabits areas near or in depressions where sodium deposits sit above an argillic horizon and often are visible at the surface. Water typically ponds above the rather impervious clay boundary of such areas during rainy periods. Narrow-leaf saltbush and seepweed in particular seem well adapted to conditions intolerable to most other perennial species. On abandoned farmland, the calcic and argillic horizons (which take several centuries to form in undisturbed soils) have been broken up by tilling, and salts have been leached by deep irrigation. The special adaptations of these plants should no longer give them a competitive advantage in tilled soils, and we speculate that, if replanted on abandoned farmland as part of a restoration effort, they would eventually be outcompeted by other species. Indeed, we have never observed narrow-leaf saltbush or seepweed in abandoned farm fields, and the narrow-leaf saltbush we grew in our test plots, initially with some success, experienced high mortality within two years of establishment.

Desert saltbush grows on a variety of soils ranging from loamy sand to silty or sandy clay, but is excluded from soils with calcic horizons (deposits of calcium carbonate) within 8 cm of the surface. The outlook for desert saltbush revegetation is somewhat better, since this shrub grows on a variety of soil types and structures that are not so inconsistent with previous tillage. It appears to be thriving, moreover, in our experimental plantings after seven years (Jackson et al. 1991).

Various species of cacti are dominant or codominant within many remnant saltbush communities today. These were not described in the Shantz and Piemeisel study (1924), which considered cacti only in the context of the "giant cactus and paloverde" association typically found at somewhat higher elevations. The sizes (height and/or circumference) of individual cacti and of many of their skeletal remains indicate that these communities have existed for at least most of this century.

Abandoned farmland has no such cactus communities. Many old fields are dominated by one of two native shrub species previously infrequent or rare: burroweed and desert broom. The original shrubby natives are either missing entirely or are relatively infrequent. For instance, in one 65 ha field abandoned in 1949, a thorough census found one creosote bush amidst a monoculture of burroweed.

The composition of winter ephemeral species on native remnants has changed significantly, based on studies undertaken primarily in 1995 following above-average rainfall. The second most dominant annual in creosote bush, and a frequent annual in desert saltbush, had been the fescue. However, we have not collected it nor have we found two other species said to be common or frequent previously—*Poa bigelovii* and *Malvastrum exile*. Purple mat (*Nama demissum*), also abundant or frequent in the 1924 study, was rare in our collections. The dominant ephemerals in 1995 were *Plantago ovata* and Indian wheat, *Lepidium lasiocarpum*, filaree (*Erodium cicutarium;* nonnative, described as less frequent or rare in the earlier study), and *Schismus* spp. (not mentioned in the earlier study). In the creosote bush association, native *Pectocarya* spp. were dominant. The nonnative annuals *Lepidium* sp., *Plantago* spp., and *Schismus* spp. were dominant in associations other than creosote bush and saltbush in 1995.

In contrast, Indian wheat, previously the most important annual in creosote bush and desert saltbush communities, was not found in old fields. These fields were variously dominated by nonnatives *Bromus rubens, Hordeum* spp., and filaree, with *Schismus* spp. and *Lepidium* common codominants. Notably, although our initial winter planting of Indian wheat on abandoned cropland established and set seed the first year, new plants failed to come up in subsequent years.

Consequences for Ecosystem Recovery

The vegetation and soils of the lower Santa Cruz Valley were shaped by the periodic flooding of dynamic wash systems, which partially recharged a shallow, fluctuating groundwater table. Because of agricultural development, the valley no longer experiences those defining processes. The groundwater has been mined, the Santa Cruz River's flow was moved several kilometers west into another watershed, the other major tributary washes were channelized, and smaller waterways are now blocked by roads and the Central Arizona Project. Microtopography has been greatly altered by field leveling and irrigation ditches. Compounding these large-scale changes, soils in some areas may have increased salinity, pesticide residues, or loss of physical structure due to repeated tillage and irrigation.

There have been important biological losses and introductions as well. Seed sources for native plants are rare, while safe sites and times for seedling establishment are few and far between in this harsh climate. Natural regeneration of many of these old fields is unlikely because they average nearly 1300 m from the nearest source of native seed. Only wind-dispersed plants such as burroweed and desert broom seem to be capable of quickly traversing these distances. It is not known to what extent the loss of certain pollinators, predators, detritivores, cryptogamic crust, mycorrhizae, etc., and the addition of exotic species (e.g., *Schismus barbatus*, filaree, and *Salsola kali*) may place additional constraints on recovery of this ecosystem.

Because of these profound changes, we believe that ecosystem recovery, either by natural succession or through various attempts at ecological restoration, will be very limited. Deliberate ecological restoration of the valley is unlikely because of social and economic factors. Very few people live in the valley, and fewer have any appreciation for what has been lost. No laws require restoration of abandoned desert farmland, in contrast with strip mining or wetlands destruction. Cattle and sheep operations are not sufficiently profitable to justify costly manipulations. Most abandoned fields do not qualify for irrigation water, having lost their water rights.

Nevertheless, should ecological restoration be attempted, we recommend attention be paid to reintegration of wash systems. In our observations, the only abandoned fields to recover from farming on their own have been adjacent to never-tilled natural areas with a waterway that flows from the natural area into the field. Not only were the seed sources abundant, but the original flooding pattern appears to have been quickly reestablished at these sites (Karpiscak 1980). Sections of channelized wash running next to abandoned fields could be blocked to recreate overland flow conditions. It would be necessary to change roads and canals so that floodwater is allowed to drain across or under them at many locations, rather than at a single point. It might require resculpting old fields to recreate shallow drainage systems.

Large-scale seeding is risky and expensive. Adequate rainfall years cannot be predicted, and in years that rainfall is adequate for some seedling establishment, herbivores may destroy the new seedlings. Another strategy would be to plant small patches of native plants along the new washes, through transplanting and drip irrigation if necessary, to provide long-lived seed source islands. More research should be conducted to elucidate the biotic and abiotic processes that create safe sites for seedling establishment of various desert species. Enhancement or mimicry of these processes near seed source islands would increase the chances for reinvasion of old fields in the rare years when rainfall is adequate, herbivore pressure is low, and seeds are available.

Acknowledgments

This work was made possible by major funding from the Arizona Game and Fish Department Heritage Fund to the Desert Botanical Garden, Phoenix, Arizona. Additional financial assistance was provided by the Marshall Fund of Arizona. We thank Joe McAuliffe, Jane Cole, Bruce Hooper, Dewitt Weddle, and other staff and volunteers of the Desert Botanical Garden for valuable contributions to fieldwork and analysis, and an anonymous reviewer for improving the manuscript. Joni Ward and Ted Harris were instrumental in mapping land use. We received much needed assistance, including the use of documents and aerial photos, from the USDA Soil Conservation Service in Casa Grande, Martin Karpiscak, and the Arizona Department of Water Resources.

Literature Cited

Arizona Agricultural Statistics Service, for the following years: 1944, 1952, 1972–1974, 1976–1992. Arizona Agricultural Statistics, Bulletin s-26. 201 E. Indianola, Suite 250, Phoenix, Arizona 85012.

Arizona Crop and Livestock Reporting Service (1966) Arizona agricultural statistics historical summary of county data 1867–1965. Bulletin s-16, Phoenix.

Arizona Crop and Livestock Reporting Service (1981) Arizona agricultural statistics historical summary of county data 1965–1980. Bulletin s-16, Phoenix.

Autodesk Inc. (1992) *AutoCAD 12 Reference Manual*. Autodesk Inc., Sausalito, Calif.

Barr G. W. (1950) *Arizona Agriculture 1949: Production, Income, Costs*. University of Arizona Agricultural Experiment Station Bulletin no. 220, Tucson.

Barr G. W. (1951) *Arizona Agriculture 1950: Production, Income, Costs*. University of Arizona Agricultural Experiment Station Bulletin no. 226, Tucson.

Barr G. W. (1952) *Arizona Agriculture 1951: Production, Income, Costs*. University of Arizona Agricultural Experiment Station Bulletin no. 232, Tucson.

Barr G. W. (1955) *Arizona Agriculture 1954: Production, Income, Costs*. University of Arizona Agricultural Experiment Station Bulletin no. 252, Tucson.

Barrios F. M. (undated) *Santa Cruz Reservoir Project*. Unpublished manuscript.

Author may be contacted at 839 E. Marconi, Phoenix, Arizona, 85022. 602-993-3569

Bowden C. (1977) *Killing the hidden waters: the slow destruction of water resources in the American Southwest*. University of Texas Press, Austin.

Brown D. E. (1982) Biotic communities of the American Southwest—United States and Mexico. *Desert Plants* 4: 1–342.

Bunce R. G. H., Howard D. C. (eds) (1990) *Species dispersal in agricultural habitats*. Belhaven Press, New York.

Cox J. R., Morton H. L., LaBaume J. T., Renard K. G. (1983) Reviving Arizona's rangelands. *Journal of Soil and Water Conservation* 38: 342–345.

Crown P. L., Judge W. J. (eds) (1991) Chaco and Hohocam: prehistoric regional systems in the American Southwest. School of American Research Press, Seattle.

Dobyns H. F. (1981) *From Fire to Flood: Historic Human Destruction of Sonoran Desert Riverine Oases*. Ballena Press Anthropology Paper no. 20, New York.

Eckman E. C., Baldwin M., Carpenter E. J. (1920) Soil survey of the Middle Gila Valley Area, Arizona. In: Whitney M. (ed) *Field Operations of the Bureau of Soil, 1917*. USDA Bureau of Soils 19th Report, Washington, D.C.

Forbes R. H. (1911) *Irrigation and Agricultural Practice in Arizona*. U. S. Department of Agriculture Office of Experiment Stations Bulletin no. 235, Washington, D.C.

Green D. E. (1973) *Land of the Underground Rain: Irrigation on the Texas High Plains, 1910–1970*. University of Texas Press, Austin.

Gumerman G. J. (ed) (1991) *Exploring the Hohokam: Prehistoric Desert People of the American Southwest*. University of New Mexico Press, Albuquerque.

Hall J. F. (1991) *Soil Survey of Pinal County, Arizona, Western Part*. USDA Soil Conservation Service, Washington, D. C.

Hammet B. A. (1992) *Maps Showing Groundwater Conditions in the Eloy and Maricopa-Stanfield Sub-basins of the Pinal Active Management Area, Pinal, Pima, and Maricopa Counties, Arizona—1989*. Department of Water Resources Hydrologic Map Series Report no. 23, Phoenix.

Hastings J. R., Turner R. M. (1965) *The Changing Mile: An Ecological Study of Vegetation Change with Time in the Lower Mile of an Arid and Semiarid Region*. University of Arizona Press, Tucson.

Jackson L. L., McAuliffe J. R., Roundy B. A. (1991) Desert restoration: revegetation trials on abandoned farmland in the Sonoran Desert lowlands. *Restoration and Management Notes* 9: 71–79.

Judd B. I. (1971) The lethal decline of mesquite on the Casa Grande National Monument. *Great Basin Naturalist* 31: 153–159.

Karpiscak M. M. (1980) Secondary succession of abandoned field vegetation in southern Arizona. Ph.D. dissertation, University of Arizona, Tucson.

McAuliffe J. R. (1991a) Demographic shifts and plant succession along a late Holocene soil chronosequence in the Sonoran Desert of Baja California. *Journal of Arid Environments* 20: 165–178.

McAuliffe J. R. (1991b) A rapid survey method for the estimation of density and cover in desert plant communities. *Journal of Vegetation Science* 1: 653–656.

McAuliffe J. R. (1994) Landscape evolution, soil formation, and ecological patterns and processes in Sonoran Desert bajadas. *Ecological Monographs* 64: 111–148.

Nabhan G. P. (1986) 'Ak-ci arroyo mouth' and the environmental setting of the Papago Indian fields in the Sonoran Desert. *Applied Geography* 6: 61–75.

Nabhan G. P. (1987) *The Desert Smells Like Rain: A Naturalist in Papago Indian Country.* North Point Press, San Francisco.

Poulson E. N., Wildermuth R., Harper W. G. (1941) *Soil Survey: The Casa Grande Area, Arizona.* USDA Bureau of Plant Industry, Series 1936, no. 7, U.S. Government Printing Office, Washington, D.C.

Primack R. B. (1993) *Essentials of conservation biology.* Sinauer Associates, New York.

Sellers W. D., Hill R. H. (eds) (1974) *Arizona Climate, 1931–1972.* University of Arizona Press, Tucson.

Shantz H. L., Piemeisel R. L. (1924) Indicator significance of the natural vegetation of the southwestern desert region. *Journal of Agricultural Research* 28: 721–801.

Shapiro E. A. (1989) Cotton in Arizona: a historical geography. Master's thesis, University of Arizona, Tucson.

Smith G. E. P. (1940) *The groundwater supply of the Eloy district in Pinal County, Arizona.* University of Arizona Agricultural Experiment Station Technical Bulletin no. 87, Tucson.

Stromberg J. C., Patton D. T., Richter B. D. (1991) Flood flows and dynamics of Sonoran riparian forests. *Rivers* 2: 221–235.

Turner R. M. (1974a) *Map showing vegetation in the Tucson Area, Arizona.* U.S. Geological Survey, Map 1-844-H.

Turner R. M. (1974b) *Map showing vegetation in the Phoenix Area, Arizona.* U.S. Geological Survey. Map 1-845-H.

University of Arizona (1963) *Map of irrigated regions of Arizona.* Departments of Agricultural Engineering and Agricultural Economics, Cooperative Extension Service, Agricultural Experiment Station, University of Arizona, Tucson.

U.S. Bureau of the Census (1956) *U.S. Census of Agriculture 1954.* Vol. 1, Counties and State Economic Areas, Part 30. U.S. Government Printing Office, Washington, D.C.

U.S. Bureau of the Census (1961) *U.S. Census of Agriculture 1959.* Vol. 1, Counties, Part 43, Arizona. U.S. Government Printing Office, Washington, D.C.

U.S. Bureau of the Census (1966) *U.S. Census of Agriculture 1964. Vol. 1, Counties, Part 43, Arizona*. U.S. Government Printing Office, Washington D.C.
U.S. Bureau of the Census (1972) *Census of Agriculture 1969. Vol. 1. Area Reports Part 43, Arizona*. U.S. Government Printing Office, Washington, D.C.
U.S. Bureau of the Census (1977) *U.S. Census of Agriculture 1974. Vol. 1, State Reports, Part 43, Arizona*. U.S. Government Printing Office, Washington D.C.
U.S. Bureau of the Census (1981) *U.S. Census of Agriculture 1978. Vol. 1, State and County Data, Part 3, Arizona*. U.S. Government Printing Office, Washington D.C.
U.S. Bureau of the Census (1984) *U.S. Census of Agriculture 1982. Vol. 1, Geographic Area Series, Part 3, Arizona State and County Data*. U.S. Government Printing Office, Washington, D.C.
U.S. Bureau of the Census (1989) *U.S. Census of Agriculture 1987 Vol. 1, Geographic Area Series, Part 3, Arizona State and County Data*. U.S. Government Printing Office, Washington D.C.
U.S. Department of Agriculture (1977) *Santa Cruz—San Pedro River Basin, Arizona. Main Report.* USDA Soil Conservation Service, Portland, Oreg.
U.S. Department of Agriculture Soil Survey Staff (1975) *Soil Taxonomy: A Basic System of Soil Classification for Making and Interpreting Soil Surveys*. United States Department of Agriculture Handbook 436. U.S. Government Printing Office, Washington, D.C.
Valentine K.A., Gerard J. B. (1968) *Life-history characteristics of the creosotebush*, Larrea tridentata. New Mexico State University Agricultural Experiment Station Bulletin no. 526: 1–32.
Van Cleve Associates (1963) *Pinal Profiles: Data for Planning Report Number Three: Population, Residential Environment, Economics*. University of Arizona Bureau of Business and Public Research, Tucson.
Vos C. C., Opdam P. (eds) (1993) *Landscape Ecology of a Stressed Environment*. Chapman and Hall, London.
White R. H. (1990) *Tribal Assets: The Rebirth of Native America*. Henry Holt, New York.
Worster D. 1985. *Rivers of Empire: Water, Aridity and the Growth of the American West*. Oxford University Press, Oxford.

8 Deep History and a Wilder West

Paul S. Martin

At the start of the twentieth century, botanists of the Carnegie Institution of Washington (CIW) initiated desert research. They liked what they saw in Tucson, Arizona, and established a Desert Laboratory on Tumamoc Hill, two miles west of downtown. They began to investigate saguaros (*Carnegiea gigantea*), paloverde (*Cercidium* spp.), creosote bush (*Larrea tridentata*), and hundreds of other desert species (McGinnies 1981; Bowers 1988). Much would be learned about the distribution and movements of desert plants. While there was no fossil record of desert plants to help matters, D. T. MacDougal, Director of the Laboratory, anticipated the discovery of Pleistocene climatic changes. He believed that the extinction of mammoths (*Mammuthus*), camels (*Camelops*), horses (*Equus*), and other large animals of this region was driven by postglacial aridity (MacDougal 1908, 66–67). In later years several generations of geologists invoked drought as the cause of the megafaunal extinctions. Tests of this idea required better fossils and improved methods of dating.

Presenting the Desert's Past

By the end of the century, radiocarbon assay had vastly improved dating of fossil faunas and the times of extinction. New sources of fossil evidence, including fossil packrat middens, revealed much about vegetation change both before and after the megafaunal extinction episode.

One result of the fossil and geochemical discoveries was a new approach to conservation and land use. The Quaternary (the ice age of the

Table 8.1 Large (> 40 kg) mammals of the late Quaternary in southern California, Nevada, Sonora, Chihuahua, Arizona, New Mexico, and western Texas.

Scientific Name	Common Name
Antilocapra americana	pronghorn
† *Arctodus simus*	giant short-faced bear
† *Bison antiquus*	ancient bison
Bison bison	bison
† *Bootherium bombifrons*	woodland muskox
† *Camelops hesternus*	western camel
† *Canis dirus*	dire wolf
Canis lupus	gray wolf
Cervus elaphus	elk, wapiti
† *Equus*, four or more species	horses, onager
† *Euceratherium collinum*	shrub-ox
Felis concolor	mountain lion
† *Glossotherium harlani*	big-tongued ground sloth
† *Glyptotherium floridanum*	Florida glyptodont
† *Hemiauchenia macrocephala*	long-legged llama
† *Mammuthus imperator*	imperial mammoth
† *Mammuthus columbi*	Columbian mammoth
† *Mammut americanum*	American mastodon
† *Megalonyx jeffersonii*	Jefferson's ground sloth
† *Mylohyus nasutus*	long-nosed peccary
† *Navahoceros fricki*	mountain deer
† *Nothrotheriops shastensis*	Shasta ground sloth
Odocoileus hemionus	mule deer
Odocoileus virginianus	Virginia deer
† *Oreamnos harringtoni*	Harrington's extinct mountain goat
Ovis canadensis	bighorn
† *Panthera leo atrox*	American lion
Panthera onca	jaguar
† *Platygonus compressus*	flat-headed peccary
† *Smilodon fatalis*	saber-cat
† *Tapirus californicus*	California tapir
† *Tremarctos floridanus*	Florida cave bear
Ursus americanus	black bear
Ursus arctos	grizzly bear

† = extinct.

last two million years) including "deep history" (the last 40 000 years measured by radiocarbon dating) is no longer regarded as too remote to be worth considering in "negotiations with nature." For example, archaeological and paleoecological data are crucial in approaches to managing big game in national parks such as Yellowstone (Wagner and Kay 1993), the livestock on the western range (Johnson and Mayeux 1992; Tausch et al. 1993), and the conservation of "red book" species (Caughley and Gunn 1996).

At stake is the granting of usufruct, the rights of using and enjoying all the advantages and profits of property, private or public, without altering or damaging the substance. Species richness of large herbivores now found in the "home where the buffalo roam, where the deer and the antelope play" is impoverished. The fossil record of the arid west indicates that until only 11 000 to 13 000 calendar years ago, the common large herbivores native to the region included proboscideans, camelids, equids, and more than one bovid; some 37 species in all. Two thirds were lost (see table 8.1).

Surrogates for these exist elsewhere; some are currently endangered in their native haunts. Potential experiments with "Pleistocene Parks" offer international as well as regional benefits. Both managers and visionaries may find their choices are greatly enriched by acknowledging a history that looks not only beyond Columbus but also beyond Clovis colonization in considering large animals suitable for the western range. The fossil record seems clear on one point. Over evolutionary time (for millions of years) a rich assemblage of large herbivores and carnivores occupied the land, until catastrophic extinction 11 000 radiocarbon years ago, roughly equivalent to 13 000 calendar years ago.

The Last Cold Stage: Nature's Norm

In scaling evolutionary time for modern species, New World conservationists need a running mean embracing a greater range than the few hundred years of written history. If one consults proxy data from ice cores or deep sea sediments of the last 10^5 to 10^6 years, the results are sobering. Marine isotopes, ice core chemistry and gas content, and marine and terrestrial fossils all indicate that the climatic and biotic conditions known to us historically and in the Holocene (last 10 000 years) are geologically aberrant. Interglacial climates have been thought to embrace 11 000 to

14 000 years, less than 10% of the mid to late Pleistocene (Imbrie and Imbrie 1979), recently reevaluated at perhaps twice that length (Winograd et al. 1997). But in either case, most of the last 6×10^5 years were cool, dry, and dusty with an atmosphere reduced in carbon dioxide and methane. Sea level dropped by 100 m; glaciers expanded, albeit not to their maximum position. Continental fossils show that many plants and animals lived elsewhere in patterns that seem strange to biogeographers and naturalists. According to Porter (1989), the geological norm last occurred in the last cold stage only 14 000 years ago. In addition, recent findings from research on ice cores reveal rapid switching into, back out of, and then again into the Holocene (Dansgaard et al. 1993; Taylor et al. 1993).

To illustrate the perceptional problem, we may consider by analogy the dilemma of an imaginary cohort of fruit flies emerging from pupae in mid-June. Their life span is less than a month. The flies experience part of summer before all die of old age. They will not experience the cooler temperatures of spring and fall, much less the extremes of winter. In understanding the pulse of the Pleistocene, the human consciousness is similarly handicapped. We suffer an overwhelming bias toward "summer," the Holocene, the interglacial of our written history, not to mention our own lifetime. We have no experience of cool stadial or full glacial climates and their biotas, the normal evolutionary theater for our own and most other modern species. Our understanding of the norm is further hampered by the unbalanced extinctions of megafauna.

Biogeographers standardize on the time frame they know personally. They think of ice age "displacements" or "refugia" as anomalies. To the contrary, the anomaly is not what was but what is. The plant and animal ranges and assemblages that we "know" are short-term epiphenomena. Conversely, continental "refugia" occupied by animals and plants in the hundred thousand years of the last cold stage are normal, not exceptional. It may take some time to learn to adjust for the "fruit fly problem," learning to discount the Holocene and adopt the last cold stage, extinct megafauna included, as normal.

Through a wealth of new data, following a paleontological "gold rush," the arid West recently contributed to better understanding of the ice age norm. Rock shelters and caves of dry regions from eastern California to western Texas and from southern Canada to northern Mexico, have attracted interest for the fossils found in ancient packrat (*Neotoma*) middens. Plant remains characterize most of the midden deposits accumulated by prehistoric packrats. Fossil middens also yield small animal bones, insects, arachnids, and, occasionally, a bone scrap of a large mammal. Radiocarbon dating of the plant and animal remains in ancient middens enables ecologists to envision some of the changes that took place and to interpret the meaning of these changes (summarized in Betancourt et al. 1990). While the last cold stage saw many plants and animals inhabiting lower latitudes and/or lower elevations than they occupy today, others were less vagile. In geographic location and species composition the ice age communities challenge our understanding. By the start of the Holocene many species were on the move to where they are now. Others endured the climatic shift and persisted in place.

Insight into what the normal ("displaced") communities of the western dry lands were like can be gained by fossil pollen analysis, by the analysis of ancient packrat middens, and by analysis of extinct animal remains, including dung from caves. Since Wells and Jorgensen (1964) pioneered the method, over 1000 fossil middens have been radiocarbon dated, many selected for dating because of their abundance of extralocal plant remains (Webb and Betancourt 1990). The bias yielded a large number of old records. Over half the dated middens exceed 8000 radiocarbon years before present (BP). Beyond 8000 BP, sample size decreases exponentially with age. There are fewer than 100 middens that predate the Wisconsin glacial maximum; that is, middens that are older than 22 000 BP (Webb and Betancourt 1990). These differ in only minor ways from those of the full glacial (22 000 to 13 000 BP). According to Sorensen's index, the similarity of full glacial fossil assemblages to modern plants in the same area is 0.25 to 0.30. They were quite unlike. The similarity index of plants in modern middens to those in a modern releve is 0.75 to 0.85. A perfect 1.00 is not obtained, since investigators do not find all the species around the site of modern nests as the packrats found themselves.

In the arid west of both the United States and northern Mexico, fossil middens of the last 40 000 years reveal biotic communities that were previously unknown and unimagined (reviewed in Betancourt et al. 1990). In the Grand Canyon, Cole (1985) found that ponderosa pine (*Pinus ponderosa*) and other woody plants now present did not arrive until after the departure of most extralocal species, including spruce and limber pine (*Pinus flexilis*). The turnovers indicate steady modernization, not only between 13 000 and 8000 years ago, but even afterward. What is now dry

Figure 8.1 Some fossil localities mentioned in text.

Utah juniper (*Juniperus osteosperma*) woodland in northern New Mexico was occupied 12 000 years earlier by limber pine and Douglas fir (*Pseudotsuga menziesii*). Only within the last few thousand years have Chihuahuan Desert communities of creosote bush and lechuguilla (*Agave lechuguilla*) invaded the Hueco Mountains of western Texas, a region occupied in the last cold stage by pinyon, juniper, oak, and winterfat (*Ceratoides lanata*; Van Devender 1990a).

At present the pinyon to the north of the Huecos is *Pinus edulis,* a conifer common in the Four Corners region (border of Colorado, Utah, Arizona and New Mexico; fig. 8.1). Thus, one might expect the fossil woodland communities, which lie due south of modern *P. edulis*, to harbor that species. Instead, in the last cold stage, the pinyon commonly brought by rats to their middens in the Hueco Mountains was *P. remota*. The modern range (the Holocene "refuge") of *P. remota* is to the east rather than to the north of western Texas. In the last cold stage, the range of *P. edulis* constricted while *P. remota* expanded, forming a unique association with the oak *Quercus hinckleyi* (Betancourt et al. 1990).

In the Snake Range on the Nevada-Utah border, Thompson (1988) found fossil middens yielding limber pine and bristlecone pine (*Pinus longaeva*) along with sagebrush (*Artemisia*). The area is now a treeless desert scrub of sagebrush and shadscale (*Atriplex confertifolia*). In the last few thousand years, *P. edulis* invaded higher elevations of the Snake Range. In the northwestern Great Basin, many of the plant species now found in the region were also present 12 000 to 30 000 years ago (Nowak et al. 1994). While species could and did move upward in elevation, northward emigration was not an alternative for cold desert species faced with Holocene warming. Since no desert climates developed in western Canada, no appreciable northern range expansion of plants of the Great Basin Desert could take place. Unlike warm desert shrubs, which could and did spread north in the Holocene, the cold desert shrubs were trapped essentially where they are now (Nowak et al. 1994).

During the last cold stage in the Mojave Desert region, juniper (*Juniperus*) occupied elevations now inhabited by both creosote bush and white bursage (*Ambrosia dumosa*). Rising some 1800 m above the Mojave Desert (600 to 800 m) in southern Nevada northeast of Las Vegas is a mountain island, the Sheep Range (fig. 8.1). During the last cold stage its woodland evidently was impoverished. The Sheep Range harbored only five conifers: white fir (*Abies concolor*), limber pine, bristlecone, singleleaf pinyon (*Pinus monophylla*), and Utah juniper. In the warmer and perhaps wetter Holocene, the five conifers moved upward in elevation and were joined by ponderosa pine and two junipers (*J. communis, J. scopulorum*). Unlike the Grand Canyon and other parts of the southwest, there apparently were no departures of cold-tolerant species. Holocene changes saw only enrichment of the Sheep Range by thermophilous shrubs (Spaulding 1990).

Although late glacial spruce (*Picea engelmannii*) descended in elevation in Arizona and occupied localities to the south of its present range, such as the Guadalupe Mountains of western Texas, it has yet to be found in the fossil record of the Sheep Range or higher mountains of Nevada or southern California, which should have been cold enough for spruce. In the Holocene, ponderosa pine not only migrated into the Sheep Range but also spread north some 2000 km from Arizona into western Canada.

In the Sonoran Desert, the glacial-age middens disclose oak (*Quercus*), juniper, occasional pinyon (*Pinus falax*), sagebrush, and Joshua tree (*Yucca brevifolia*) in regions where the modern assemblage of brittlebush (*Encelia farinosa*), ironwood (*Olneya tesota*), mesquite (*Prosopis*), and blue paloverde (*Cercidium floridum*) first appears in middens of the early

Holocene. At Organ Pipe Cactus National Monument (NM), the organ pipe cactus (*Stenocereus thurberi*), foothill paloverde (*C. microphyllum*), and Mexican jumping bean (*Sapium biloculare*) arrived from northwestern Mexico only within the last 5000 years (Van Devender 1990b). At an earlier time, the arborescent plants in the Puerto Blanco Mountains at 600 m in Organ Pipe Cactus NM (fig. 8.1) would have looked much more like those to be found now around Joshua Tree National Park, complete with rabbit brush (*Chrysothamnus nauseosus*), juniper, Whipple yucca, Whipple cholla, and Joshua trees themselves. Higher elevations in Organ Pipe Cactus NM harbored other Mojave elements: shadscale, singleleaf pinyon, and snowberry (*Symphoricarpos* spp.).

Evidence of late glacial-age enrichment compared with modern vegetation is seen in the southern Chihuahuan Desert in Mexico. Near Torreon, Coahuila, a constellation of species from various compass points, including the south, were identified in middens 12 500 years old (Van Devender and Burgess 1985). Insects from fossil middens in the Chihuahuan Desert region presently occupy ranges scattered in various directions from the midden sites (Elias 1992). It would appear that at least locally the Mexican Plateau, famous for high biotic diversity, was even more enriched in the last cold stage.

In the western United States since the last cold stage, the range of some woody plants expanded, while that of others shrank. Many moved north and a few moved south. Some moved meridonally; some altitudinally (usually up, in one case down); a few may not have moved at all. This is inferred when displaced plants co-occurred with species that persist in the region, such as Turk's head barrel cactus (*Echinocactus horizonthalonius nicholii*) in calcic rocks in the Waterman Mountains outside Tucson. During glacial times packrats incorporated this ground-hugging cactus into middens of the Watermans along with fragments of juniper, oak, and sagebrush (Van Devender 1990b). Today its associates include paloverde, ironwood, saguaro, and other species representative of the Sonoran Desert. The anomaly, if there is one, is the modern and not the fossil assemblage. This example illustrates the fluidity of desert plant "associations." These and other records of range changes in plants and plant communities probed by radiocarbon dating, including the relatively late arrival of foothill paloverde and organ pipe cactus, might well have surprised some of the early botanists at the Desert Laboratory on Tumamoc Hill. The discoveries certainly surprised those of us who came afterward.

Table 8.2 Trees, shrubs, or succulents found together in fossil packrat middens of the last cold stage and not known to coexist now.

	In Range	Displaced
Sonoran Desert		
Waterman Mts., 795 m 12 000 to 22 000 BP (Van Devender 1990a)	*Echinocactus horizonthalonius nicolii*	*Pinus monophylla* *Artemisia tridentata* *Yucca brevifolia*
Ajo Mts. 13 500 to 22 000 BP (Van Devender 1990a)	*Quercus ajoensis*	*Atriplex confertifolia* *Symphoricarpus*
Kofa Mts. 9750 BP (Van Devender 1990a)	*Castela emoryi*	*Juniperus californica* *Yucca brevifolia* *Nolina bigelovia*
Tinajas Altas Mts. 13 330 to 15 800 BP (Van Devender 1990a)	*Ambrosia ilicifolia*	*Pinus monophylla* *Juniperus californica* *Rhus trilobata*
Mohave Desert (Spaulding 1990)	*Atriplex confertifolia*	*Pinus flexilis*
Great Basin Desert (Nowak et al. 1994)	*Atriplex confertifolia*	*Pinus albicaulis*
Colorado Plateau (Betancourt et al. 1990)	*Atriplex canescens* *Ephedra* spp. *Yucca* spp. *Opuntia* spp.	*Picea pungens* *Juniperus communis*

In the last cold stage in unglaciated North America, the ranges of some mammals that now are divergent were then locally sympatric. Some mammalogists account for the pulling apart by postulating colder winters and warmer summers (for a review of the small mammal records, see Semken 1988). Glacial climates, they propose, were more equable. Western plant assemblages of the last cold stage include certain cases in which unusual or unexpected overlaps in range have been noted in fossil middens (table 8.2, see also Van Devender 1990b, 156–157). The plant overlaps have been interpreted by some authors as the mammal overlaps were; that is, as the result of more equable conditions (cooler summers and warmer winters) during the last cold stage (Van Devender 1990b). Other explanations are possible (Cole 1995), including the view that edaphically restricted species like the Turk's head barrel cactus are inherently less vagile than those species able to grow on many soil types. The "stay-at-homes" simply do not respond to the climatic changes of the Holocene by changing their range. By not responding, they help generate the "hind-sighted" view of anomalous communities of species not supposed to mix, which were in fact the norm.

Shadscale in the Sheep Range of southern Nevada is another case in point. In glacial times it ranged slightly *above* its present upper elevational limit to meet bristlecone and limber pine, displaced downward; presently shadscale is altitudinally separated from the pines by 600 m (Spaulding 1990). Shadscale, bristlecone, and limber pine are strictly continental in range; none occur in equable climates. Equability is not a plausible explanation for their past overlap; more likely is a change in soil chemistry and depth. Mid-valley deflation of sodium carbonate deposits with deposition on slopes would favor shadscale, a salt-tolerant shrub in the saltbush family. It is also possible that CO_2 depletion during the last cold stage (Van de Water et al. 1994) altered the ecological opportunity for some species. The result should have favored saltbush species. Their C4 metabolic pathway is better adapted to low concentrations of CO_2. Other examples of unexpected range overlaps from fossil middens are indicated in table 8.2. Most are full or late glacial; a few are early Holocene in age. While all invite independent radiocarbon assay to verify the assumption of contemporaneity (see Van Devender et al. 1985), none should be held anomalous or "disharmonious."

As noted previously, what *is* anomalous is the expanded nature of interglacial and Holocene ranges. As continental ice sheets melted away, exposing new lands, we find that some species moved north a great deal, others moved less, some moved not at all, and a very few moved south. The changing climate and melting ice opened higher latitudes and higher elevations to colonization; species responded in different ways to this change. No climatic explanation, no change in "equability" or "continentality," is required to account for the apparent reduction in range overlaps in the Holocene. Unless all march exactly in step with the drumbeat of climatic change, species will naturally pull apart during their differential expansion, after ice sheets vanish and the cold stage ends.

The sudden loss of megafaunal diversity also seen in North America near the end of the last cold stage (Martin and Klein 1984; Martin 1990; Stuart 1991) is another matter. Understanding the extinction of dozens of large animals is no less important in understanding the deep history of western North America than is understanding the profound changes in plant ranges mentioned above. While the two events have been linked, since both occurred at a time of climatic change, there are reasons to treat them independently.

Tracking an Extinct Megafauna

Fossils reveal a Quaternary megafauna far more diverse in species than that known historically. Prior to the extinction of large herbivores, Arizona and adjacent states supported more than twice the number of species than are present in the Holocene (table 8.1). The many bones of mammoth, camel, horse, and extinct species of bison found in alluvial or lacustrine deposits (Lundelius et al. 1983) suggest that these animals were the most common species in western North America prior to extinction. Less common fossils of the west include remains of mastodons, tapir, ground sloths, dwarf antelope, shrub oxen, giant peccary, machairodonts (sabercats), and others (table 8.1). While there is no rigorous method of aliquots on which to base an objective estimate of numbers of the native North American megafauna, Mosimann and Martin (1975) suggested 100 to 150 million animal units (one unit = 1000 pounds). They based their estimate on African analogues and on the modern carrying capacity of domestic livestock and wild game in the United States. Their estimate may be excessive if glacial age CO_2 depletion had a significant effect in reducing the productivity of forage plants, as mentioned above, or if regulation of megaherbivore populations by disease or predation was severe.

The foraging activity of extinct proboscidea, camelids, equids, bovids, and other megafauna played a role in seed dispersal (Janzen and Martin 1982). Hooked seeds of beggars ticks (*Bidens*) and adhesive seeds of tick trefoil (*Desmodium*) attached themselves to shaggy coats of grazers. Sweet pods attractive to large herbivores contain beans within a hard woody coat or mesocarp. The coat protects the seeds within from chewing and digestion. They benefit by dung dispersal, while the large herbivore benefits from the fruit, which ripens at a season of drought stress. According to Janzen and Martin (1982), cattle, horses, and burros replaced extinct bovids, equids, and proboscideans, the original dispersers of the seed pods of legumes, such as mesquite, huisache (*Acacia*), and ebano (*Pithecollobium*), or the tunas (fruits) of cacti, such as prickly pear and cholla (*Opuntia* spp.).

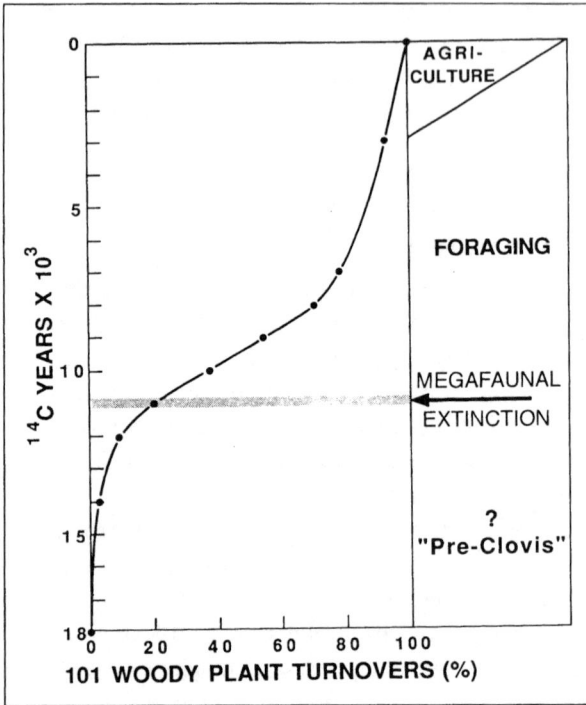

Figure 8.2 North American megafaunal extinction models related to cultural events. Radiocarbon age plotted on ordinate; percent loss of species on abscissa. The "spontaneous" curve approximates my interpretation of rate and time of species extinction. The "gradual" curve approximates Grayson's interpretation (1993) in which undated or poorly dated extinct species are thought to vanish before Clovis colonization.

Perhaps the most dramatic example of passive dispersal in the Sonoran region is seen in the case of the unicorn plant family, Martyniaceae. *Martynia*, an erect annual of disturbed sites in southern Sonora, has a dangling pod the size of a large cockroach and armed with two short (3 mm) stout hooks waiting to attach themselves to legs of any passing forager. Its close relative, the well-known Devil's claw (*Proboscidea parviflora*), whose vines drape soft alluvium at the end of the summer rainy season, has wiry hooks 10 to 15 cm in length extending beyond a fusiform fruit. The hooks attach themselves to any hoof (or boot) they intercept. The seed pod lodges underfoot and in soft ground the hoof itself serves as both soil imprinter and planter for seeds of Devil's claw.

Thanks to unusual fossil preservation in dry caves, an unusual opportunity both to date faunal extinction by radiocarbon and to investigate paleodiet of extinct herbivores is available, especially in and around the Grand Canyon of the Colorado River in Arizona and the Canyonlands of

Utah. Suites of radiocarbon dates on the youngest deposits of extinct fauna provide an unequaled opportunity to analyze tempo and mode of extinction.

The radiocarbon record may be subject to stretch if some extinctions did in fact occur closer to 10 000 than 11 000 BP. A recently discovered plateau in the radiocarbon dating calibration curve occurs in the Younger Dryas, hard on the heels of the extinctions. Based on thorium/uranium dating and tree-ring dates, 10 000 BP (radiocarbon years) will calibrate to an absolute date of about 11 000 years ago (BP) or 9000 BC. A radiocarbon date of 11 000 BP calibrates to roughly 13 000 BP or 11 000 BC. The offset is apparently driven by an adjustment in the oceanic reservoir of carbonate (Edwards et al. 1993). It may complicate efforts to determine rate of change of various deglacial events, megafaunal extinction included. Nevertheless, terminal radiocarbon dates obtained for New World genera of large mammals converge on 11 000 BP (Martin 1987b, 1990; Stuart 1991; Stafford 1995, personal correspondence). Unlike the gradual extinction of large mammals in the Old World, which played out over the last 50 000 years (Stuart 1991), New World losses were spontaneous, taking place within 2000 years or perhaps much less (fig. 8.2).

Extinct Goats, Ground Sloths, and Condors in the Grand Canyon

Dry caves of the Grand Canyon have yielded mats of dung, hair, and bone with tissue or keratin attached. Robbins et al. (1984) attribute cough-drop size pellets to Harrington's extinct goat (*Oreamnos harringtoni*). Shasta ground sloths (*Nothrotheriops shastensis*) deposited baseball-size boluses. The Canyon caves have also yielded remains of other large mammals and birds, such as extinct species of *Bison, Mammuthus, Equus, Camelops,* and the California condor (*Gymnogyps californianus*). The modern fauna of large animals native along the Colorado River in the Grand Canyon is limited to scattered records of bighorn sheep and mule deer. Mule deer descended to river level only in this century (Webb 1995). Unknown there previously in both the historic and the fossil record, elk (*Cervus*) have begun to invade Grand Canyon National Park in the last two decades.

Both dung and bones of extinct goats, condors, and ground sloths are found on or close to the surface of caves. Little or no sedimentation or

Figure 8.3 Cranium of Harrington's extinct goat with keratinous horn sheaths removed from floor of Stevens Cave, Grand Canyon National Park.

infilling occurred in the Grand Canyon caves to bury their remains. The pellets of the fossil goats are larger than those of living mountain goats, mule deer, or mountain sheep and are smaller than those of mature elk. At Rampart Cave, Bida Cave, Stanton's Cave, Stevens Cave, and elsewhere the pellets are commonly associated with bones and horn sheaths of Harrington's extinct goat, evidently their source (Robbins et al. 1984; Mead et al., Dung and diet, 1986). The pellets, horn sheaths, and bones supporting dry tissue are so fresh in appearance and so abundant in the surface litter of certain caves that one cannot escape a haunting feeling that the animals must still be alive (fig. 8.3).

Nevertheless, contrary to appearances, the perishable fossil material is much too old to be modern. Some 25 samples of either horn sheaths or goat pellets yielded radiocarbon age estimates from 10 200 to over 30 000 BP (Mead et al., Extinction, 1986; Mead et al., Dung and diet, 1986). In Stanton's Cave, extinct goat pellets were overlain by a 10 cm deposit of smaller pellets attributed to living bighorn sheep. The sheep samples were dated at 8000 BP or less (Robbins et al. 1984), verifying their survival into the Holocene.

None of the extinct goat samples proved to be younger than 10 000 BP (Mead et al., Extinction, 1986), and one surface sample of fecal pellets of the extinct goats was 39 000 years old by radiocarbon dating (Emslie et al. 1995). A recent study by Emslie, Mead, and Coats (1995) indicates that despite virtual inaccessibility, certain caves rich in fossils attracted prehistoric human visitors—a desert culture (western Archaic) people that arrived thousands of years after megafaunal extinction. The ancient visitors built low shrines of cave spall or blocks of packrat middens and left split sticks, grass bundles, or twig figurines beneath rocks on or near remains of the extinct goats. They left few other unmistakable artifacts and no hearths. They did not inhabit the shrine caves. Without radiocarbon dating it would have been virtually impossible for archaeologists to escape the erroneous conclusion that the desert culture people were contemporaries of the extinct goats.

In the years before radiocarbon dating, Harrington (1933) made exactly that mistake, interpreting archaic cultural material from Gypsum Cave, Nevada (outside Las Vegas), as associated with extinct fauna, by virtue of its being interbedded with ground sloth dung. Dates by Heizer and Berger (1970) revealed the error. Thanks to the dates on the shrine caves, it is possible to imagine that the cave fossils of an extinct species (a mountain goat), discovered by prehistoric foragers within the Inner Gorge of the Grand Canyon, triggered some ritual, spiritual, or magical response. The evidence of prehistoric people venerating fossils is not trivial. They demonstrated more interest in and concern for the extinct goat caves than any one since, at least until the arrival of research teams led by Euler, Mead, and Emslie.

In the eastern Grand Canyon, late glacial departures of the trees and shrubs mentioned above preceded the arrival of Holocene (modern) shrub species by 2000 years (Cole 1985). Presumably, climatic change forced the change in vegetation and, some suggest, the shift from glacial age mountain goats to Holocene mountain sheep. There is a difficulty with such a view. According to cuticle analysis, grasses served as forage for Harrington's extinct mountain goat (Robbins et al. 1984; Mead et al., Dung and diet, 1986). Between elevations of 1500 and 2100 m, cool, dry, grassy woodlands persist in southern or central Utah and western Colorado. One must ask why the goats did not simply relocate.

Furthermore, if the regional climate was becoming too warm and thus unsuitable for the mountain goat lineage, which evolved in boreal mountains and may have been adapted to cooler climates, it is unclear why ground sloths should have vanished from Grand Canyon caves at about the same time as the goats (Martin et al. 1985). Evolving in South America, the ground sloths were widely distributed through tropical and temperate regions. They are related to the tropical tree sloths. Concurrent extinction of both species, one of boreal ancestry, the other of tropical ancestry, suggests a forcing function independent of climatic change.

In the 1930s, two dry caves, Rampart and Muav, in the western Grand Canyon were discovered to contain fossil remains of the extinct Shasta ground sloth. One also yielded bones and perishable remains of extinct goats (Martin et al. 1961). The ground sloth fossils, including dung boluses and fragments to the depth of a meter or more, were remarkably fresh in appearance. They were so fresh that, as in the case of the goat caves mentioned above, one might well imagine the animals were still alive. Nevertheless, 26 radiocarbon measurements on the distinctive softball-size lumps of sloth dung indicated that the animals survived no later than 10 000 BP. It is more likely that extinction occurred closer to 11 000 BP (Martin et al. 1985).

While it is impossible to determine by current dating methods whether all populations of an extinct animal crashed simultaneously within a decade or less, or whether they sputtered out more slowly over hundreds or even thousands of years, Shasta ground sloth extinction was rapid. Samples of ground sloth dung from six caves in southwestern North America from southern Nevada to the Guadalupe Mountains of western Texas revealed no trace of ground sloths in those regions after 10 500 years. Within the limits of radiocarbon error, New Mexican, Texan, and Arizonan populations vanished concurrently (Martin et al. 1985).

From dietary analysis (Hansen 1978 and references therein), it appears that the more common forage plants eaten by sloths inhabiting Rampart and Muav Caves in the western Grand Canyon were globemallow (*Sphaeralcea,* Malvaceae) and Indian or Mormon tea (*Ephedra,* Ephedraceae). Other forage plants included *Atriplex* (Chenopodiaceae), Cactaceae, and *Yucca* (Liliaceae). Presently all are common and widespread at various elevations in the region. Guthrie (1984) claimed the increase in *Ephedra* in the sloth dung portended dietary stress leading to extinction. However, *Ephedra* appears in the diet of various megaherbivores, including livestock, wild burros, and mountain sheep as well as ground sloths (Hansen and Martin 1973). Even if forage supply failed locally in the western

Grand Canyon and triggered a die-off of ground sloths, it is not clear why a region as topographically and environmentally variable as Arizona and adjacent states, including northern and central Mexico, could not have sustained ground sloth populations at higher elevations or within more productive regions such as central Texas and the Mexican Plateau.

At the end of their tenure in the western Grand Canyon, Shasta ground sloths coexisted with juniper, ash (*Fraxinus*), desert almond (*Prunus*), and a variety of other shrubs and herbs to be expected in dry woodland (midden analysis of Phillips, summarized by Spaulding 1990, 191). Juniper and other extralocal plant species persisted for at least 2000 years after the extinction of both the Shasta ground sloth and Harrington's extinct goat. The packrat middens reveal no major disruption of regional vegetation coinciding with the time of the extinctions (Spaulding 1990).

Grand Canyon caves disclosed the extinction of a third species that offers another glimpse of Arizona's megafauna prior to its impoverishment, as well as a further test of the extinction chronology. Bones, egg shell, and feathers of California condors had been found in various caves from the Southwest in circumstances that, just as in the case of Harrington's mountain goat and the Shasta ground sloth, strongly suggested youthfulness. Some bones were found with potsherds or other artifacts of late prehistoric people (Emslie 1987 and references therein), an apparent association that Emslie has shown by radiocarbon dates to be spurious. Stevens Cave harbored a remarkably fresh condor cranium; the bird's keratinous beak was still attached (fig. 8.4). It assayed $12\,540 \pm 790$ BP. The result is further testimony to the remarkable ability of Grand Canyon caves to preserve perishable fossils over 10 000 years old.

Other Grand Canyon caves yielded new records of condors, including condor bones rich in collagen and a condor nest with juvenile bird bones and food scraps. Emslie (1987) dated samples of bone collagen or tissue of 16 condor fossils from three states: Texas, New Mexico, and Arizona. The oldest date was 22 000 BP; the youngest radiocarbon date, from Sandblast Cave in the Grand Canyon, was 9580 BP. Condors disappeared from the Grand Canyon around the same time or perhaps slightly later than the Shasta ground sloth and Harrington's extinct goat. None of the radiocarbon measurements support the conclusion that condors lived as late as the time of the Anasazi or even as late as the much older Archaic culture. Given the elegant preservation of the fossils, their apparent association with late prehistoric artifacts, and a historic sight record of a California

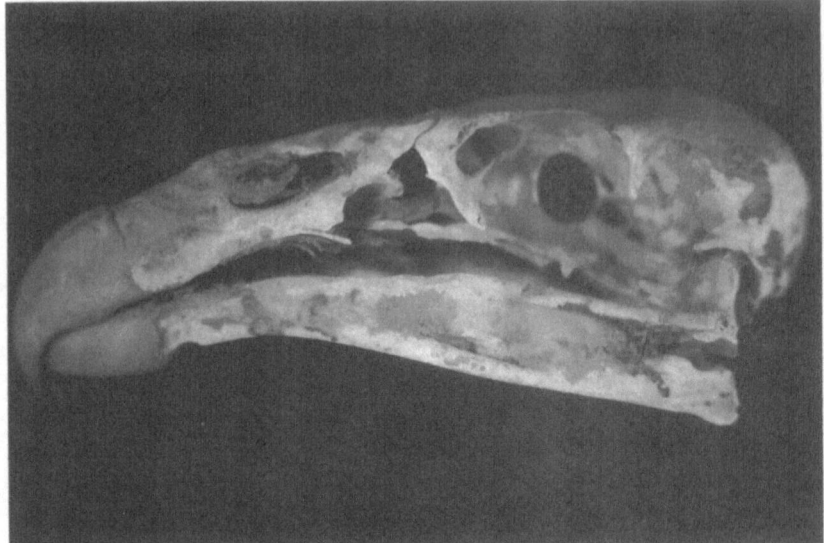

Figure 8.4 Condor (*Gymnogyps*) beak, keratinous sheath and soft tissue, from Stevens Cave, Grand Canyon National Park, 12 500 years old.

condor near Pierce Ferry, Mojave County, Arizona, it would be hard without radiocarbon dating to dispute the proposition that the birds lived in the Canyon prehistorically and actually lingered into the time of Anglo settlement.

Unlike black and turkey vultures, condors provide their fledglings with bone scraps. Bone fragments identified by Emslie from a condor nest at Sandblast Cave included those of extinct megafauna: horse (*Equus* sp.), bison (*Bison* sp.), mammoth (*Mammuthus* sp.), and camel (*Camelops* sp.). In addition, there were bones of extinct goats, which inhabited many caves in the Inner Gorge as we have seen. Grand Canyon localities yielding these animals along with condor bones include Stanton's (Euler 1984) and Vulture Caves (Mead and Phillips 1981). It is possible that most, if not all, records of extinct horse, bison, mammoth, and camel from caves of the Inner Gorge represent condor transport of bones from carcasses of large animals that died on the Coconino and Kaibab Plateaus. Some may be derived from carcasses washed down the Colorado River, deposited on beaches, and brought to the caves by condors or other scavengers.

With their food supply in jeopardy during the time of megaherbivore extinction, the condors, along with other scavenging predators of the region (presumably dire wolves [*Canis dirus*], lions [*Felis leo atrox*], giant short-faced bears [*Arctodus simus*], and perhaps the saber-cat [although there are no Arizona records of them at the time]), vanished as well.

Mammoth Dung in Utah

For many years researchers at the Desert Laboratory sought without success to locate more caves containing ground sloth dung. As indicated above, Ken Cole, Steve Emslie, and Jim I. Mead found horn sheaths and other perishable remains of extinct goats in caves where such material had been unknown previously, and Art Phillips had recovered the hoof of an extinct small horse from Rampart Cave. No one imagined that the region harbored caves or rock shelters that would yield trophies as extraordinary as dung of mammoth. While bones, tooth plates, or tusks of Pleistocene mammoths are often found in Southwestern fluviatile or lacustrine sediments, caves rarely yield such fossils. Even those few caves large enough to receive elephants would not necessarily attract them. The idea of caves yielding dung of extinct mammoth seemed preposterous.

Richard Hansen was the first to identify mammoth dung from a North American cave (Hansen 1980). The material, excavated by Jess Jennings from beneath rich archaeological remains in Cowboy Cave, Utah, was much more fibrous in texture than abundant associated bison dung dated at 11 000 BP and older. The fragments were not directly dated by radiocarbon and Hansen had not seen them in place.

Late in 1982 the Desert Laboratory received a fragment of unknown dung collected from a cave in southern Utah by Loren Haury and Steve Carothers, along with managers of the Glen Canyon National Recreation Area of the National Parks Service. The team was conducting a natural resources survey. The fragment resembled the coarse-textured manure of a monogastric (ground sloths, equids, mammoths, and other elephants are monogastrics) rather than the finely divided excreta of a ruminant such as a cow or cervid. The sample was much more fibrous than any of our reference material of *Nothrotheriops* or *Equus*. Accompanied by National Park Service personnel, Jim Mead and Larry Agenbroad (geologists from Northern Arizona University), visited the site to be known as Bechan Cave. A test pit in the floor of the cave yielded two large dung balls: one measured 230 by 170 by 85 mm and the other, subspherical, was roughly

225 mm in diameter. In the reference collection of modern and fossil mammalian megafaunal dung at the Desert Laboratory, these extraordinary fossils uniquely matched the size and texture of the manure of living elephants (Davis et al. 1984; Mead et al., *Mammuthus*, 1986).

The deposit within Cowboy Cave is mainly of bison manure; that from Bechan Cave is mainly of the newly discovered dung. While Bechan Cave is much larger, both caves are high enough to admit proboscideans. In both, trampling is extensive and entire boluses are exceptional. Trampling by ground sloths evidently disaggregated and compacted much of the ground sloth dung in Rampart Cave.

A trench was dug in Bechan Cave and entire samples or bolus fragments were dated from various depths. Initial dates of 11 670 ± 300 and 13 505 ± 580 BP on the two dung balls mentioned above suggested the possibility of a significant difference in age (Davis et al. 1984). Instead, 13 subsequent dates on graminoid remains from individual samples of dung averaged 12 400 BP, all within the range established above (Martin 1987b). Apparently the manure blanket in Bechan Cave was deposited rapidly within less than a few years, and perhaps within less than a month (Martin 1987b).

The fibrous plant remains of mammoth dung are mainly graminoids (grasses and sedges) with individual culms up to 60 mm in length and 4.5 mm in width. Macrofossils identified by O. K. Davis include seeds of Indian rice grass (*Oryzopsis*), sedges (*Cyperus, Carex*), and prickly pear (*Opuntia*). Other fecal pellets found in Bechan and adjacent caves or shelters, which approach those of elk in size, may, according to Mead and Agenbroad (1992), represent ejecta of shrub ox (*Euceratherium*). A pollen profile from Bechan Cave indicates that when mammals visited the cave, conditions were cooler, with spruce displaced some 900 m below its present lower limit in the adjacent Henry Mountains (Davis et al. 1984; Mead et al., *Mammuthus*, 1986).

The Bechan Cave mammoth dung deposit is over 1000 years older than the youngest dates averaged on ground sloths, condors, and extinct goats from Grand Canyon caves (Martin 1987b) and older than the upper part of the dung deposit of *Bison* from Cowboy Cave (Spaulding and Petersen 1980). Bechan and Cowboy Caves are not the only Utah caves to yield mammoth dung; boluses attributed to mammoth have been obtained in adjacent rock shelters (Mead and Agenbroad 1992). Elsewhere in the region, mammoth survived to 11 000 BP as shown both by mammoth bones within Clovis sites (C. V. Haynes 1993) and without them (Gillette and Madsen 1993). The Bechan Cave mammoth dung is 1000 years too old to coincide with the time of Clovis arrival and mammoth extinction in adjacent parts of the Southwest.

A virtually complete skeleton of a Columbian mammoth from Huntington Canyon in the Wasatch Mountains, central Utah, represents a record elevation for mammoths, 2700 m. The site sheds further insight on diet and on the importance of direct dating to establish the age of an individual that might well be contemporary with Clovis colonization. A senescent animal, it suffered from fused vertebrae and worn molars (Gillette and Madsen 1993). Determining the age of the deposit proved to be complicated. Radiocarbon dates on organic material above and below the animal (including dates on insect larvae) suggested survival to 10 000 years or less, as did an associated Goshen point. This would make the Huntington Canyon mammoth 1000 years younger than those from Clovis sites (C. V. Haynes 1993; Taylor et al. 1996), thought by many to be the last mammoths alive in the west.

Fortunately, bones of both the Huntington Canyon mammoth and an associated giant bear, *Arctodus* (Gillette and Madsen 1992), yielded ample bone collagen for accelerator dating (Stafford 1995, personal communication). The hydroxyproline amino acid fraction of purified collagen from mammoth bone (sample AA-4936) was $11\ 220 \pm 110$ BP. That on *Arctodus* collagen was $10\ 870 \pm 75$ BP (D. B. Madsen 1992, personal communication). The animals could be contemporary. Arguably, the Goshen point and embedded organic material dated around 10 000 radiocarbon years is a postmortem deposit. If so, it indicates the hazard of assuming contemporaneity of organic material intimately associated with fossil bones in a lacustrine environment. Well-dated mammoths in North America younger than Clovis sites remain unknown.

Organic material from the visceral region of the mammoth yielded the following results, which seemed baffling at first: Beta 29808, $11\ 020 \pm 110$ BP on bulk bolus with mammoth skeleton; AA-4934, $11,420 \pm 110$ BP on gut contents (a fir needle); and Pitt 543, $12,340 \pm 85$ BP on a bulk bolus associated with the mammoth skeleton (Gillette and Madson 1993). AA-4934 and Beta 29808 are in agreement with the bone dates and pose no problems. Pitt 543 is significantly older. How could dung dates differ in age?

Drought-stressed African elephants are known to ingest mud (G. Haynes 1991), and the same is suspected of the Hillsborough, New Brunswick, mastodon (Harington et al. 1993). Unlike the haylike or fibrous texture of dung from Bechan Cave, the muddy appearance of the gut contents of the Huntington Canyon mammoth suggests a moribund animal swallowing earth (D. D. Gillette and D. B. Madsen 1990, personal communication). In this peculiar case, ancient organic material in the soil or lake muds consumed by the animal before death could have "time averaged" Pitt 543. This would account for the discord between it and samples AA-4934 and Beta 29808. Whether or not I am right, the case of the Huntington Canyon mammoth illustrates the critical value of accelerator dating in investigating context.

Certain extinct genera of the late Pleistocene are as yet undated to within a thousand years of 11 000 BP, and some authors are not persuaded that the extinctions were spontaneous. They propose that some animals (those poorly dated) vanished at some time before 11 000 BP (Grayson 1989, 1991, 1993), perhaps similar to the gradual extinction of large animals in Eurasia (Stuart 1991; Martin and Stuart 1995). They reject a catastrophic extinction chronology implied by the line shown in figure 8.2.

There are many reasons to believe late Quaternary megafaunal extinction in the West was catastrophic. Except for apparent "manuports" (prehistoric trophies placed in a "rock shop" and other examples), no extinct genera have been found in archaeological sites after Clovis. The radiocarbon dates for all species that have been repeatedly dated terminate at 11 000 radiocarbon years, or perhaps slightly later (Stuart 1991). The dates come from populations at various elevations and latitudes (Martin 1987b, 1990). On stratigraphic or other geological evidence, no one has claimed extinction prior to 11 000 years ago for any of the undated genera listed in table 8.1. The harvest of dates (see Stuart 1991), with the more common genera being the ones more reliably dated, suggests a statistical effect, where "even sudden extinctions may appear gradual as sampling efficiency declines near the boundary" (Flessa 1990). Thus uncommon or rare genera will inevitably be more difficult to date and dated less often.

A final reason for doubting gradual loss of genera is the fossil record of scavengers. These include the California condors in the Grand Canyon, the saber-cat at Rancho La Brea, the dire wolf at the Murray Springs site in Arizona, and the giant bear in the Wasatch Mountains. All are dated to 11 000 BP, the doomsday millennium. Had the environment gradually lost biomass over the last few thousands of years prior to doomsday, one might see one or more predators or scavengers predecease the rest. No gradual loss of predators or scavengers is evident.

Clovis Colonization

The earliest robust evidence of human colonization in the western drylands is occasionally revealed in stratified alluvial or lacustrine deposits found in the states of Washington, Wyoming, Colorado, New Mexico, Oklahoma, Texas, and Arizona. Here the Clovis culture (Llano Complex) left stone tools—scrapers, knives, burins, and distinctive fluted points known as Clovis points, which ranged from 4 to 8 cm in length, with superbly crafted ceremonial blades over 20 cm long at East Wenatchee, Washington (Mehringer and Foit 1990). Nine stratified and critically dated Clovis sites fell within 400 years of 11 000 BP (uncorrected radiocarbon age estimates, C. V. Haynes 1993, corrected in Taylor et al. 1996). The deeply fluted points of the Folsom culture are barely younger than Clovis; at Blackwater Draw, where both occur together, Folsom points lie stratigraphically above Clovis (Taylor et al. 1996).

Claims for human occupation predating Clovis appear regularly. They have yet to be verified or replicated by independent investigators and, in fact, no established protocol exists for site verification. Ecologically attractive sites that might be expected to yield lithic evidence of pre-Clovis colonists include caves, lake shores, floodplains, and spring deposits. Well known to paleontologists, such sites are often the source of bones of extinct megafauna, especially camel, horse, bison, and mammoth. While they may be overlain by sediments yielding abundant Holocene lithics and ceramics, the pre-Holocene sediments have yet to disclose the long sought "holy grail"; that is, cultural remains demonstrably older than Clovis (Frison 1990; C. V. Haynes 1993). Even if we accept unverified claims, there is nothing in the New World before 12 000 BP to approach the rich Paleolithic record in the Old. Given the wealth of resources available to those who first gained a foothold in America (Martin 1987a), it would seem inevitable (to ecologists at least) that soon after the first Americans arrived south of the Laurentide ice sheet, they would have experienced a population explosion. The artifact scatter of such a population should be easily detected and, like the discovery of the Old World sites of Middle and Upper Paleolithic age, should have been made and abundantly verified at least a century ago. At this late date the intensity of the hunt for pre-Clovis sites, a discovery now long overdue if such sites actually exist, suggests something less than serious science, akin to the ever popular search for "Big Foot" or the "Loch Ness Monster."

Undaunted by skeptics or arguments, some archaeologists regularly advance claims of pre-Clovis sites. The existence of such sites is seriously entertained by others (see Dixon 1993). Nevertheless, the oldest *verifiable* archaeological record begins with Clovis. Clovis foragers hunted or, possibly, scavenged American mammoths. Eight of nine high quality sites are tightly dated around 11 000 BP (C. V. Haynes 1993); the ninth, the Aubrey site of central Texas, is dated around 11 500 BP. Nothing much older has been verified by independent investigators.

Mammoth butchering at Clovis sites was rather casual compared with intense butchering of African elephants hunted historically (G. Haynes 1991). As mentioned, the number of known stratified Clovis sites is small. The greatest concentration is in southeastern Arizona with four sites along the San Pedro River (Colorado River drainage) in Cochise County; they are Naco, Murray Springs, Lehner, and Escapule. The geochronology of the sites has been studied by C. V. Haynes (1991, 1993) and the paleontology by Saunders (1992). Here and elsewhere in the region, no diagnostic archaeological remains occur in bone-rich fossil units beneath or older than Clovis.

Possibly a few of the extinct North American mammals lasted beyond Clovis into Folsom or later time. Folsom association with western camel (*Camelops*) has been claimed in a few cases (Haynes and Stanford 1984). Secondary associations are known, as in the case of the shrine caves in the Grand Canyon mentioned above. More can be expected. Late Holocene records of mammoth molars at the prehistoric pueblo of Paquime, Casas Grandes, Chihuahua, are considered to be manuports (Martin 1990). While anticipated since the days of Jefferson, there is no robust evidence for Holocene survival of mammoth or other extinct megafauna in the New World to match the mid-Holocene survival of woolly mammoth recently reported on Wrangel Island in the Arctic Ocean off the Arctic coast of Siberia (Vartanyan et al. 1993). More to the point, the Eurasian sequence of extinction of large mammals reviewed in detail by Stuart (1991) is gradual, both within and between species, starting around 40 000 years ago or earlier. By comparison, the megafaunal extinction chronology of North America is spontaneous (Stuart 1991). The circumstance of Clovis colonization in America coinciding with megafaunal extinction looms large. In rate, if not in scale, it resembles extinctions of smaller animals on oceanic islands, also in step with human colonization (Burney 1993; Steadman 1995). While not established beyond doubt, the model of spontaneous extinction by prehistoric overkill continues to win favor among ecologists (Diamond 1992; Flannery 1995; Wilson 1992). What are some of the consequences?

The Incumbents

Some 40 000 years of change in climate, vegetation, megafauna, and ancient cultures sketched above left a major imprint on the modern biota of American drylands. Increasingly conservationists, managers, and futurists are coming to grips with the ways the fossils affect choice in planning. Bighorn (*Ovis*), collared peccary (*Tayassu*), elk, pronghorn (*Antilocapra*), two species of deer (*Odocoileus*), black and brown (grizzly) bear (*Ursus*), wolves (*Canis*), mountain lions (*Felis*), and jaguar (*Panthera*) that inhabit the U.S.-Mexican borderlands are not the full complement of natives, but the extinction-ravaged survivors of a much richer patrimony (table 8.1). Some of these (collared peccary, elk, and grizzly or brown bear [*Ursus arctos*]) have such poor fossil records in the southwest (see Harris 1993) that they must be quite recent invaders, as well as survivors of late Quaternary extinction. Traditionally only the incumbent animals, the survivors, won serious consideration by managers and planners. Such a shortsighted approach leaves much to be desired.

As noted previously, the more common fossils of the late Pleistocene deposits in western North America are those of large bovids, equids, camelids, and proboscideans. Vast amounts of genetic and behavioral information vanished with the extinction of these taxa. Fortunately, the loss is by no means totally irredeemable. Among the bovids, many breeds of *Bos* occupy the western range, while species of *Bison* and *Ovibos* survived. In parts of the west, wild horses and wild burros may, arguably, reoccupy fragments of the extinct equid niche. Afro-Asian camels (*Camelus*) and South American llamas (*Lama*) can restore the camelid way of life to the Sonoran Desert and the western range, at least at the familial if not the generic level. Llamas are increasingly popular as pack animals and in guarding sheep. Data on activity, forage consumption, and adaptability is available for the equids, bovids, and camelids. Only foraging behavior of elephants in the New World remains entirely unknown beyond the record of fossil dung mentioned earlier. The opportunity is at hand to initiate new research on the sustainability of the western rangelands based on substituting close living representatives for the megaherbivores that provided most of the late Quaternary fossils. Some may well prove to be keystone species whose absence seriously distorts our understanding of New World herbivory.

Ignoring deep history, planners have typically turned to 1492 for some sort of baseline (Leopold et al. 1963). This practice attributes important change to European influence and discounts any significant impacts by Native Americans (Wagner and Kay 1993). Historically, there has been a tradition of viewing European contact and European hunting practice as sounding the death knell to pristine America. Historical losses of the "nobler animals," including the extermination in New England of cougar, lynx, wolverine, wolf, bear, moose, and other wildlife, are mourned by Thoreau in his widely quoted "wish to know an entire heaven and an entire earth" (1984). There is no question that increasing numbers of European settlers blighted the native faunas. On the other hand, if disease and other traumas caused severe depopulation of Native Americans following European contact as Dobyns (1983), Crosby (1986), Reff (1991), and most anthropologists conclude, the immediate effect of contact, prior to settlement, may have been to release hunting pressure on wildlife (Rostlund 1960; Martin and Szuter, unpublished manuscript).

Today in the northern Rockies, elk occur locally in greater numbers than other big game. Zooarchaeological data are discordant. In Wyoming and six other adjacent states, 52 624 identified bones of megaherbivores from archaeological sites yielded the following counts: 29 585 (56%) deer, 12 301 (23%) bighorn, 4865 (9%) antelope, 4362 (8%) bison, 1511 (3%) elk, and one moose (Kay 1990, 325). From this Kay concludes that elk and bison were scarce prehistorically. In and near Yellowstone National Park, Kay (1990, 28) summarizes historical big game observations from 1835 to 1876 including sightings, sign, and animals killed. The proportions are 38% elk, 27% deer, 15% antelope, 13% bighorn, 5% bison, and 2% moose. Based on 1989 population estimates, elk now comprise 79% of total ungulates in the Yellowstone ecosystem, an elk herd Kay (1990, 1994) considers unprecedented in size. The shift in numbers of animals from mainly deer to mainly elk invites an explanation. Kay invokes relaxation of human predation. Smallpox in particular decimated Native American societies, and the arrival of the horse, guns, and alcohol caused widespread dislocations. In the larger sense, the increase in game animals on the land 300 years after Columbus was all part of the recovery from the late Quaternary faunal extinctions.

One may ask what the borderlands might have been like had its native fauna been totally free ranging and not subjected to any human hunting or anthropogenic habitat alteration before European settlement. The late prehistoric record suggests heavy utilization of animal resources by the Hohokam (Szuter 1991). While nineteenth century Arizona was not particularly rich in large game, in the time of the Hohokam large animals may have been even less numerous. Szuter's sample of 50 000 identified items yielded no javelina. In any case, the last unhunted landscape in the region (and in America) disappeared when the Clovis culture arrived 11 000 years earlier. Had hunting by Native Americans been inconsequential when Columbus arrived, the American range might well have seen buffalo ranging coast to coast, as in the late Pleistocene. In the absence of mammoths, mastodons, extinct horses, camelids, ground sloths, etc., *and in the absence of human hunters,* the survivors — bison, elk, deer, antelope, mountain goats, bighorn, and javelina and their associated wolves, bears, and jaguar — would have ranged throughout the west. Numbers of game animals — deer, elk, antelope, or bighorn — would have been considerably in excess of those present historically. Bones of these animals would be found much more commonly in all natural deposits of the Holocene, such as in caves, sink holes, alluvial, and lacustrine beds and springs, than is the case. In defining the range of native large mammals, investigators have ignored the role of human hunters as keystone predators. In the total absence of humans, the above scenario is at least plausible.

Prospect: Winning Usufruct

The late Quaternary record of biotic communities in Arizona and adjacent states includes ecologically informative, perishable organic deposits in dry sites. Rock shelters and caves have yielded over 1000 radiocarbon dates on fossil packrat middens and over 100 on tissue and dung of the megafauna of the late Quaternary. The regional data indicate that during the first 30 000 of the last 40 000 years, species distribution and, presumably, vegetation were unlike what we find now, with montane conifers typically occupying lower latitudes and lower elevations. Many Sonoran and Mojave Desert communities that now support desert shrubs were then occupied by juniper, pinyon, sagebrush, and other woodland species, indicative of a cooler climate. Evidently in the last cold stage, the structure and composition of biotic communities differed appreciably from the peculiar interglacial vegetation pattern seen at present. If fossil packrat middens and marine isotopic data are a reliable guide to the terrestrial climate, the types of desert communities with which we are familiar were not the norm during the last 600 000 years (Porter 1989).

The fossil record reveals sudden extinction of two thirds of the region's species of large mammals around 11 000 radiocarbon years ago (table 8.1). While the modern climate and vegetation is distinctly different from that of 12 000 years ago, the lost mammals had experienced and endured comparable (*interglacial*) conditions earlier in their evolutionary history, most notably in the Sangamon. No unique climatic pulse or biomass collapse has been detected that might explain when and why megafaunal extinctions occurred.

During most of the Quaternary, Arizona and adjacent states supported a rich megafauna reminiscent of the diversity of big game to be found in protected parts of Africa. The most common species in the fossil record of alluvial deposits are mammoth and extinct species of horse, camel, and bison. Grand Canyon caves are rich in dung deposits of an extinct mountain goat often associated with bones of the California condor. At least nine caves in the western drylands harbor dung of the extinct Shasta ground sloth, while two caves have yielded fossil dung of extinct mammoth. The perishable material and collagen-rich bones are ideally suited for direct radiocarbon measurement. Suites of dates available for the more common species and some rare ones indicate last occurrences at or around 11 000 years ago, coincidental with late glacial shifts in plants and with invasion of Clovis foragers. Whatever caused extinction (to me cultural forcing seems much more plausible than climatic change), the Quaternary fossils make it possible to envision a much more diversified array of species of large animals than are now considered native to the western range. If diversity of megafauna is a measure of "wildness," no wilder west has been seen since.

In considering options for restoration of western rangelands, conservation biologists search for a baseline. Until recently, they have standardized on what was on the land five centuries ago (before Euroamerican "first contact" with Native Americans). Inevitably this approach showcased the Holocene survivors exclusively. Lewis and Clark's accounts of abundant and tame bison, deer, elk, wolves, and bears on the Upper Missouri were taken as entirely natural and idealized. This "top down" view rarely descended into evolutionary time with its extinct mammoths and other megafauna of the Quaternary.

The view "bottom up" is through an unknown biomass of high diversity, dozens of genera of large mammals including proboscideans, camelids, equids, bovids, tayassuids, gravigrade edentates, machairodonts, giant bears, and lions, not to mention teratorns, condors, and other large scavenging or commensal birds. While these animals vanished under climates unknown to us historically, they had also experienced interglacial climates akin to ours and endured whatever climatic switching occurred in earlier times. In the southwest of the United States, only one third of the species of large mammals native to the region in the late Quaternary managed to survive (table 8.1), and even some of these escape mention in early historical accounts.

For managers, deep history (the "bottom up" view) provides an opportunity for presenting the aboriginal American patrimony to the public; one focused not only on a richer fauna incorporating extinct species but also on a much earlier colonization of America, out of Asia. The extinct megafauna prompts new thoughts on how to design with nature in the broadest sense. For example, are cattle, as a surrogate species for native *Bison* and the other bovids on the western range, the only or the best choice? More than that, are bovids alone to be granted usufruct? Why not equids, camelids, and even proboscideans? What was the role on the Western range of native species of proboscideans—the mammoths, mastodons, and gomphotheres? Before riparian ecologists can claim to understand the dynamics of streamside and riverside vegetation in the arid West, it might help to examine the impact on a New World floodplain of a free-ranging experimental herd of Indian or African elephants. They are potential vicars for the extinct Columbian mammoths that left so many bones along the ice age waterways of the West.

Having returned to extinct proboscidea in the arid West, a theme initiated by D.T. MacDougal when he was director of the Desert Laboratory, it goes without saying that deep history and its ramifications are vastly enriched by radiocarbon dating of the perishable fossils of this region. By considering the changes disclosed by radiocarbon age estimates, we can approach conservation biology from its roots. We can offer much more choice to managers to consider in their dilemma of granting usufruct. We can assist ecologists in their effort at determining the biotic potential of the land. A wilder west is at stake.

Acknowledgments

My views and visions of ice age biogeography in and around Arizona, on the distribution and extinction of western megafauna in the last ice age, and the meaning of these events were shaped, often at lunch, over the last 40 years by students, staff, professors, ranchers, Earth Firsters, and other friends and neighbors of the Desert Laboratory on Tumamoc Hill, overlooking the city of Tucson. Some may even share some of them. I especially thank Julio Betancourt, Janice Bowers, Jim Brown, Martha Ames Burgess, Tony Burgess, Wayne Burkhardt, Ken Cole, Russell Davis, Betty Fink, Karl Flessa, Vance Haynes, Charles Kay, Mary Ellen Morbeck, James Mosimann, Gary Nabhan, Mary Kay O'Rourke, Betsy Pierson, Jay Quade, Jennifer Shopland, Geof Spaulding, Tom Stafford, David Steadman, Chris Szuter, Robin Tausch, Thomas Van Devender, Peter Van de Water, Fred Wagner, and Robert Webb. Jo Ann Overs patiently processed innumerable drafts of a cantankerous manuscript.

Literature Cited

Betancourt J. L., Van Devender T. R., Martin P. S. (1990) *Packrat Middens: The Last 40,000 Years of Biotic Change*. University of Arizona Press, Tucson.

Bowers J. E. (1988) *A Sense of Place: The Life and Work of Forrest Shreve*. University of Arizona Press, Tucson.

Burney D. A. (1993) Recent animal extinctions: recipes for disaster. *American Scientist* 81: 530–541.

Caughley G., Gunn A. (1996) *Conservation Biology in Theory and Practice*. Blackwell Science, Cambridge, Mass.

Cole K. L. (1985) Past rates of change, species richness, and a model of vegetational inertia in the Grand Canyon, Arizona. *American Naturalist* 125: 289–303.

Cole K. L. (1995) Equable climates, mixed assemblages and the regression fallacy. In: Mead J. I., Steadman D. W. (eds) *Late Quaternary Environments and Deep History*, pp 131–138. The Mammoth Site of Hot Springs, South Dakota, Inc. Scientific Papers, Vol. 3.

Crosby A. W. (1986) *Ecological Imperialism, the Biological Expansion of Europe, 900–1900*. Cambridge University Press, New York.

Dansgaard W., Johnsen S. J., Clausen H. B., Dahl-Jensen D., Gundestrup N. S., Hammer C. U., Hvidberg C. S., Steffensen J. P., Sveinbjornsdottir A. E., Jouzel J., Bond G. (1993) Evidence for general instability from a 250-kyr ice-core record. *Nature* 364: 218–220.

Davis O. K., Agenbroad L. D., Martin P. S. (1984) The Pleistocene dung blanket of Bechan Cave, Utah. In: Genoways H. H., Dawson M. R. (eds) *Contributions in Quaternary Vertebrate Paleontology: A Volume in Memorial of John E. Guilday*, pp 267–282. Carnegie Museum of Natural History Special Publication Vol. 8, Pittsburgh.

Diamond J. (1992) *The Third Chimpanzee: The Evolution and Future of the Human Animal*. Harper-Collins, New York.

Dixon E. J. (1993) *Quest for the Origins of the First American*. University of New Mexico Press, Albuquerque.

Dobyns H. F. (1983) *Their Number Became Thinned*. University of Tennessee Press, Knoxville.

Edwards R. L., Beck J. W., Burr G. S., Donahue D. J., Chappell J. M. A., Bloom A. L., Druffel E. R. M., Taylor F. W. (1993) A large drop in atmospheric $^{14}C/^{12}C$ and reduced melting in the Younger Dryas, documented with ^{230}Th ages of corals. *Science* 260: 962–968.

Elias S. A. (1992) Late Quaternary zoogeography of the Chihuahuan Desert insect fauna, based on fossil records from packrat middens. *Journal of Biogeography* 19: 285–297.

Emslie S. D. (1987) Age and diet of fossil California condors in Grand Canyon, Arizona. *Science* 237: 768–770.

Emslie S. D., Mead J. I., Coats L. (1995) Split-twig figurines in Grand Canyon, Arizona: new discoveries and interpretations. *Kiva* 61: 145–173.

Euler R. C. (ed) (1984) *The Archaeology, Geology and Paleobiology of Stanton's Cave*. Grand Canyon Natural History Association Monograph No. 6, Grand Canyon, Ariz.

Flannery T. F. (1995) *The Future Eaters: An Ecological History of the Australian Lands and People*. Reed Books, Chatswood, NSW, Australia.

Flessa K. W. (1990) The "facts" of mass extinctions. *Geological Society of America Special Paper* 247: 1–7.

Frison E. C. (1990) Clovis, Goshen and Folsom: lifeways and cultural relationships. In: Agenbroad L. D., Mead J. I., Nelson L. W. (eds) *Megafauna and Man: Discovery of America's Heartland*. The Mammoth Site of Hot Springs, South Dakota, Inc. Scientific Papers, Vol. 1.

Gillette D. D., Madsen D. B. (1992) The short-faced bear *Arctodus simus* from the Late Quaternary in the Wasatch Mountains of central Utah. *Journal of Vertebrate Paleontology* 12: 107–112.

Gillette D. D., Madsen D. B. (1993) The Columbian mammoth, *Mammuthus columbi*, from the Wasatch Mountains of central Utah. *Journal of Paleontology* 87: 669–680.

Grayson D. K. (1989) The chronology of North American late Pleistocene extinctions. *Journal of Archaeological Science* 16: 153–165.

Grayson D. K. (1991) Late Pleistocene mammalian extinctions in North America: taxonomy, chronology, and explanations. *Journal of World Prehistory* 5: 193–231.

Grayson D. K. (1993) *The Desert's Past: A Natural Prehistory of the Great Basin.* Smithsonian Institution Press, Washington, D.C.

Guthrie R. D. (1984) Mosaics, allelochemics, and nutrients: an ecological theory of late Pleistocene megafaunal extinction. In: Martin P. S., Klein R. G. (eds) *Quaternary Extinctions: A Prehistoric Revolution,* pp 259–298. University of Arizona Press, Tucson.

Hansen R. M. (1978) Shasta ground sloth food habits, Rampart Cave, Arizona. *Paleobiology* 4: 302–319.

Hansen R. M. (1980) Late Pleistocene plant fragments in the dungs of herbivores at Cowboy Cave. In: Jennings J. D. (ed) *Cowboy Cave,* pp 179–189. University of Utah Anthropological Papers 104, Salt Lake City.

Hansen R. M., Martin P. S. (1973) Ungulate diets in the lower Grand Canyon. *Journal of Range Management* 26: 380–381.

Harington C. R., Grant D. R., Mott R. J. (1993) The Hillsborough, New Brunswick, mastodon and comments on other Pleistocene mastodon fossils from Nova Scotia. *Canadian Journal of Earth Science* 30: 1242–1253.

Harrington M. R. (1933) *Gypsum Cave, Nevada.* Southwest Museum Papers, No. 8, Los Angeles.

Harris A. H. (1993) Quaternary vertebrates of New Mexico. In: Lucas S. G., Zidek J. (eds) *New Mexico Museum of Natural History and Science Bulletin 2,* pp 179–191. Albuquerque.

Haynes C. V. Jr (1991) Geoarchaeological and paleohydrological evidence for a Clovis-age drought in North America and its bearing on extinction. *Quaternary Research* 35: 438–450.

Haynes C. V. Jr (1993) Contributions of radiocarbon dating to the geochronology of the peopling of the New World. In: Taylor R. E., Long A., Kra R. S. (eds) *Radiocarbon Dating After Four Decades,* pp 355–374. Springer-Verlag, New York.

Haynes G. (1991) *Mammoths, Mastodonts and Elephants.* Cambridge University Press, Cambridge.

Haynes G., Stanford D. (1984) On the possible utilization of *Camelops* by early man in North America. *Quaternary Research* 22: 216–230.

Heizer R. F., Berger R. (1970) *Radiocarbon Age of Gypsum Cave Culture.* Contributions of the University of California (Berkeley) Archaeological Research Facility 7: 13–18.

Imbrie J., Imbrie H. P. (1979) *Ice Ages: Solving the Mystery.* Enslow Publishers, Holland, N.J.

Janzen D., Martin P. S. (1982) Neotropical anachronisms: the fruits the gomphotheres ate. *Science* 215: 19–27.

Johnson H. B., Mayeux H. S. (1992) Viewpoint: a view on species additions and deletions and the balance of nature. *Journal of Range Management* 45: 322–333.

Kay C. (1990) Yellowstone's northern elk herd: a critical evaluation of the "natural regulation" paradigm. Ph.D. dissertation, Utah State University, Logan.

Kay C. (1994) Aboriginal overkill: the role of native Americans in structuring western ecosystems. *Human Nature* 5: 359–398.

Leopold A. S., Cain S. A., Cottam C. M., Gabrielson I. N., Kimball T. L. (1963) Wildlife management in the national parks. *Transactions of the North American Wildlife and Natural Resources Conference* 24: 29–44.

Lundelius E. L., Graham R. W., Anderson E., Guilday J., Holman J. A., Steadman D., Webb D. (1983) Terrestrial vertebrate faunas. In: Porter S. C. (ed) *Late Quaternary Environments of the United States*, Vol. 1, *The Late Pleistocene*, pp 311–353. University of Minnesota Press, Minneapolis.

MacDougal D. T. (1908) *Botanical Features of North American Deserts*. Carnegie Institution of Washington Publication no. 99, Washington, D.C.

Martin P. S. (1987a) Clovisia the beautiful. *Natural History* 96: 10–13.

Martin P. S. (1987b) Late Quaternary extinctions: the promise of TAMS C-14 dating. *Nuclear Instruments and Methods in Physics Research* B29: 179–186.

Martin P. S. (1990) Who or what destroyed our mammoths? In: Agenbroad L. D., Mead J. I., Nelson L. W. (eds) *Megafauna and Man: Discovery of America's Heartland*, pp. 109–117. The Mammoth Site of Hot Springs, South Dakota, Inc. Scientific Papers Vol. 1.

Martin P. S., Klein R. G. (eds) (1984) *Quaternary Extinctions: A Prehistoric Revolution*. University of Arizona Press, Tucson.

Martin P. S., Sabels B., Shutler D. Jr (1961) Rampart Cave coprolite and ecology of the Shasta ground sloth. *American Journal of Science* 259: 102–127.

Martin P. S., Stuart A. J. (1995) Mammoth extinction: two continents and Wrangel Island. *Radiocarbon* 37: 7–10.

Martin P. S., Thompson R. S., Long A. (1985) Shasta ground sloth extinction: a test of the blitzkrieg model. In: Mead J. I., Meltzer D. J. (eds) *Environments and Extinctions: Man in Late Glacial North America*, pp 5–14. Center for the Study of Early Man, Orono, Maine.

McGinnies W. G. (1981) *Discovering the Desert: Legacy of the Carnegie Desert Botanical Laboratory*. University of Arizona Press, Tucson.

Mead J. I., Agenbroad L. D. (1992) Isotope dating of Pleistocene dung deposits from the Colorado Plateau, Arizona and Utah. *Radiocarbon* 34: 1–19.

Mead J. I., Agenbroad L. D., Davis O. K., Martin P. S. (1986) Dung of *Mammuthus* in the arid southwest, North America. *Quaternary Research* 25: 121–127.

Mead J. I., Martin P. S., Euler R. C., Long A., Jull A.J.T., Toolin L. J., Donahue

O. J., Linick T. W. (1986) Extinction of Harrington's mountain goat. *Proceedings of the National Academy of Sciences* 83: 836–839.

Mead J. I., O'Rourke M. K., Foppe T. M. (1986) Dung and diet of the extinct Harrington's mountain goat (*Oreamnos harringtoni*). *Journal of Mammalogy* 67: 284–293.

Mead J. I., Phillips A. M. III (1981) The late Pleistocene and Holocene fauna and flora of Vulture Cave, Grand Canyon, Arizona. *The Southwestern Naturalist* 26(3): 257–288.

Mehringer P. J., Foit F. F. (1990) Volcanic ash dating of the Clovis cache at East Wenatchee, Washington. *National Geographic Research* 6(4): 495–503.

Mosimann J., Martin P. S. (1975) Simulating overkill by Paleoindians. *American Scientist* 63: 304–313.

Nowak C. L., Nowak R. S., Tausch R. J., Wigand P. E. (1994) A 30 000 year record of vegetation dynamics at a semi-arid locale in the Great Basin. *Journal of Vegetation Science* 5: 579–590.

Porter S. C. (1989) Some geological implications of average Quaternary conditions. *Quaternary Research* 32: 245–261.

Reff D. T. (1991) *Disease, Depopulation, and Culture Change in Northwestern New Spain, 1518–1764*. University of Utah Press, Salt Lake City.

Robbins E. I., Martin P. S., Long A. (1984) Paleoecology of Stanton's Cave, Grand Canyon, Arizona. In: Euler R. C. (ed) *The Archaeology, Geology, and Paleobiology of Stanton's Cave,* pp 117–130. Grand Canyon Natural History Association Monograph no. 6, Grand Canyon, Ariz.

Rostlund E. (1960) The geographic range of the historic bison in the southeast. *Annals Association American Geographers* 50(4): 395–407.

Saunders J. J. (1992) Blackwater draws: mammoth and mammoth hunters in the terminal Pleistocene. In: Fox J. W., Smith C. B., Wilkins K. T. (eds) *Proboscidean and Paleoindian Interactions,* pp. 123–147. Baylor University Press, Waco, Tex.

Semken H. A. (1988) Environmental interpretations of the "disharmonious" Late Wisconsinan biome of southeastern North America. In: Laub R. S., Miller N. G., Steadman D. W. (eds) *Late Pleistocene and Early Holocene Paleoecology and Archeology of the Eastern Great Lakes Region,* pp 185–194. Bulletin of the Buffalo Society of Natural Sciences, vol. 33. Buffalo, N.Y.

Spaulding W. G. (1990) Vegetational and climatic development of the Mojave Desert: the last glacial maximum to the present. In: Betancourt J. L., Van Devender T. R., Martin P. S. (eds) *Packrat Middens: The Last 40,000 Years of Biotic Change,* pp 166–199. University of Arizona Press, Tucson.

Spaulding W. G., Petersen K. L. (1980) Late Pleistocene and early Holocene paleoecology of Cowboy Cave. *University of Utah Anthropology Papers* 104: 163–177.

Steadman D. W. (1995) Prehistoric extinctions of Pacific island birds: biodiversity meets zooarchaeology. *Science* 267: 1123–1131.
Stuart A. J. (1991) Mammalian extinctions in the late Pleistocene of northern Eurasia and North America. *Biological Reviews* 66: 453–562.
Szuter C. R. (1991) *Hunting by Prehistoric Horticulturists in the American Southwest*. Garland, New York and London.
Tausch R. J., Wigand P. E., Burkhardt J. W. (1993) Viewpoint: plant community thresholds, multiple steady states, and multiple successional pathways: legacy of the Quaternary? *Journal of Range Management* 46: 439–447.
Taylor K. C., Lamorey G. W., Doyle G. A., Alley R. B., Grootes P. M., Mayewski P. A., White J.W.C., Barlow K. L. (1993) The "flickering switch" of late Pleistocene climate change. *Nature* 361: 432–436.
Taylor R. E., Haynes C. V., Stuiver M. (1996) Calibration of the Late Pleistocene radiocarbon time scale: Clovis and Folsom age estimates. *Antiquity* 70: 515–525.
Thompson R. S. (1988) Western North America: vegetation dynamics in the western United States; modes of response to climatic fluctuations. In: Huntley B., Webb T. (eds) *Vegetation History*, pp 415–458. Kluwer Academic Publishers, Dordrecht, Boston.
Thoreau H. D. (1984) *The Journal of Henry David Thoreau*, Vol.8. Edited by B.Torrey and F.H. Allen, reprinted by Peregrine Smith Books, Salt Lake City.
Van Devender T. R. (1990a) Late Quaternary vegetation and climate of the Chihuahuan Desert, United States and Mexico. In: Betancourt J. L., Van Devender T. R., Martin P. S. (eds) *Packrat Middens: The Last 40,000 Years of Biotic Change*, pp 104–133. University of Arizona Press, Tucson.
Van Devender T. R. (1990b) Late Quaternary vegetation and climate of the Sonoran Desert, United States and Mexico. In: Betancourt J. L., Van Devender T. R., Martin P. S. (eds) *Packrat Middens: The Last 40,000 Years of Biotic Change*, pp 134–165. University of Arizona Press, Tucson.
Van Devender T. R., Burgess T. L. (1985) Late Pleistocene woodlands in the Bolson de Mapimi: a refugium for the Chihuahuan desert biota? *Quaternary Research* 24: 346–353.
Van Devender T. R., Martin P. S., Thompson R. S., Cole K. L., Jull A.J.T., Long A., Toolin L. J., Donahue D. J. (1985) Fossil packrat middens and the tandem accelerator mass spectrometer. *Nature* 317: 610–613.
Van de Water P. K., Leavitt S. W., Betancourt J. L. (1994) Trends in stomatal density and $^{13}C/^{12}C$ ratios of *Pinus flexilis* needles during last glacial-interglacial cycle. *Science* 264: 239–243.
Vartanyan S. L., Garrutt V. E., Sher A. V. (1993) Holocene dwarf mammoths from Wrangel Island in the Siberian Arctic. *Nature* 362: 337–340.
Wagner F. H., Kay C. E. (1993) "Natural" or "healthy" ecosystems: are U.S.

national parks providing them? In: McDonnell M. J., Pickett S. T. (eds) *Humans as Components of Ecosystems,* pp 257–270. Springer-Verlag, New York.

Webb R. H. (1995) *A Century of Environmental Change in Grand Canyon: Repeat Photography of the 1889–1890 Stanton Expedition.* University of Arizona Press, Tucson.

Webb R. H., Betancourt J. L. (1990) The spatial and temporal distribution of radiocarbon ages from packrat middens. In: Betancourt J. L., Van Devender T. R., Martin P. S. (eds) *Pack Rat Middens: The Last 40,000 Years of Biotic Change,* pp 85–102. University of Arizona Press, Tucson.

Wells P. V., Jorgensen C. D. (1964) Pleistocene wood rat middens and climatic change in Mojave Desert — a record of juniper woodlands. *Science* 143: 1171–1174.

Wilson E. O. (1992) *The Diversity of Life.* W.W. Norton, New York.

Winograd I. J., Landwehr J. M., Ludwig K. R., Copen T. B., Riggs A. C. (1997) Duration and structure of the past four interglaciations. *Quaternary Research* 48; 141–154.

Contributors

Janice E. Bowers, U.S. Geological Survey, Water Resources Division, 1675 W. Anklam Rd., Tucson, AZ 85745

Alberto Búrquez, Instituto de Ecología, Universidad Nacional Autónoma de México, Apartado Postal 1354, Hermosillo, Sonora 83000, México

Patricia W. Comus, Desert Botanical Garden, 1201 N. Galvin Pkwy., Phoenix AZ 85008

Garry A. Duncan, Department of Biology, Nebraska Wesleyan University, Lincoln, NE 68504

William J. Etges, Department of Biological Sciences, University of Arkansas, Fayetteville, AR 72701

Richard S. Felger, Drylands Institute, 2509 N. Campbell Ave., #405, Tucson, AZ 85719

William B. Heed, Department of Ecology & Evolutionary Biology, University of Arizona, Tucson, AZ 85721

Greg Huckins, Department of Ecology & Evolutionary Biology, University of Arizona, Tucson, AZ 85721

Laura L. Jackson, Department of Biology, University of Northern Iowa, Cedar Falls, IA 50614

William R. Johnson, Department of Ecology & Evolutionary Biology, University of Arizona, Tucson, AZ 85721

Contributors

Michael E. Loik, Department of Environmental Studies, University of California, Santa Cruz, CA 95064

Paul S. Martin, Desert Laboratory, Department of Geosciences, University of Arizona, Tucson, AZ 85721

Angelina Martínez-Yrízar, Instituto de Ecología, Universidad Nacional Autónoma de México, Apartado Postal 1354, Hermosillo, Sonora 83000, México

Joseph R. McAuliffe, Desert Botanical Garden, 1201 N. Galvin Pkwy., Phoenix AZ 85008

Steven P. McLaughlin, Bioresources Research Facility, 250 E. Valencia Rd., Tucson, AZ 85706

Park S. Nobel, Department of Organismic Biology, Ecology, and Evolution, University of California, Los Angeles, CA 90095

Catherine E. Pake, Portland Community College, Rock Creek Campus, Rm. 7202, P.O. Box 19000, Portland, OR 97280

D. Lawrence Venable, Department of Ecology & Evolutionary Biology, University of Arizona, Tucson, AZ 85721

David Yetman, Southwest Center, University of Arizona, 1052 N. Highland Ave., Tucson, AZ 85721

About the Editor

Robert H. Robichaux is an associate professor in the Department of Ecology and Evolutionary Biology, University of Arizona, Tucson. His research focuses on plant diversity, evolution, and conservation, with an emphasis on plants of the Sonoran Desert and the Hawaiian Islands.

Index

Acacia spp., 45, 97
Agave spp., 27, 146, 158–59, 260
agriculture, 10, 215; and groundwater pumping, 216–17, 230–33; in lower Santa Cruz Valley, 218, 223–24, 227–38
Ak Chin Reservation, 221
alluvial fans, 73, 76, 90
Ambrosia (bursage): *deltoidea* (triangle-leaf), 49, 84, 93, 94(fig.), 95(fig.), 96, 97, 99, 100, 119; *dumosa* (white), 5, 84, 261
Analysis of Molecular Variance (AMOVA), 177, 186, 187, 190, 200, 202–3n. 4
Angel de la Guarda, Isla, 60
animals, 117; and cacti, 153–54; and plants, 9–10, 130, 164; Quaternary, 255, 256(table), 257; seed dispersal by, 265–66. *See also* megafauna
annuals, 3, 6, 9, 27, 42, 49; bet hedging in, 136–37; lower Santa Cruz, 220–21; population dynamics of, 117–18, 119–23; seed dispersal of, 130–36; seed size of, 125–27; Sonoran Desert, 115, 116(fig.); species coexistence in, 127–30; study plots of, 118–19, 124(fig.)
apical region, 144, 146, 150, 158
archaeology, 81–82, 216, 228, 269, 274, 275, 277–78, 281
Archaic tradition, 269
argillic horizons, 88, 89, 90, 91; plant associations with, 93, 95(fig.), 96, 97, 99, 100, 101, 102(fig.), 103, 248

Arizona, 5, 21, 22, 147, 153, 216, 261, 265; *Drosophila* in, 171, 175, 177, 201; floristic research in, 15, 16, 19. *See also various features; regions; towns*
Arizona City, 245(fig.), 247
Arizona Upland District, 22, 23, 24, 27, 30, 49
Arizona Upland subdivision, 8, 9, 20, 25, 193; mountainside terrain of, 99–101; physiographic units in, 6, 7; soils in, 88–89, 93–99; as thornscrub, 53–54
Artemisia spp. (sagebrush), 261, 262
Asteraceae, 27, 119
Astragalus spp., 27
Atriplex spp. (saltbush), 27, 57, 218, 220, 225, 226(fig.), 227, 247, 248, 249, 270; *confertifolia* (shadscale), 261, 262, 264

Baccharis sarothroides (desert broom), 227, 243, 248, 249, 250
Bacoachi, Río, 41
Baja California, 21, 31, 39, 50, 51, 68, 85, 144; columnar cacti in, 191–92; creosote bush in, 79, 81(fig.); *Drosophila* in, 168, 169, 171, 173, 174–75, 177, 181, 187, 188, 190–91, 193, 197, 200; geology in, 82, 84, 101–3; Sonoran Floristic Province in, 12, 16, 20; species diversity in, 25, 26, 28
Baja California Norte, 5, 16
Baja California Sur, 5, 16, 37, 53

Index

bajadas. *See* piedmonts, alluvial
barrel cactus. *See Ferocactus*
Basin and Range Province, 21, 39–40
Baucarit Formation, 40
Bechan Cave, mammoth dung from, 273–75
bet hedging, 125, 137
Bida Cave, 268
biomass productivity, 48–50
Bison spp. (bison), 265, 267, 272, 274, 279, 280
Black Mountains (Death Valley), 23
Blackwater Draw, 277
boojum tree. *See Fouquieria columnaris*
Boraginaceae, 27, 119
Brahea elegans (palm), 3, 46
Brassicaceae, 27
Brickellia spp., 27
brittlebush. *See Encelia farinosa*
Bromus rubens, 131, 249
burrobush. *See Hymenoclea salsola*
bursage. *See Ambrosia*
Bursera spp., 3, 6, 58, 59

Cabeza Prieta National Wildlife Refuge, 30
Cabo San Lucas, 188, 189, 197
Cactaceae, cacti, 27, 96, 143, 145(fig.), 248, 265, 270; and animal-plant associations, 9–10; columnar, 3, 45, 54, 58; decay of, 164–65; and nurse plants, 46, 157–59; ribs and tubercles of, 148–50; roots of, 154–57; spines and pubescence on, 150–54; stem morphology and orientation of, 144, 146–48. *See also various genera; species*
calcic horizons, 93, 100; development of, 88–90; and erosion, 90–91
calcium carbonate, 89–90, 103, 248
California, 5, 14, 15, 16, 21, 27, 50, 147, 216, 261; desert subdivisions in, 19–20; *Drosophila* in, 190, 201; Sonoran Floristic Province in, 12, 23, 24, 25. *See also various features; regions; towns*
California, Gulf of, 20, 36, 169
California Desert District, 22, 23, 24, 27, 30
California Floristic Province, 30
CAM. *See* Crassulacean acid metabolism
Camelops hesternus (western camel), 255, 265, 267, 272, 278

Camissonia spp., 27
Canal del Infiernillo, 57
Canis spp., 279; *dirus* (dire wolf), 273, 276
Canyonlands, 266–67
Cape Region, 25, 26, 177, 188, 189–90
carbon dioxide (CO_2) uptake, 144, 146, 147, 148, 149, 150, 155, 159–60, 264
cardón. *See Pachycereus pringlei*
Carnegiea gigantea (saguaro), 3, 45, 46, 58, 143, 144, 145(fig.), 148, 155, 158, 255, 262; on mountainsides, 99, 100; ribs of, 149, 150; spines and pubescence on, 151, 152(fig.)
Carnegie Institution of Washington, 3, 255. *See also* Desert Laboratory
Casa Grande, 225, 228
Casa Grande Mountain, 221, 223
Castle Dome Mountains, 25, 30
caves in Grand Canyon, 267–73
Central Arizona Project, 237, 238, 249
Central Gulf Coast subdivision, 6, 7, 48, 58, 175, 176, 179, 187, 190
Cercidium spp., 46, 47, 58, 59, 255, 262; *floridum* (blue paloverde), 14, 158, 261; *microphyllum* (foothills paloverde), 45, 93, 95(fig.), 97, 157–58
Chamaesyce spp., 27
channelization, 235, 237
Chihuahuan Desert, 21, 87, 260, 262
cholla. *See Opuntia*
chorology: and numerical studies, 21–25; and phytogeography, 19–21
chromosomes, 171; *Drosophila*, 173–87, 209–14(tables); polytene, 165, 167–68
Chuckwalla Mountains, 19
Chuichu, 221
cladodes, 147–48, 159–60
climate, 43, 218; and inverse polymorphism, 192–94; and plant associations, 93, 96; Quaternary, 257–58; Sonoran Desert, 37–39; and vegetation, 51–53, 68. *See also* climate change; weather
climate change, 10, 255, 264, 269, 281–82
Clovis culture, 274, 275, 276, 277, 278–79, 282
Coahuila, 262
coast, 57
coastal plain, 40–41
Coastal Thornscrub, 53, 54–55

Colorado Desert, 19–20, 22, 28, 30, 31
Colorado Plateau, 21
Colorado River, 20, 24, 37
competition, 84, 127, 129–30, 137, 144
Coolidge Dam, 224, 228
Cordilleran Province, 14, 23
Cowboy Cave, 273, 274
Crassulacean acid metabolism (CAM), 143, 146
creosote bush. *See Larrea tridentata*
Cretaceous geology, 40
Cryptantha spp., 27
cryptophytes, 42

Dalea s.s. spp., 27
dams, 216
Death Valley, 15, 19, 20, 21, 23, 82, 84
deep history, 10–11, 257, 280, 283
Desemboque, 169, 175, 200
desert broom. *See Baccharis sarothroides*
Desert Laboratory, 3, 15, 16, 69, 76, 119, 137, 148–49, 255
desert pavement, 103, 104(fig.), 105, 106(fig.), 107
desert remnants: in lower Santa Cruz Valley, 240–41, 242(fig.), 243; vegetation on, 247–48
desert scrub, 58, 218
Desert Soil–Geomorphology Project, 87, 88
distance trees, 10; *Drosophila,* 195–99
diversity, 7–8, 15, 129–30; gamma, 27–30
Drosophila spp. (pomace flies), 10, 202n. 1; *arizonae,* 197; and climate, 192–94; *mojavensis,* 165, 168, 169, 171, 173, 175, 177, 178(fig.), 179, 180(fig.), 181, 182(table), 183, 186, 187, 190–92, 193, 197, 199, 200, 202–3nn. 3, 4, 209–11(table); *pachea,* 165, 168, 169, 173, 174–75, 176(fig.), 177, 179, 181, 183, 184(table), 186, 187–90, 193, 199, 200, 201, 202–3n. 4, 212–14(table); *pseudoobscura* and *subobscura,* 168, 179, 185, 186, 199
drought, 255, 275–76
dunes, coastal, 57

Eastern Imperial County, 23, 25, 27
Echinocactus horizonthalonius (Turk's head barrel cactus), 262, 264

Echinocereus spp., 100, 144, 148, 155, 159
ecoclines in Sonora, 43–47
El Aguaje, Sierra, 41, 45, 46, 53
El Bacatete, Sierra, 41, 45, 53
elephant tree. *See Pachycormus discolor. See also Bursera*
Eloy, 221, 229, 230
Encelia farinosa (brittlebush), 45, 243, 261
ephemerals, 42, 44, 49, 227
epiphytes, 54
Eriastrum diffusum (miniature wool star), 131
Eriogonum spp., 27, 101
Eriophyllum lanosum (wooly daisy), 119
Erodium spp., 119, 227, 249
erosion, 90–91, 94(fig.), 96–97, 99, 229
estuaries, 39, 41, 57
Euphorbiaceae, 27
Evax multicaulis, 119
exotic species, 115, 117(fig.), 119, 249, 250
extinctions, Quaternary, 10, 255, 257, 264–65, 266(fig.), 267, 269–73, 276–77, 278–79

Fabaceae, 27
farmland, abandoned, 217, 218–20, 226(fig.), 238–40, 241–47, 248–49, 250
Ferocactus spp. (barrel cactus), 146, 148, 153, 154, 172(fig.), 226(fig.); *acanthodes,* 146, 149(fig.), 150, 151, 155, 157, 158; *cylindraeous* [= *acanthodes*] (California), 171, 179
Ficus spp. (figs), 46, 69
fitness and seed size, 125–27
flies, 164; pomace, *see Drosophila*
flooding: in lower Santa Cruz, 224, 229, 237, 238, 249, 250; piedmont, 79, 81–82
floodwater farming, 215–16, 221, 223
floras, 12; local and regional, 14–15, 16–19; numerical studies of, 21–25; phytogeography of, 19–21
Florence–Casa Grande Canal, 221, 223, 228–29, 241
floristic areas, 12, 14
Folsom culture, 277, 278
Foothills of Sonora subdivision, 6, 7, 43
Foothills Thornscrub, 53, 55
forbs, 57, 131, 132, 133(fig.), 134. *See also various species*

fossils, 257; and biotic change, 255, 258–62, 281–82; in Grand Canyon, 267–73
Fouquieria: columnaris (boojum tree), 60, 101, 85, 86(fig.); *splendens* (ocotillo), 58, 97, 119
freezes, 39, 50, 150, 158

gene mapping, 165, 167–68
genetic variation, 10, 164–65
geology, 8, 40–41; and vegetation distribution, 45–46, 58–59, 99–101
geomorphology: and plant ecology, 73–78, 79; of southern Sonoran Desert, 39–42
Geraniaceae, 119
germination, 158; delayed, 131–32, 133(fig.), 134, 137; and reproductive success, 128–29; and seed size, 126–27; of winter annuals, 119, 121, 123, 125
Giganta, Sierra, 53, 175, 177
Gila Bend, 121
Gila River, 23, 216, 218, 221, 224, 225, 228, 233
Gila Valley, 217
Gilia spp., 27
Grand Canyon, 266; *Drosophila* in, 171, 175, 190; fossil evidence from, 259, 267–73, 282
Gran Desierto, 37, 41, 42, 45, 57
Granite Mountains, 20, 147
grasses, 3, 131, 134
grassland and grazing, 223, 227
Great Basin, 31, 261
Greene's Canal, 221, 223, 229, 237, 241
Greene's Reservoir, 223, 224, 229, 237
Greene's Wash, 221, 223, 237
groundwater: in lower Santa Cruz, 224–25, 241, 243; pumping of, 216–17, 223–24, 230–33
Guaymas, 39, 43, 53; *Drosophila* in, 183, 187, 188, 193, 200; geology near, 40–41
Gymnogyps californianus (California condor), 267, 271–73, 276, 282
Gypsum Cave, 269

habitat diversity and scale, 43–47. See also diversity
halophytic shrubs, 57
Hermosillo, 44, 58–59, 187, 217
Hohokam culture, 81–82, 216, 228, 281
Holocene, 282–83; alluvial deposits of, 76, 78, 79, 82, 84, 88, 90, 94(fig.), 95(fig.), 101, 107–8; climate of, 257–58, 264, 269. See also Quaternary
host plants for *Drosophila*, 167, 169–71, 187–92, 200
Huachuca Mountains, 16
Hueco Mountains, 260
Huntington Canyon, 275, 276
hydrology, 41; of lower Santa Cruz Valley, 221–25, 228, 235–37
Hymenoclea salsola (burrobush), 79, 81, 82–84

immigration. See seed dispersal
Intermountain Province, 23
introduced species. See exotic species
inversion: and climate, 192–94; in *Drosophila*, 174–75, 177, 185–87, 195–200; genetic, 168–69
Inyo subdivision, 20, 22
ironwood. See *Olneya tesota*
irrigation, 216, 223–24, 235, 237
islands, 36, 39, 60, 169
Isocoma acradenia (burroweed), 225, 226(fig.), 227, 243, 248, 249, 250

Jatropha spp., 6, 45, 58, 97
Joshua tree. See *Yucca brevifolia*
Joshua Tree National Park, 23, 30, 262
Juniperus spp. (juniper), 262, 271; *communis*, 261; *osteosperma* (Utah), 260, 261; *scopulorum*, 261

karyotypic variation: in *Drosophila*, 10, 174–75, 176(fig.), 177–79, 199, 200–201, 202–3n. 4, 209–14(tables); and vegetational subdivisions, 179–85
Kingston Range, 24
Kino, Bahía, 41, 55, 57, 169, 187, 189
Krameria spp. (ratany), 58, 97, 119

lagoons, 50
La Laguna, Sierra, 53
land leveling, 233–35
landscape and species diversity, 27–30
La Paz, 177, 188, 193, 200
Larrea tridentata (creosote bush), 5, 14, 49, 58, 80(fig.), 98(fig.), 119, 159, 255, 260,

261; on abandoned farmland, 243, 246, 247, 249; in Lower Santa Cruz, 220, 225; piedmonts, 79, 81, 82, 83(fig.), 84, 94(fig.), 95(fig.), 97; and soil, 70, 96, 101, 103, 107–8
Lasthenia coulteri (goldfields), 227
Lepidium lasiocarpum, 125, 249
Libre, Sierra, 41, 45, 53
littoral scrub, 57
Lophocereus: gatesii, 169; *schottii* (senita), 58, 144, 150, 151, 169, 170(fig.), 171, 179, 187, 188–92, 200
Lower Colorado Valley subdivision, 6, 7, 8, 20, 22, 25, 42, 48, 68, 121; *Drosophila* in, 171, 175, 181, 193; soils, 103–8
Lupinus spp., 27
Lycium spp. (wolfberry), 97, 220, 225, 226(fig.)
Lysiloma spp., 59

macroalgae communities, 57
Madrean Floristic Province, 14, 16, 25
Magdalena, Bahía, 37
Magdalena, Río, 41
Magdalena Plains, 175, 177, 188
Magdalena Region, 7
mammals, Quaternary, 256(table), 263, 264–77, 282
Mammillaria spp., 27, 100, 144, 148, 159(fig.)
Mammuthus spp. (mammoth), 255, 265, 267, 272, 276, 278; dung of, 273–75
mangroves, 39, 57
Mátape, Río, 41
Mayo, Río, 16, 43
Mazatán, Sierra, 41, 53
McClellan Wash, 221, 223, 237
McDowell Mountain Regional Park, 30
megafauna: extinctions of, 264–65, 276–77, 278–79; Quaternary, 255, 256(table), 265–67, 268–77, 282–83
Mentzelia spp., 27
meristem, 144, 146, 151, 158
mesic habitats, 45–46
mesquite forests, 49–50, 58. See also *Prosopis*
Mexican Plateau, 262
microphyllous desert, 5
Miocene period, 40

Mitchell Caverns State Nature Preserve, 147
Mohavian Biotic Province, 20
Mojave Desert, 19, 20, 21, 22, 31, 36, 82, 121, 147, 171, 261; and creosote bush, 14, 79; species diversity in, 25, 27, 28, 30
Monoptilon bellioides (Mohave desert star), 119
"monsoon" season, 37–38
mortality, winter annual, 119–21
mountains, Arizona Upland, 99–101
Muav Cave, 270
Muddy Mountains, 23
Murray Springs, 276, 278

Nacapule, Cañón del, 46
Neotropical species, 25, 26, 27
Nevada, 12, 16, 19, 21, 22, 23–24, 261
Newberry Mountains, 23
Niño, El, 96, 115, 121
Northern Mojave Desert District, 22, 23, 25, 27
Nothrotheriops shastensis (Shasta ground sloth), 267, 270, 271
numerical studies of vegetation distribution, 21–25
nurse plants, 46–47, 157–59

oak woodlands, 60
oases, 50
ocotillo. See *Fouquieria splendens*
Olneya tesota (ironwood), 14, 49, 261, 262; distribution of, 44–45, 58, 59; as nurse plant, 46–47, 48(table)
Opuntia spp., 27, 93, 94(fig.), 147, 148, 149(fig.), 151, 155, 157, 160, 171, 262, 265; *bigelovii* (teddy-bear cholla), 100, 144, 145(fig.), 153–54; *Drosophila* in, 169, 179; *fulgida* (jumping cholla), 119, 153–54, 159; *leptocaulis* (pencil cholla), 144, 159; *phaeacantha* (prickly pear), 95(fig.), 119, 144, 145(fig.), 153, 154, 274; *versicolor* (staghorn cholla), 95(fig.), 154
Oreamnos harringtoni (Harrington's extinct goat), 267, 268, 269–70, 272, 282
organ pipe cactus. See *Stenocereus thurberi*
Organ Pipe Cactus National Monument, 30, 80(fig.), 151, 262

Index

Pachycereus: pecten-aboriginum (etcho), 54, 58, 59; *pringlei* (cardón), 3, 85, 86(fig.), 101
Pachycormus discolor (elephant tree), 3
Pacific coast, 43
Pacific Ocean, 36
packrat (*Neotoma*) middens, 255, 282; fossil data in, 258–62, 271
Paleozoic geology, 40
palms, 3, 46
paloverde. *See Cercidium*
Pectocarya recurvata, 119, 129–30, 131, 132
Pennisetum ciliare (buffelgrass), 58
Penstemon spp., 27
perennials, 3, 6, 27, 28, 43–44, 116(fig.), 117(fig.), 220
performance hierarchies, 129–30
Phacelia spp., 27
photosynthetic photon flux density (PPFD), 144, 153; and nurse plants, 158–59; and ribs and tubercles, 149, 150; and stem orientation, 146–48, 159–60
physical geography, 28–29
phytogeography of deserts, 19–21
Picacho Mountains, 223, 228
Picacho Reservoir, 221, 223, 228, 241
piedmonts, alluvial, 8, 73; cross-sections of, 74–75(figs.); development of, 76–77; and plant communities, 69–70, 72, 78–79, 81–85, 93–99, 107–8; rock types in, 77–78; soil development on, 88–90
Pinacate, El, 41, 45
Pinal County, 232(fig.); abandoned farmland in, 218–20, 238–39; agriculture in, 230, 231(fig.), 233
pine-oak forests, 60
Pinus spp. (pine), 259, 260, 261, 262, 264
pitahaya agria. *See Stenocereus gummosus*
Plains of Sonora subdivision, 7, 42, 44–45, 49, 58–60
Plantago spp., 119, 249; *patagonica* (Indian wheat), 129–30, 131, 132, 227
plant communities, 5–6, 7, 70, 87, 109; competition in, 84–85; local, 68–69; lower Santa Cruz, 225, 227; piedmont, 78–79, 81–84; Quaternary, 259–64, 281–82; and soils, 68–69, 93–108

plant succession, 70, 85, 87
Pleistocene, 267; alluvial fans of, 76, 78, 96–97, 107; climate of, 255; and packrat middens, 258–60; soils of, 42, 89–90, 93, 94(fig.), 95(fig.), 101, 103–5, 108(fig.). *See also* Quaternary
Pleuraphis rigida (big galleta), 158–59, 220, 227
Poaceae, 27, 119
Polemoniaceae, 27
Polygonaceae, 27
population dynamics of annuals, 117–18, 119–23
population sources of seeds, 131–34, 137
population structure: of *Drosophila*, 177–200; and gene inversion, 168–69
PPFD. *See* photosynthetic photon flux density
precipitation, 43–44; and soils, 91, 93; Sonoran Desert, 5, 37–39, 51–52, 68; and winter annuals, 121–23, 131–32
prickly pear. *See Opuntia*
productivity, biomass, 48–50
Prosopis spp. (mesquite), 58; at Bahía Kino, 47, 233, 248, 261, 265; *glandulosa,* 50, 58; *velutina* (velvet), 58, 225, 226(fig.), 227
pubescence on cacti, 150–54
Puerto Lobos, 57
pumps and pumping: deep well, 216–17; in lower Santa Cruz, 223–24, 230–33
Punta Chueca, 49
Punta Prieta, 101–3

Quaternary, 40, 255; climate in, 10, 257–58; mammals of, 256(table), 264–77, 278–79, 282–83; vegetation of, 258–64, 281–82
Quercus spp. (oak), 260, 261, 262

radiocarbon dating, 255, 259; of fossil material, 266–69, 270, 274, 275, 276
Rainbow Valley, 217
rainfall. *See* precipitation
Rampart Cave, 268, 270, 273, 274
ranching, 227, 228
reproduction, *in situ,* 131, 133(fig.), 134
reproductive success, 128–30

Reserva de la Biósfera El Pinacate y El Gran Desierto de Altar, 45
ribs, cactus, 148–50
Rincon Mountains, 14
riparian canyons, 46
Riverside County, 19, 20
Robles Wash, 223, 227, 229
rocks and piedmont development, 77–78
Roosevelt Dam, 216
roots, cactus, 154–57
rot density, 191–92

saguaro. See *Carnegiea gigantea*
Saguaro National Monument, 148
saltbush. See *Atriplex*
Salton Sink, 15, 19, 50
Salt River, 216, 225
Salt River Project, 216
saltscrub, 57
Salvia spp., 27
San Agustin, 188
San Bartolo, 42, 188
San Bernardino County, 19, 20
Sandblast Cave, 271, 272
San Diego County, 16
San Ignacio, 174, 197
San Ignacio/Arivaipa, Río, 41
Sanlucan Biotic Province, 20
Sanlucan element, 26, 27
San Manuel, 76
San Pedro Martir, 175, 177, 179, 188
San Pedro Nolasco, 179
San Pedro River Valley, 76, 278
Santa Catalina Island, 171, 175, 179, 190, 201
Santa Catalina Mountains, 15, 21, 49
Santa Cruz Canal Company, 228
Santa Cruz Flats, 221, 223, 229
Santa Cruz Reservoir Company, 229
Santa Cruz River, 218, 220, 221, 223, 224, 227, 229
Santa Cruz Valley: abandoned cropland in, 238–39, 243–47; agriculture in, 10, 227–38; desert remnants in, 240–41, 242(fig.); ecosystem recovery in, 249–51; groundwater pumping in, 230–33; hydrology of, 221–25; vegetation of, 217–21, 226–27

Santa Rosa Wash, 223, 237
sarcocaulescent desert, 5–6
Sasco Flats, 223, 227, 229
Sawtooth Mountains, 221, 229
scale and flora, 14–15
scavengers, extinction of, 276–77
Schismus barbatus, 119, 129–30, 131, 249
Scrophulariaceae, 27
sea grass meadows, 57
seed banks, 6, 117–18, 123; fitness of, 125–27; persistent, 128–29
seed dispersal: of desert annuals, 130–36; by mammals, 265–66
seed size and fitness, 125–27
seepweed. See *Suaeda moquinii*
senita. See *Lophocereus schottii*
shadscale. See *Atriplex confertifolia*
Sheephole Mountains, 20
Sheep Range, 261, 264
Shopishk, 221
Shreve, Forrest, 16, 19, 85, 150; desert chorology of, 20, 21; publications of, 5–7; on soils, 68–69; on Sonoran Desert subdivisions, 5–6, 12, 14
Sierra Madre Occidental, 36, 37, 40, 41, 43
Sierra Nevada, 14, 27
Sierra Rosario, 30
Siete Cerros, 58
Silverbell, 49
Silver Bell piedmont, 70, 72, 79, 80(fig.), 84, 90, 97, 98(fig.)
Silver Reef Mountains, 221
sina. See *Stenocereus alamosensis*
Sinaloa, 21, 31, 43, 50, 54, 171, 175, 188
Snake Range, 261
soils, 8, 72, 78, 248; development of, 87–91; lower Santa Cruz, 221, 241; and plant communities, 68–70, 85, 93–108; southern Sonoran Desert, 41–42; texture of, 91–93; water potential of, 155–57, 159
Sonoita, 189
Sonora, 5, 12, 21, 28, 31, 216; climate of, 37–39; *Drosophila* in, 171, 175, 177, 187; ecosystem dynamics in, 48–50; geomorphology of, 39–42; habitat diversity in, 43–48; Sonoran Desert in, 36–37, 50;

thornscrub in, 53–55. *See also various features; regions; towns*
Sonora, Río, 39, 41, 58
Sonoran batholith, 40
Sonoran Biotic Province, 20
Sonoran Desert, 21, 22, 71(map), 147; climate of, 37–39; ecosystem dynamics of, 48–50; features of, 36–37; geographic limits of, 50–53; species diversity in, 25, 26, 27, 28; and thornscrub, 53–55; vegetational subdivisions of, 4(fig.), 5–6, 7, 12, 14, 20, 166(fig.)
Sonoran element, 23, 26–27
Sonoran Floristic Province, 12, 13(fig.), 14, 15–16, 31; numerical studies of, 21–25; phytogeography of, 19–21; species diversity in, 25–30
Sonoran Province, 21
Sonoyta, Río, 41
Southern Pacific Railroad, 224
species coexistence of annuals, 127–30, 137
species diversity: in Sonoran Floristic Province, 25–30; of winter annuals, 129–30
spines, cactus, 150–54
Sporobolus spp., 57, 227
Spring Range, 23, 24
Stanton's Cave, 268
stem morphology, 144, 146
stem orientation, 146–48, 159–60
Stenocereus: alamosensis (sina), 54, 58, 171, 190; *gummosus* (pitahaya agria), 144, 169, 171, 172(fig.), 179, 190, 191–92, 200; *thurberi* (organ pipe cactus), 3, 54, 58, 150, 151, 171, 172(fig.), 179, 190, 191–92, 262
Steven's Cave, 268, 271
Stylocline micropoides, 119, 131, 132
Suaeda moquinii (seepweed), 57, 220, 225, 248
subdivisions of Sonoran Desert, 4(fig.), 5–6, 7, 12, 14, 20

temperature(s), 144, 146, 149, 158; and *Drosophila* distribution, 192–93; and Sonoran Desert, 38(fig.), 39, 51–52; and spines and pubescence, 150–53
temporal heterogeneity, 127, 137
Tertiary geology, 40, 41

Texas, 216, 260, 261
thornscrub, 7, 49; in Sonora, 53–55, 59; transition to, 8, 42, 43
Tiburón, Isla, 53, 60, 173, 177, 179
Tohono O'odham, 215–16, 221
Toltec, 245(fig.), 247
Tonto National Monument, 23, 30
topography, 228; and farm land, 219–20; and land leveling, 233–35; and stem orientation, 147–48
Tortolita Mountains, 75(fig.), 78, 79, 81, 83(fig.)
transects, Bahía Kino-Maycoba, 55–60
tropical deciduous forest, 43, 51, 59, 60
tropics, 46, 53
tubercles, cactus, 148–50
Tucson, 3, 49, 221
Tucson Mountains, 15, 30, 89; piedmonts in, 72, 73–74, 76; vegetation in, 69–70, 93–99
Tumamoc Hill, 76, 118–19, 124(fig.), 255

University of Nevada at Las Vegas, 19
Upper Gila River Valley, 76
Upper Verde River Valley, 76
U.S./International Biological Program, Desert Biome Project, 70–71
Utah, 22, 261, 275

Vallecito, 179
Valle Montevideo, 79, 81(fig.), 82, 84, 85
vegetation, 12, 15; scales of variation in, 43–47, 68–70; transects, 56–60
vegetational units in southern Sonoran Desert, 42–43
Vekol Wash, 221
Vizcaíno Region, 7, 8, 101–3, 175

Wasatch Mountains, 275, 276
wasteland, 233
Waterman Mountains, 99–100, 262
water storage capacity of cacti, 144, 146
water table, 224–25
water uptake: of cacti, 154–55; and soils, 155–57
weather: and species performance, 129–30; and winter annuals, 119–21
wetlands, 50
White Tank Mountains, 25

Wild Burro Wash, 79, 81–82, 83(fig.)
Wildhorse Wash, 73, 76
Willcox Playa, 216, 217
winter, 123; annual growth during, 119–21
wolfberry. *See Lycium*
woodlands in Pinal County, 232(fig.), 233

Yaqui, Río, 40, 41, 43, 59–60
Yellowstone National Park, 280
Yucca spp., 3, 101, 262, 270; *brevifolia*, 14, 261
Yuma, 68
Yuma County, 25, 107